The Oxford Solid State Basics

The Oxford Solid State Basics

Steven H. Simon
University of Oxford

OXFORD
UNIVERSITY PRESS

OXFORD
UNIVERSITY PRESS
Great Clarendon Street, Oxford, OX2 6DP,
United Kingdom

Oxford University Press is a department of the University of Oxford.
It furthers the University's objective of excellence in research, scholarship,
and education by publishing worldwide. Oxford is a registered trade mark of
Oxford University Press in the UK and in certain other countries

First Edition published in 2013
Reprinted 2014 (twice), 2015 (twice), 2016 (with corrections), 2017 (twice), 2019 (with corrections)

Published in the United States of America by Oxford University Press
198 Madison Avenue, New York, NY 10016, United States of America

British Library Cataloguing in Publication Data
Data available

Library of Congress Control Number: 2013936358

ISBN 978–0–19–968076–4 (Hbk.)
978–0–19–968077–1 (Pbk.)

Printed and bound in Great Britain by
Ashford Colour Press Ltd

Preface to the 2019 Reprint

Mostly I have the same comments as for the 2016 reprint. Changes are again limited to correction of small errors and typos, and a few added comments or clarifications.

Oxford, United Kingdom
January 2019

Preface to the 2016 Reprint

The 2016 reprint of this book is a "bug-fix" release — changes are limited to corrections of small errors and typos, and occasional (always very small) added explanatory material. While there have been numerous requests to add a chapter on this-or-that subject, there has not been much concensus on what subjects need to be added, so I concluded that it is probably best not to add any new material in hopes of keeping the book as streamlined as possible. (That said, I did add one short paragraph in chapter 18 about light emitting diodes in honor of the 2014 Nobel prize).

A large number of the errors in the book were in exercises that were not properly vetted before being released into the wild. Hopefully most of the problems with these exercises are fixed in this reprint. A few additional exercises are added as well. Those that are new or have been substantially changed have been marked with the symbol "&" to indicate where things differ from the original printing.

Thanks are due to a number of people. First, to those who have chosen to use this book for teaching. Second, to those who have read this book and have posted positive reviews on the web or otherwise recommended its use to others. Third, to those who have contacted me and pointed out typos or other problems to be fixed. Finally, to those who have generally supported me and made my efforts possible. I apologize for not mentioning all of you by name — you know who you are.

Oxford, United Kingdom
January 2016

Preface after Teaching this Course

Although things were a bit bumpy the first few times I taught this course, I learned a lot from the experience, and hopefully I have now managed to smooth out many of the rough parts. The good news is that the course has been viewed mostly as a success, even by the tough measure of student reviews. I would particularly like to thank the student who wrote on his or her review that I deserve a raise—and I would like to encourage my department chair to post this review on his wall and refer to it frequently.

If you can think of ways that this book could be further improved (correction of errors or whatnot) please let me know. The next generation of students will certainly appreciate it and that will improve your Karma. ☺

Oxford, United Kingdom
April 2013

Preface

When I was an undergraduate I thought solid state physics (a sub-genre of condensed matter physics) was perhaps the worst subject that any undergraduate could be forced to learn—boring and tedious, "squalid state" as it was commonly called.[1] How much would I really learn about the universe by studying the properties of crystals? I managed to avoid taking this course altogether. My opinion at the time was not a reflection of the subject matter, but rather was a reflection of how solid state physics was taught.

[1] This gibe against solid state physics can be traced back to the Nobel Laureate Murray Gell-Mann, discoverer of the quark, who famously believed that there was nothing interesting in any endeavor but particle physics. Interestingly he now studies complexity—a field that mostly arose from condensed matter.

Given my opinion as an undergraduate, it is a bit ironic that I have become a condensed matter physicist. But once I was introduced to the subject properly, I found that condensed matter was my favorite subject in all of physics—full of variety, excitement, and deep ideas. Sadly, a first introduction to the topic can barely scratch the surface of what constitutes the broad field of condensed matter.

Last year, when I was told that a new course was being prepared to teach condensed matter physics to third year Oxford undergraduates, I jumped at the opportunity to teach it. I felt that it *must* be possible to teach a condensed matter physics course that is just as interesting and exciting as any other course that an undergraduate will ever take. It must be possible to convey the excitement of real condensed matter physics to the undergraduate audience. I hope I will succeed in this task. You can judge for yourself.

The topics I was asked to cover are not atypical for a solid state physics course. Some of these topics are covered well in standard solid state physics references that one might find online, or in other books. The reason I am writing this book (and not just telling students to go read a standard reference) is because condensed matter/solid state is an enormous subject—worth many years of lectures—and one needs a guide to decide what subset of topics are most important (at least in the eyes of an Oxford examination committee). The material contained here gives depth in some topics, and glosses over other topics, so as to reflect the particular topics that are deemed important at Oxford as well as to reflect the subjects mandated by the UK Institute of Physics.

I cannot emphasize enough that there are many many extremely good books on solid state and condensed matter physics already in existence. There are also many good resources online (including the rather infamous "Britney Spears' guide to semiconductor physics"—which is tongue-in-cheek about Britney Spears, but is actually a very good reference about semiconductors). Throughout this book, I will try to point you to other good references appropriately.

So now we begin our journey through condensed matter. Let us go then, you and I...

Oxford, United Kingdom
January 2011

About this Book

This book is meant to be a first introduction to solid state and condensed matter physics for advanced undergraduate students. There are several main prerequisites for this course. First, the students should be familiar with basic quantum mechanics (we will sometimes use bra and ket notation). Secondly, the students should know something about thermodynamics and statistical mechanics. Basic mechanics and basic electromagnetism are also assumed. A very strong student might be capable of handling the material without all of the prerequisites, but the student would have to be willing to do some extra work on the side.

At the end of each chapter I give useful references to other books. A full list of all the books cited, along with proper reference and commentary, is provided in Appendix B.

Most chapters also have exercises included at the end. The exercises are marked with * if they are harder (with multiple *s if they are much harder). Exercises marked with ‡ are considered to be fundamental to the core syllabus (at least at Oxford).

A sample exam is provided (with solutions) in Appendix A. The current Oxford syllabus covers this entire book with the exception of Chapter 18 on device physics and Chapter 23 on the Hubbard model (interactions and magnetism).

Acknowledgments

Needless to say, I pilfered a fair fraction of the intellectual content of this book from parts of other books (mostly mentioned in Appendix B). The authors of these books put great thought and effort into their writing. I am deeply indebted to these giants who have come before me. Additionally, I have stolen many ideas about how this book should be structured from the people who have taught the condensed matter courses at Oxford in years past. Most recently this includes Mike Glazer, Andrew Boothroyd, and Robin Nicholas. I also confess to having stolen (with permission) many examples and exercises from the Oxford course or from old Oxford exams.

I am also very thankful to all the people who have helped me proofread, correct, and otherwise tweak this book. Among others, this includes Mike Glazer, Alex Hearmon, Simon Davenport, Till Hackler, Paul Stubley, Stephanie Simmons, Katherine Dunn, Joost Slingerland, Radu Coldea, Stanislav Zavjalov, Nathaniel Jowitt, Thomas Elliott, Ali Khan, Andrew Boothroyd, Jeremy Dodd, Marianne Wait, Seamus O'Hagan, Simon Clark, Joel Moore, Natasha Perkins, Christiane Riedinger, Deyan Mihaylov, Philipp Karkowski, William Bennett, Francesca Mosely, Bruno Balthazar, Richard Fern, Dmitry Budker, Rafe Kennedy, Sabine Müller, Carrie Leonard-McIntyre, and Nick Jelley (and I apologize if I have left anyone's name off this list). I am also very grateful for the hospitality of the Aspen Center for Physics, the Nordic Institute for Theoretical Physics, National University of Ireland Maynooth, the Galileo Galilei Institute for Theoretical Physics, 139 Edgeview Lane, and the Economy Section of United Airlines Transatlantic where major parts of this book were written.

Finally, I thank my father for helping proofread and improve these notes... and for a million other things.

Contents

About Condensed Matter Physics

1

This chapter is my personal take on why this topic is interesting. You might want to read it to figure out why you should think this book is interesting if that isn't otherwise obvious.

1.1 What Is Condensed Matter Physics

Quoting Wikipedia:

> Condensed matter physics is the field of physics that deals with the macroscopic and microscopic physical properties of matter. In particular, it is concerned with the "condensed" phases that appear whenever the number of constituents in a system is extremely large and the interactions between the constituents are strong. The most familiar examples of condensed phases are solids and liquids, which arise from the electromagnetic forces between atoms.

The use of the term "condensed matter", being more general than just the study of solids, was coined and promoted by Nobel laureate Philip W. Anderson.

Condensed matter physics is by far the largest single subfield of physics. The annual meeting of condensed matter physicists in the United States attracts over 6000 physicists each year! Topics included in this field range from the very practical to the absurdly abstract, from down-to-earth engineering to mathematical topics that border on string theory. The commonality is that all of these topics relate to the fundamental properties of matter.

1.2 Why Do We Study Condensed Matter Physics?

There are several very good answers to this question

(1) **Because it is the world around us**
Almost all of the physical world that we see is in fact condensed matter. We might ask questions such as

- why are metals shiny and why do they feel cold?

- why is glass transparent?
- why is water a fluid, and why does fluid feel wet?
- why is rubber soft and stretchy?

These questions are all in the domain of condensed matter physics. In fact almost every question you might ask about the world around you, short of asking about the sun or stars, is probably related to condensed matter physics in some way.

(2) **Because it is useful**
Over the last century our command of condensed matter physics has enabled us humans to do remarkable things. We have used our knowledge of physics to engineer new materials and exploit their properties to change our world and our society completely. Perhaps the most remarkable example is how our understanding of solids enabled new inventions exploiting semiconductor technology, which enabled the electronics industry, which enabled computers, iPhones, and everything else we now take for granted.

(3) **Because it is deep**
The questions that arise in condensed matter physics are as deep as those you might find anywhere. In fact, many of the ideas that are now used in other fields of physics can trace their origins to condensed matter physics.
A few examples for fun:

- The famous Higgs boson, which was recently observed at CERN, is no different from a phenomenon that occurs in superconductors (the domain of condensed matter physicists). The Higgs mechanism, which gives mass to elementary particles is frequently called the "Anderson–Higgs" mechanism, after the condensed matter physicist Phil Anderson[1] who described much of the same physics before Peter Higgs, the high-energy theorist.
- The ideas of the renormalization group (Nobel Prize to Kenneth Wilson in 1982) was developed simultaneously in both high-energy and condensed matter physics.
- The ideas of topological quantum field theories, while invented by string theorists as theories of quantum gravity, have been discovered in the laboratory by condensed matter physicists!
- In the last few years there has been a mass exodus of string theorists applying black-hole physics (in N-dimensions!) to phase transitions in real materials. The very same structures exist in the lab that are (maybe!) somewhere out in the cosmos!

That this type of physics is deep is not just my opinion. The Nobel committee agrees with me. During this course we will discuss the work of no fewer than fifty Nobel laureates! (See the index of scientists at the end of this book.)

[1] The same guy who coined the term "condensed matter".

(4) **Because reductionism doesn't work**
`begin{rant}` People frequently have the feeling that if you continually ask "what is it made of" you learn more about something. This approach to knowledge is known as *reductionism*. For example, asking what water is made of, someone may tell you it is made from molecules, then molecules are made of atoms, atoms of electrons and protons, protons of quarks, and quarks are made of who-knows-what. But none of this information tells you anything about why water is wet, about why protons and neutrons bind to form nuclei, why the atoms bind to form water, and so forth.[2] Understanding physics inevitably involves understanding how many objects all interact with each other. And this is where things get difficult very quickly. We understand the Schroedinger equation extremely well for one particle, but the Schroedinger equations for four or more particles, while in principle solvable, in practice are never solved because they are too difficult—even for the world's biggest computers. Physics involves figuring out what to do then. How are we to understand how quarks form a nucleus, or how electrons and protons form an atom if we cannot solve the many particle Schroedinger equation?
Even more interesting is the possibility that we understand very well the microscopic theory of a system, but then we discover that macroscopic properties *emerge* from the system that we did not expect. My personal favorite example is when one puts together many electrons (each with charge $-e$) one can sometimes find new particles emerging, each having one third the charge of an electron![3] Reductionism would never uncover this—it misses the point completely. `end{rant}`

[2] Phil Anderson provocatively wrote "The ability to reduce everything to simple fundamental laws does not imply the ability to start from those laws and reconstruct the universe. In fact, the more elementary particle physicists tell us about the fundamental laws of the universe, the less relevance they seem to have..."

[3] Yes, this really happens. The Nobel Prize in 1998 was awarded to Dan Tsui, Horst Stormer, and Bob Laughlin, for discovery of this phenomenon known as the fractional quantum Hall effect.

(5) **Because it is a laboratory**
Condensed matter physics is perhaps the best laboratory we have for studying quantum physics and statistical physics. Those of us who are fascinated by what quantum mechanics and statistical mechanics can do often end up studying condensed matter physics which is deeply grounded in both of these topics. Condensed matter is an infinitely varied playground for physicists to test strange quantum and statistical effects.
I view this entire book as an extension of what you have already learned in quantum and statistical physics. If you enjoyed those courses, you will likely enjoy this as well. If you did not do well in those courses, you might want to go back and study them again, because many of the same ideas will arise here.

1.3 Why Solid State Physics?

Being that condensed matter physics is so huge, we cannot possibly study all of it in one book. Instead we will focus on just one particular subfield, known as "solid state physics". As the name suggests, this is the study of matter in its solid state (as compared to being in a liquid state, a gas state, a superfluid state, or some other state of matter). There are several reasons why we choose to focus on the solid state. First of all, solid state physics is by far the biggest single subfield of condensed matter physics.[4] Secondly, solid state physics is the most successful and most technologically useful subfield of condensed matter physics. Not only do we know far more about solids than we know about other types of matter, but also solids are far more useful than other types of matter. Almost all materials that have found their way to industrial application are in their solid state. Paramount among these materials are the solids known as semiconductors which are the basis of the entire electronics industry. Indeed, frequently the electronics industry is even called the "solid state" industry.[5] More importantly, however, the physics of solids provides a paradigm for learning other topics in physics. The things we learn in our study of solids will form a foundation for study of other topics both inside the field of condensed matter, and outside of it.

[4] Perhaps this is not surprising considering how many solid objects there are in the world.

[5] This stems from the term "solid state electronics" which describes any electronic device where electrons travel within a solid. This is in comparison to the old vacuum tube-based electronic systems where the electrons actually traveled *in vacuo*. The old-style tubes have been replaced in almost every application—with very few exceptions. One interesting exception is that many audiophiles and musicians prefer sound amplification using tubes rather than solid state electronics. What they prefer is that the tubes amplify sound with a characteristic distortion that the musicians somehow find appealing. For a pure amplification without distortion, solid state devices are far better.

Part I

Physics of Solids without Considering Microscopic Structure: The Early Days of Solid State

Specific Heat of Solids: Boltzmann, Einstein, and Debye

2

Our story of condensed matter physics starts around the turn of the last century. It was well known (and you should remember from your prior study of statistical physics) that the heat capacity[1] of a monatomic (ideal) gas is $C_v = 3k_B/2$ per atom, with k_B being Boltzmann's constant. The statistical theory of gases described why this is so.

As far back as 1819, however, it had also been known that for many solids the heat capacity is given by[2]

$$\begin{aligned} C &= 3k_B \quad \text{per atom} \\ \text{or} \quad C &= 3R \end{aligned}$$

which is known as the *law of Dulong–Petit,*[3] where R is the ideal gas constant. While this law is not always correct, it frequently is close to true. For example, see Table 2.1 of heat capacities at room temperature and pressure. With the exception of diamond, the law $C/R = 3$ seems to hold extremely well at room temperature, although at lower temperatures all materials start to deviate from this law, and typically C drops rapidly below some temperature (and for diamond when the temperature is raised, the heat capacity increases towards $3R$ as well, see Fig. 2.2).

In 1896 Boltzmann constructed a model that accounted for this law fairly well. In his model, each atom in the solid is bound to neighboring atoms. Focusing on a single particular atom, we imagine that atom as

[2]Here I do not distinguish between C_p (at constant pressure) and C_v (at constant volume) because they are very close to the same. Recall that $C_p - C_v = VT\alpha^2/\beta_T$, where β_T is the isothermal compressibility and α is the coefficient of thermal expansion. For a solid, α is relatively small.

[3]Both Pierre Dulong and Alexis Petit were French chemists. Neither is remembered for much else besides this law.

Table 2.1 Heat capacities of some solids at room temperature and pressure.

Material	C/R
Aluminum (Al)	2.91
Antimony (Sb)	3.03
Copper (Cu)	2.94
Gold (Au)	3.05
Silver (Ag)	2.99
Diamond (C)	0.735

[1]We will almost always be concerned with the heat capacity C per atom of a material. Multiplying by Avogadro's number gives the *molar heat capacity* or heat capacity per mole. The *specific heat* (denoted often as c rather than C) is the heat capacity per unit mass. However, the phrase "specific heat" is also used loosely to describe the molar heat capacity, since they are both intensive quantities (as compared to the total heat capacity which is extensive—i.e., proportional to the amount of mass in the system). We will try to be precise with our language, but one should be aware that frequently things are written in non-precise ways and you are left to figure out what is meant. For example, really we should say C_v per atom $= 3k_B/2$ rather than $C_v = 3k_B/2$ per atom, and similarly we should say C per mole $= 3R$. To be more precise I really would have liked to title this chapter "Heat Capacity per Atom of Solids" rather than "Specific Heat of Solids". However, for over a century people have talked about the "Einstein Theory of Specific Heat" and "Debye Theory of Specific Heat", and it would have been almost scandalous to not use this wording.

being in a harmonic well formed by the interaction with its neighbors. In such a classical statistical mechanical model, the heat capacity of the vibration of the atom is $3k_B$ per atom, in agreement with Dulong–Petit. (You should be able to show this with your knowledge of statistical mechanics and/or the equipartition theorem; see Exercise 2.1).

Several years later in 1907, Einstein started wondering about why this law does not hold at low temperatures (for diamond, "low" temperature appears to be room temperature!). What he realized is that quantum mechanics is important!

Einstein's assumption was similar to that of Boltzmann. He assumed that every atom is in a harmonic well created by the interaction with its neighbors. Further, he assumed that every atom is in an identical harmonic well and has an oscillation frequency ω (known as the "Einstein" frequency).

The quantum-mechanical problem of a simple harmonic oscillator is one whose solution we know. We will now use that knowledge to determine the heat capacity of a single one-dimensional harmonic oscillator. This entire calculation should look familiar from your statistical physics course.

2.1 Einstein's Calculation

In one dimension, the energies of the eigenstates of a single harmonic oscillator are

$$E_n = \hbar\omega(n + 1/2) \tag{2.1}$$

with ω the frequency of the harmonic oscillator (the "Einstein frequency"). The partition function is then[4]

$$\begin{aligned} Z_{1D} &= \sum_{n\geqslant 0} e^{-\beta\hbar\omega(n+1/2)} \\ &= \frac{e^{-\beta\hbar\omega/2}}{1-e^{-\beta\hbar\omega}} = \frac{1}{2\sinh(\beta\hbar\omega/2)}. \end{aligned}$$

The expectation of energy is then (compare to Eq. 2.1)

$$\langle E \rangle = -\frac{1}{Z_{1D}}\frac{\partial Z_{1D}}{\partial \beta} = \frac{\hbar\omega}{2}\coth\left(\frac{\beta\hbar\omega}{2}\right) = \hbar\omega\left(n_B(\beta\hbar\omega) + \frac{1}{2}\right) \tag{2.2}$$

where n_B is the Bose[5] occupation factor

$$n_B(x) = \frac{1}{e^x - 1}.$$

This result is easy to interpret. The mode ω is an excitation that is excited on average up to the n_B^{th} level, or equivalently there is a "boson" orbital which is "occupied" by n_B bosons.

Differentiating the expression for energy we obtain the heat capacity for a single oscillator,

$$C = \frac{\partial\langle E\rangle}{\partial T} = k_B(\beta\hbar\omega)^2 \frac{e^{\beta\hbar\omega}}{(e^{\beta\hbar\omega}-1)^2}.$$

[4] We will very frequently use the standard notation $\beta = 1/(k_B T)$.

[5] Satyendra Bose worked out the idea of Bose statistics in 1924, but could not get it published until Einstein lent his support to the idea.

Note that the high-temperature limit of this expression gives $C = k_B$ (check this if it is not obvious!).

Generalizing to the three-dimensional case,

$$E_{n_x,n_y,n_z} = \hbar\omega[(n_x + 1/2) + (n_y + 1/2) + (n_z + 1/2)]$$

and

$$Z_{3D} = \sum_{n_x,n_y,n_z \geqslant 0} e^{-\beta E_{n_x,n_y,n_z}} = [Z_{1D}]^3$$

resulting in $\langle E_{3D} \rangle = 3\langle E_{1D} \rangle$, so correspondingly we obtain

$$C = 3k_B(\beta\hbar\omega)^2 \frac{e^{\beta\hbar\omega}}{(e^{\beta\hbar\omega} - 1)^2}.$$

Plotted, this looks like Fig. 2.1.

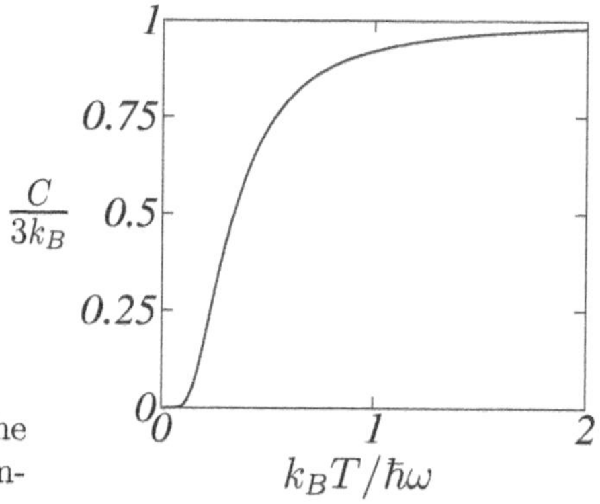

Fig. 2.1 Einstein heat capacity per atom in three dimensions.

Note that in the high-temperature limit $k_BT \gg \hbar\omega$ we recover the law of Dulong–Petit: $3k_B$ heat capacity per atom. However, at low temperature ($T \ll \hbar\omega/k_B$) the degrees of freedom "freeze out", the system gets stuck in only the ground-state eigenstate, and the heat capacity vanishes rapidly.

Einstein's theory reasonably accurately explained the behavior of the heat capacity as a function of temperature with only a single fitting parameter, the Einstein frequency ω (sometimes this frequency is quoted in terms of the Einstein temperature $\hbar\omega = k_BT_{Einstein}$). In Fig. 2.2 we show Einstein's original comparison to the heat capacity of diamond.

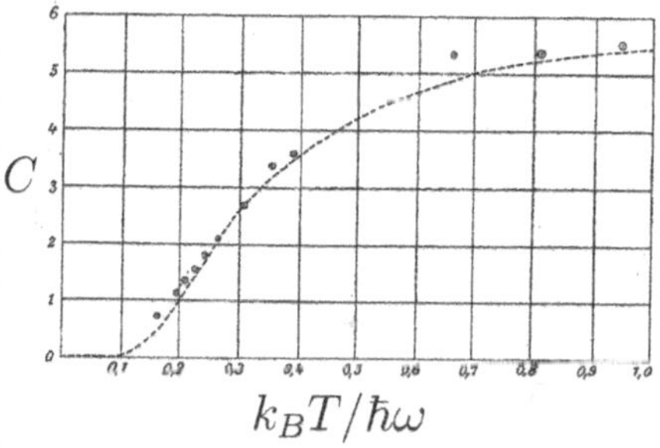

Fig. 2.2 Plot of molar heat capacity of diamond from Einstein's original paper. The fit is to the Einstein theory. The y axis is C in units of cal/(K-mol). In these units, $3R \approx 5.96$. The fitting parameter $T_{Einstein} = \hbar\omega/k_B$ is roughly 1320K. Figure from A. Einstein, *Ann. Phys.*, **22**, 180, (1907), Copyright Wiley-VCH Verlag GmbH & Co. KGaA. Reproduced with permission.

For most materials, the Einstein frequency ω is low compared to room temperature, so the Dulong–Petit law holds fairly well (being relatively high temperature compared to the Einstein frequency). However, for diamond, ω is high compared to room temperature, so the heat capacity is lower than $3R$ at room temperature. The reason diamond has such a high Einstein frequency is that the bonding between atoms in diamond is very strong and the atomic mass of the carbon atoms that comprise diamond is relatively low, hence a high $\omega = \sqrt{\kappa/m}$ oscillation frequency, with κ a spring constant and m the mass. These strong bonds also result in diamond being an exceptionally hard material.

Einstein's result was remarkable, not only in that it explained the temperature dependence of the heat capacity, but more importantly it told us something fundamental about quantum mechanics. Keep in mind that Einstein obtained this result 19 years before the Schroedinger equation was discovered![6]

[6]Einstein was a pretty smart guy.

2.2 Debye's Calculation

Einstein's theory of specific heat was extremely successful, but still there were clear deviations from the predicted equation. Even in the plot in his first paper (Fig. 2.2) one can see that at low temperature the experimental data lie above the theoretical curve.[7] This result turns out to be rather important! In fact, it was known that at low temperatures

[7]Although perhaps not obvious, this deviation turns out to be real, and not just experimental error.

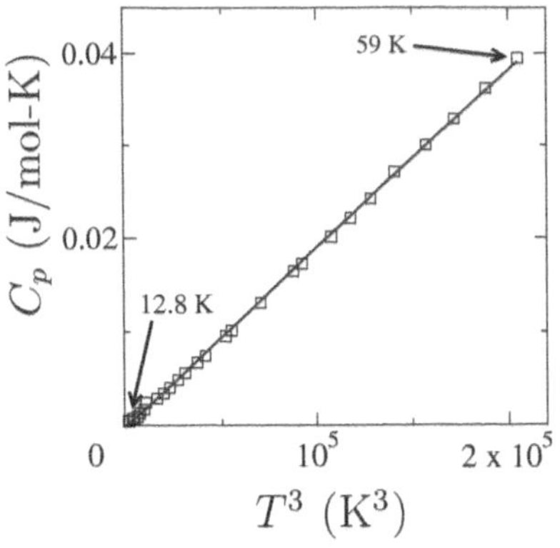

Fig. 2.3 Heat capacity of diamond is proportional to T^3 at low temperature. Note that the temperatures shown in this plot are far far below the Einstein temperature and therefore correspond to the very bottom left corner of Fig. 2.2. Data from Desnoyehs and Morrison, *Phil. Mag.*, **3**, 42 (1958).

most materials have a heat capacity that is proportional to T^3. See for example, Fig. 2.3. (Metals also have a very small additional term proportional to T which we will discuss later in Section 4.2. Magnetic materials may have other additional terms as well.[8] Non-magnetic insulators have only the T^3 behavior). At any rate, Einstein's formula at low temperature is exponentially small in T, not agreeing at all with the actual experiments.

In 1912 Peter Debye[9] discovered how to better treat the quantum mechanics of oscillations of atoms, and managed to explain the T^3 dependance of the specific heat. Debye realized that oscillation of atoms is the same thing as sound, and sound is a wave, so it should be quantized the same way as Planck[10] had quantized light waves in 1900. Besides the fact that the speed of light is much faster than that of sound, there is only one minor difference between light and sound: for light, there are two polarizations for each wavevector $\mathbf{k}$, whereas for sound there are three modes for each $\mathbf{k}$ (a longitudinal mode, where the atomic motion is in the same direction as $\mathbf{k}$ and two transverse modes where the motion is perpendicular to $\mathbf{k}$; light has only the transverse modes[11]). For simplicity of presentation here we will assume that the transverse and longitudinal modes have the same velocity, although in truth the longitudinal velocity is usually somewhat greater than the transverse velocity.[12]

We now repeat essentially what was Planck's calculation for light. This calculation should also look familiar from your statistical physics course. First, however, we need some preliminary information about waves:

2.2.1 Periodic (Born–von Karman) Boundary Conditions

Many times in this course we will consider waves with periodic or "Born–von Karman" boundary conditions. It is easiest to describe this first in one dimension. Here, instead of having a one-dimensional sample of length L with actual ends, we imagine that the two ends are connected together making the sample into a circle. The periodic boundary condition means that, any wave in this sample e^{ikr} is required to have the same value for a position r as it has for $r+L$ (we have gone all the way around the circle). This then restricts the possible values of k to be

$$k = \frac{2\pi n}{L}$$

for n an integer. If we are ever required to sum over all possible values of k, for large enough L we can replace the sum with an integral obtaining

$$\sum_k \rightarrow \frac{L}{2\pi}\int_{-\infty}^{\infty} dk.$$

A way to understand this mapping is to note that the spacing between allowed points in k space is $2\pi/L$, so the integral $\int dk$ can be replaced by a sum over k points times the spacing between the points.[13]

[8] We will discuss magnetism in part VII.

[9] Peter Debye later won a Nobel Prize in chemistry for something completely different.

[10] Max Planck did not like his own calculation of the quantization of light. He later referred to it as "an act of desperation". It seems that he viewed it mostly as a way to fudge the calculation to get an answer in agreement with experiment rather than being the revolutionary beginning of the new field of quantum physics.

[11] Sound in fluids is longitudinal only.

[12] It is not too hard to keep track of the fact that the transverse and longitudinal velocities are different. Note also that we assume the sound velocity to be the same in every direction, which need not be true in real materials. It is not too hard to include such anisotropy in Debye's theory as well. See Exercise 2.6.

In three dimensions, the story is extremely similar. For a sample of size L^3, we identify opposite ends of the sample (wrapping the sample up into a hypertorus!) so that if you go a distance L in the x, y or z direction, you get back to where you started.[14] As a result, our $\mathbf{k}$ values can only take values

$$\mathbf{k} = \frac{2\pi}{L}(n_1, n_2, n_3)$$

for integer values of n_i, so here each $\mathbf{k}$ point now occupies a volume of $(2\pi/L)^3$. Because of this discretization of values of $\mathbf{k}$, whenever we have a sum over all possible $\mathbf{k}$ values we obtain

$$\sum_{\mathbf{k}} \to \frac{L^3}{(2\pi)^3} \int \mathbf{dk}$$

with the integral over all three dimensions of $\mathbf{k}$-space (this is what we mean by the bold $\mathbf{dk}$). One might think that wrapping the sample up into a hypertorus is very unnatural compared to considering a system with real boundary conditions. However, these boundary conditions tend to simplify calculations quite a bit, and most physical quantities you might measure could be measured far from the boundaries of the sample anyway and would then be independent of what you do with the boundary conditions.

[13] In your previous courses you may have used particle-in-a-box boundary conditions where instead of plane waves $e^{i2\pi nr/L}$ you used particle in a box wavefunctions of the form $\sin(n\pi r/L)$. This gives you instead

$$\sum_k \to \frac{L}{\pi}\int_0^\infty dk$$

which will inevitably result in the same physical answers as for the periodic boundary condition case. All calculations can be done either way, but periodic Born–von Karman boundary conditions are almost always simpler.

[14] Such boundary conditions are very popular in video games, such as the classic time-wasting game of my youth, *Asteroids* (you can find it online). It may also be possible that our universe has such boundary conditions—a notion known as the *doughnut universe*. Data collected by Cosmic Microwave Background Explorer (led by Nobel Laureates John Mather and George Smoot) and its successor the Wilkinson Microwave Anisotropy Probe appear consistent with this structure.

2.2.2 Debye's Calculation Following Planck

Debye decided that the oscillation modes of a solid were waves with frequencies $\omega(\mathbf{k}) = v|\mathbf{k}|$ with v the sound velocity—and for each $\mathbf{k}$ there should be three possible oscillation modes, one for each direction of motion. Thus he wrote an expression entirely analogous to Einstein's expression (compare to Eq. 2.2)

$$\begin{aligned}\langle E\rangle &= 3\sum_{\mathbf{k}} \hbar\omega(\mathbf{k})\left(n_B(\beta\hbar\omega(\mathbf{k})) + \frac{1}{2}\right) \\ &= 3\frac{L^3}{(2\pi)^3}\int \mathbf{dk}\,\hbar\omega(\mathbf{k})\left(n_B(\beta\hbar\omega(\mathbf{k})) + \frac{1}{2}\right).\end{aligned}$$

Each excitation mode is a boson of frequency $\omega(\mathbf{k})$ and is occupied on average $n_B(\beta\hbar\omega(\mathbf{k}))$ times.

By spherical symmetry, we may convert the three-dimensional integral to a one-dimensional integral

$$\int \mathbf{dk} \to 4\pi\int_0^\infty k^2 dk$$

(recall that $4\pi k^2$ is the area of the surface of a sphere[15] of radius k) and we also use $k = \omega/v$ to obtain

$$\langle E\rangle = 3\frac{4\pi L^3}{(2\pi)^3}\int_0^\infty \omega^2 d\omega(1/v^3)(\hbar\omega)\left(n_B(\beta\hbar\omega) + \frac{1}{2}\right). \tag{2.3}$$

[15] Or to be pedantic, $\int \mathbf{dk} \to \int_0^{2\pi} d\phi \int_0^\pi d\theta \sin\theta \int_0^\infty k^2 dk$ and performing the angular integrals gives 4π.

It is convenient to replace $nL^3 = N$ where n is the density of atoms. We then obtain

$$\langle E \rangle = \int_0^\infty d\omega \, g(\omega)(\hbar\omega)\left(n_B(\beta\hbar\omega) + \frac{1}{2}\right) \tag{2.4}$$

where the *density of states* is given by[16]

$$g(\omega) = L^3\left[\frac{12\pi\omega^2}{(2\pi)^3 v^3}\right] = N\left[\frac{12\pi\omega^2}{(2\pi)^3 n v^3}\right] = N\frac{9\omega^2}{\omega_d^3} \tag{2.5}$$

where

$$\omega_d^3 = 6\pi^2 n v^3 \tag{2.6}$$

This frequency will be known as the *Debye frequency*, and in the next section we will see why we chose to define it this way with the factor of 9 removed.

The meaning of the density of states[17] here is that the total number of oscillation modes with frequencies between ω and $\omega + d\omega$ is given by $g(\omega)d\omega$. Thus the interpretation of Eq. 2.4 is simply that we should count how many modes there are per frequency (given by g), then multiply by the expected energy per mode (compare to Eq. 2.2), and finally we integrate over all frequencies. This result, Eq. 2.3, for the quantum energy of the sound waves is strikingly similar to Planck's result for the quantum energy of light waves, only we have replaced $2/c^3$ by $3/v^3$ (replacing the two light modes by three sound modes). The other change from Planck's classic result is the $+1/2$ that we obtain as the zero-point energy of each oscillator.[18] At any rate, this zero-point energy gives us a contribution which is temperature independent.[19] Since we are concerned with $C = \partial\langle E \rangle/\partial T$ this term will not contribute and we will separate it out. We thus obtain

$$\langle E \rangle = \frac{9N\hbar}{\omega_d^3}\int_0^\infty d\omega \frac{\omega^3}{e^{\beta\hbar\omega} - 1} \quad + \quad T \text{ independent constant.}$$

By defining a variable $x = \beta\hbar\omega$ this becomes

$$\langle E \rangle = \frac{9N\hbar}{\omega_d^3(\beta\hbar)^4}\int_0^\infty dx \, \frac{x^3}{e^x - 1} \quad + \quad T \text{ independent constant.}$$

The nasty integral just gives some number[20]—in fact the number is $\pi^4/15$. Thus we obtain

$$\langle E \rangle = 9N\frac{(k_B T)^4}{(\hbar\omega_d)^3}\frac{\pi^4}{15} \quad + \quad T \text{ independent constant.}$$

[16] Although it now appears that the number of atoms N and the atomic density n are relevant parameters of the problem, in fact, these two factors cancel and only the original L^3 matters for our results in this section! The reason we have introduced such canceling factors here is because writing our results this way prepares us for the next section (Sec. 2.2.3) where N becomes an important physical parameter different from L^3!

[17] We will encounter the concept of density of states many times, so it is a good idea to become comfortable with it!

[18] Planck should have gotten this energy as well, but he didn't know about zero-point energy—in fact, since it was long before quantum mechanics was fully understood, Debye didn't actually have this term either.

[19] The contribution of the zero-point energy is temperature independent and also infinite. Handling infinities like this is something that gives mathematicians nightmares, but physicists do it happily when they know that the infinity is not really physical. We will see in Section 2.2.3 how this infinity gets properly cut off by the Debye frequency.

[20] If you wanted to evaluate the nasty integral, the strategy is to reduce it to the famous Riemann zeta function. We start by writing

$$\int_0^\infty dx \frac{x^3}{e^x - 1} = \int_0^\infty dx \frac{x^3 e^{-x}}{1 - e^{-x}} = \int_0^\infty dx \, x^3 e^{-x}\sum_{n=0}^\infty e^{-nx} = \sum_{n=1}^\infty \int_0^\infty dx \, x^3 e^{-nx}$$

The integral can be evaluated and the expression can then be written as $3!\sum_{n=1}^\infty n^{-4}$. The resultant sum is a special case of the famous Riemann zeta function defined as $\zeta(p) = \sum_{n=1}^\infty n^{-p}$, where here we are concerned with the value of $\zeta(4)$. Since the zeta function is one of the most important functions in all of mathematics (see margin note 24 of this chapter), one can just look up its value on a table to find that $\zeta(4) = \pi^4/90$, thus giving us the stated result that the nasty integral is $\pi^4/15$. However, in the unlikely event that you were stranded on a desert island and did not have access to a table, you could even evaluate this sum explicitly, which we do in the appendix to this chapter.

Notice the similarity to Planck's derivation of the T^4 energy of photons. As a result, the heat capacity is

$$C = \frac{\partial \langle E \rangle}{\partial T} = N k_B \frac{(k_B T)^3}{(\hbar \omega_d)^3} \frac{12\pi^4}{5} \sim T^3.$$

This correctly obtains the desired T^3 specific heat. Furthermore, the prefactor of T^3 can be calculated in terms of known quantities such as the sound velocity. Note that the Debye frequency in this equation is sometimes replaced by a temperature.

$$k_B T_{Debye} = \hbar \omega_d$$

known as the *Debye temperature* (see Table 2.2), so that this equation reads

$$C = \frac{\partial \langle E \rangle}{\partial T} = N k_B \frac{(T)^3}{(T_{Debye})^3} \frac{12\pi^4}{5} .$$

Table 2.2 Some Debye temperatures.

Material	T_{Debye} (K)
Diamond (C)	1850
Beryllium (Be)	1000
Silicon (Si)	625
Copper (Cu)	315
Silver (Ag)	215
Lead (Pb)	88

Note that hard materials like diamond have high Debye temperatures, whereas soft materials like lead have low Debye temperatures. These data are measured at standard temperature and pressure (meaning the speed of sound and density are measured at this temperature and pressure). Since real materials change depending on the environment (expand with temperature, etc.) the Debye temperature is actually a very weak function of ambient conditions.

2.2.3 Debye's "Interpolation"

Unfortunately, now Debye has a problem. In the expression just derived, the heat capacity is proportional to T^3 up to arbitrarily high temperature. We know however, that the heat capacity should level off to $3k_B N$ at high T. Debye understood that the problem with his approximation is that it allowed an infinite number of sound wave modes—up to arbitrarily large $\mathbf{k}$. This would imply more sound wave modes than there are atoms in the entire system. Debye guessed (correctly) that really there should be only as many modes as there are degrees of freedom in the system. We will see in Chapters 9–13 that this is an important general principle. To fix this problem, Debye decided to not consider sound waves above some maximum frequency ω_{cutoff}, with this frequency chosen such that there are exactly $3N$ sound wave modes in the system (three dimensions of motion times N particles). We thus define ω_{cutoff} via

$$3N = \int_0^{\omega_{cutoff}} d\omega \, g(\omega). \tag{2.7}$$

We correspondingly rewrite Eq. 2.4 for the energy (dropping the zero-point contribution) as[21]

$$\langle E \rangle = \int_0^{\omega_{cutoff}} d\omega \, g(\omega) \, \hbar\omega \, n_B(\beta\hbar\omega). \tag{2.8}$$

[21] Here, since the integral is now cut off, had we kept the zero-point energy, its contribution would now be finite (and temperature independent still).

Note that at very low temperature, this cutoff does not matter at all, since for large β the Bose factor n_B will very rapidly go to zero at frequencies well below the cutoff frequency anyway.

Let us now check that this cutoff gives us the correct high-temperature limit. For high temperature

$$n_B(\beta\hbar\omega) = \frac{1}{e^{\beta\hbar\omega} - 1} \rightarrow \frac{k_B T}{\hbar\omega} .$$

Thus in the high-temperature limit, invoking Eqs. 2.7 and 2.8 we obtain

$$\langle E \rangle = k_B T \int_0^{\omega_{cutoff}} d\omega g(\omega) \; = 3 k_B T N$$

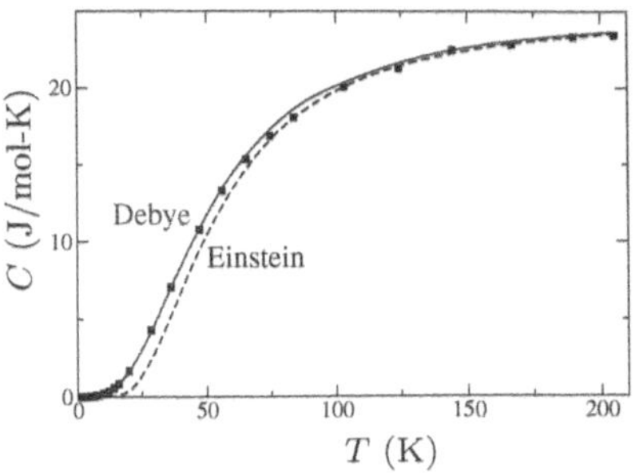

Fig. 2.4 Heat capacity of silver compared to the Debye and Einstein models. The high-temperature asymptote is given by $C = 3R = 24.945$ J/(mol-K). Over the entire experimental range, the fit to the Debye theory is excellent. At low T it correctly recovers the T^3 dependence, and at high T it converges to the law of Dulong–Petit. The Einstein theory clearly is incorrect at very low temperatures. The Debye temperature is roughly 215 K, whereas the Einstein temperature roughly 151 K. Data is taken from C. Kittel, *Solid State Physics,* 2ed Wiley (1956).

yielding the Dulong–Petit high-temperature heat capacity $C = \partial \langle E \rangle / \partial T = 3 k_B N = 3 k_B$ per atom. For completeness, let us now evaluate our cutoff frequency,

$$3N = \int_0^{\omega_{cutoff}} d\omega g(\omega) = 9N \int_0^{\omega_{cutoff}} d\omega \frac{\omega^2}{\omega_d^3} = 3N \frac{\omega_{cutoff}^3}{\omega_d^3} .$$

We thus see that the correct cutoff frequency is exactly the Debye frequency ω_d. Note that $k = \omega_d / v = (6\pi^2 n)^{1/3}$ (from Eq. 2.6) is on the order of the inverse interatomic spacing of the solid.

More generally (in the neither high- nor low-temperature limit) one has to evaluate the integral (Eq. 2.8), which cannot be done analytically. Nonetheless it can be done numerically and then can be compared to actual experimental data as shown in Fig. 2.4. It should be emphasized that the Debye theory makes predictions without any free parameters, as compared to the Einstein theory which had the unknown Einstein frequency ω as a free fitting parameter.

2.2.4 Some Shortcomings of the Debye Theory

While Debye's theory is remarkably successful, it does have a few shortcomings.

- The introduction of the cutoff seems very *ad hoc.* This seems like a successful cheat rather than real physics.
- We have assumed sound waves follow the law $\omega = vk$ even for very very large values of k (on the order of the inverse lattice spacing), whereas the entire idea of sound is a long-wavelength idea, which doesn't seem to make sense for high enough frequency and short enough wavelength. At any rate, it is known that at high enough frequency the law $\omega = vk$ no longer holds.
- Experimentally, the Debye theory is very accurate, but it is not exact at intermediate temperatures.
- Metals also have a term in the heat capacity that is proportional to T, so the overall heat capacity is $C = \gamma T + \alpha T^3$ and at low enough T the linear term will dominate.[22] You can't see this contribution on the plot Fig. 2.4, but at very low T it becomes evident, as shown in Fig. 2.5.

[22] In magnetic materials there may be still other contributions to the heat capacity reflecting the energy stored in magnetic degrees of freedom. See Part VII, and in particular Exercise 20.3, below.

Of these shortcomings, the first three can be handled more properly by treating the details of the crystal structure of materials accurately (which we will do starting in Chapter 9). The final issue requires us to carefully study the behavior of electrons in metals to discover the origin of this linear T term (see Section 4.2).

Nonetheless, despite these problems, Debye's theory was a substantial improvement over Einstein's.[23]

[23] Debye was pretty smart too... even though he was a chemist.

Chapter Summary

- (Much of the) heat capacity (specific heat) of materials is due to atomic vibrations.
- Boltzmann and Einstein models consider these vibrations as N simple harmonic oscillators.
- Boltzmann classical analysis obtains law of Dulong–Petit $C = 3Nk_B = 3R$.
- Einstein quantum analysis shows that at temperatures below the oscillator frequency, degrees of freedom freeze out, and heat capacity drops exponentially. Einstein frequency is a fitting parameter.
- Debye Model treats oscillations as sound waves with no fitting parameters.
  - $\omega = v|k|$, similar to light (but three polarizations not two)
  - quantization similar to Planck quantization of light
  - maximum frequency cutoff ($\hbar\omega_{Debye} = k_B T_{Debye}$) necessary to obtain a total of only $3N$ degrees of freedom
  - obtains Dulong–Petit at high T and $C \sim T^3$ at low T.
- Metals have an additional (albeit small) linear T term in the heat capacity.

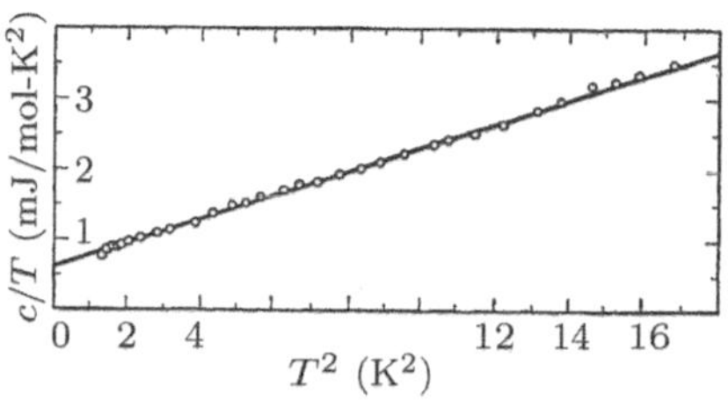

Fig. 2.5 Heat capacity divided by temperature of silver at very low temperature plotted against temperature squared. At low enough temperature one can see that the heat capacity is actually of the form $C = \gamma T + \alpha T^3$. If the dependence were purely T^3, the curve would have a zero intercept. The cubic term is from the Debye theory of specific heat. The linear term is special to metals and will be discussed in Section 4.2. Figure from Corak *et al.*, *Phys. Rev.* **98** 1699 (1955), http://prola.aps.org/abstract/PR/v98/i6/p1699_1, copyright American Physical Society. Used by permission.

References

Almost every solid state physics book covers the material introduced in this chapter, but frequently it is done late in the book only after the idea of phonons is introduced. We will get to phonons in Chapter 9. Before we get there the following references cover this material without discussion of phonons:

- Goodstein, sections 3.1–3.2
- Rosenberg, sections 5.1–5.13
- Burns, sections 11.3–11.5

Once we get to phonons, we can look back at this material again. Discussions are then given also by

- Dove, sections 9.1–9.2
- Ashcroft and Mermin, chapter 23
- Hook and Hall, section 2.6
- Kittel, beginning of chapter 5

2.3 Appendix to this Chapter: $\zeta(4)$

The Riemann zeta function is defined as[24]

$$\zeta(p) = \sum_{n=1}^{\infty} n^{-p}.$$

[24] One of the most important unproven conjectures in all of mathematics is known as the Riemann hypothesis and is concerned with determining for which values of p does $\zeta(p) = 0$. The hypothesis was written down in 1869 by Bernard Riemann (the same guy who invented Riemannian geometry, crucial to general relativity) and has defied proof ever since. The Clay Mathematics Institute has offered one million dollars for a successful proof.

This function occurs frequently in physics, not only in the Debye theory of solids, but also in the Sommerfeld theory of electrons in metals (see Chapter 4), as well as in the study of Bose condensation.

In this appendix we are concerned with the value of $\zeta(4)$. To evaluate this we write a Fourier series for the function x^2 on the interval $[-\pi, \pi]$. The series is given by

$$x^2 = \frac{a_0}{2} + \sum_{n>0} a_n \cos(nx) \tag{2.9}$$

with coefficients given by

$$a_n = \frac{1}{\pi} \int_{-\pi}^{\pi} dx\, x^2 \cos(nx).$$

These can be calculated straightforwardly to give

$$a_n = \begin{cases} 2\pi^2/3 & n = 0 \\ 4(-1)^n/n^2 & n > 0. \end{cases}$$

We now calculate an integral in two different ways. First we can directly evaluate

$$\int_{-\pi}^{\pi} dx (x^2)^2 = \frac{2\pi^5}{5}.$$

On the other hand, using the Fourier decomposition of x^2 (Eq. 2.9) we can write the same integral as

$$\begin{aligned}\int_{-\pi}^{\pi} dx (x^2)^2 &= \int_{-\pi}^{\pi} dx \left(\frac{a_0}{2} + \sum_{n>0} a_n \cos(nx)\right)\left(\frac{a_0}{2} + \sum_{m>0} a_m \cos(mx)\right) \\ &= \int_{-\pi}^{\pi} dx \left(\frac{a_0}{2}\right)^2 + \int_{-\pi}^{\pi} dx \sum_{n>0} (a_n \cos(nx))^2\end{aligned}$$

where we have used the orthogonality of Fourier modes to eliminate cross terms in the product. We can do these integrals to obtain

$$\int_{-\pi}^{\pi} dx (x^2)^2 = \pi\left(\frac{a_0^2}{2} + \sum_{n>0} a_n^2\right) = \frac{2\pi^5}{9} + 16\pi\zeta(4).$$

Setting this expression to $2\pi^5/5$ gives us the result $\zeta(4) = \pi^4/90$.

Exercises

(2.1) **Einstein Solid**

(a) *Classical Einstein (or "Boltzmann") Solid:*
Consider a three-dimensional simple harmonic oscillator with mass m and spring constant k (i.e., the mass is attracted to the origin with the same spring constant in all three directions). The Hamiltonian is given in the usual way by

$$H = \frac{\mathbf{p}^2}{2m} + \frac{k}{2}\mathbf{x}^2.$$

▷ Calculate the classical partition function

$$Z = \int \frac{\mathbf{dp}}{(2\pi\hbar)^3} \int \mathbf{dx}\, e^{-\beta H(\mathbf{p},\mathbf{x})} \,.$$

Note: in this exercise $\mathbf{p}$ and $\mathbf{x}$ are three-dimensional vectors.

▷ Using the partition function, calculate the heat capacity $3k_B$.

▷ Conclude that if you can consider a solid to consist of N atoms all in harmonic wells, then the heat capacity should be $3Nk_B = 3R$, in agreement with the law of Dulong and Petit.

(b) *Quantum Einstein Solid:*
Now consider the same Hamiltonian quantum-mechanically.

▷ Calculate the quantum partition function

$$Z = \sum_j e^{-\beta E_j}$$

where the sum over j is a sum over all eigenstates.

▷ Explain the relationship with Bose statistics.

▷ Find an expression for the heat capacity.

▷ Show that the high-temperature limit agrees with the law of Dulong and Petit.

▷ Sketch the heat capacity as a function of temperature.

(See also Exercise 2.7 for more on the same topic)

(2.2) **Debye Theory I**

(a)‡ State the assumptions of the Debye model of heat capacity of a solid.

▷ Derive the Debye heat capacity as a function of temperature (you will have to leave the final result in terms of an integral that cannot be done analytically).

▷ From the final result, obtain the high- and low-temperature limits of the heat capacity analytically.

You may find the following integral to be useful

$$\int_0^\infty dx \frac{x^3}{e^x - 1} = \sum_{n=1}^{\infty} \int_0^\infty x^3 e^{-nx} = 6 \sum_{n=1}^{\infty} \frac{1}{n^4} = \frac{\pi^4}{15}.$$

By integrating by parts this can also be written as

$$\int_0^\infty dx \frac{x^4 e^x}{(e^x - 1)^2} = \frac{4\pi^4}{15}.$$

(b) The following table gives the heat capacity C for potassium iodide as a function of temperature.

T(K)	C(J K^{-1}mol^{-1})
0.1	8.5×10^{-7}
1.0	8.6×10^{-4}
5	.12
8	.59
10	1.1
15	2.8
20	6.3

▷ Discuss, with reference to the Debye theory, and make an estimate of the Debye temperature.

(2.3) **Debye Theory II**

Use the Debye approximation to determine the heat capacity of a two-dimensional solid as a function of temperature.

▷ State your assumptions.

You will need to leave your answer in terms of an integral that one cannot do analytically.

▷ At high T, show the heat capacity goes to a constant and find that constant.

▷ At low T, show that $C_v = KT^n$ Find n. Find K in terms of a definite integral.

If you are brave you can try to evaluate the integral, but you will need to leave your result in terms of the Riemann zeta function.

(2.4) **Debye Theory III**

Physicists should be good at making educated guesses. Guess the element with the highest Debye temperature. The lowest? You might not guess the ones with the absolutely highest or lowest temperatures, but you should be able to get close.

(2.5) **Debye Theory IV**

From Fig. 2.3 estimate the Debye temperature of diamond. Why does it not quite match the result listed in Table 2.2?

(2.6) **Debye Theory V***

In the text we derived the low-temperature Debye heat capacity assuming that the longitudinal and transverse sound velocities are the same and also that the sound velocity is independent of the direction the sound wave is propagating.

(a) Suppose the transverse velocity is v_t and the longitudinal velocity is v_l. How does this change the Debye result? State any assumptions you make.

(b) Instead suppose the velocity is anisotropic. For example, suppose in the $\hat{x}, \hat{y}$ and $\hat{z}$ direction, the sound velocity is v_x, v_y and v_z respectively. How might this change the Debye result?

(2.7) **Diatomic Einstein Solid***

Having studied Exercise 2.1, consider now a solid made up of diatomic molecules. We can (very crudely) model this as two particles in three dimensions, connected to each other with a spring, both in the bottom of a harmonic well.

$$H = \frac{\mathbf{p_1}^2}{2m_1} + \frac{\mathbf{p_2}^2}{2m_2} + \frac{k}{2}\mathbf{x_1}^2 + \frac{k}{2}\mathbf{x_2}^2 + \frac{K}{2}(\mathbf{x_1} - \mathbf{x_2})^2$$

where k is the spring constant holding both particles in the bottom of the well, and K is the spring constant holding the two particles together. Assume that the two particles are distinguishable atoms.

(If you find this exercise difficult, for simplicity you may assume that $m_1 = m_2$.)

(a) Analogous to Exercise 2.1, calculate the classical partition function and show that the heat capacity is again $3k_B$ per particle (i.e., $6k_B$ total).

(b) Analogous to Exercise 2.1, calculate the quantum partition function and find an expression for the heat capacity. Sketch the heat capacity as a function of temperature if $K \gg k$.

(c)** How does the result change if the atoms are indistinguishable?

(2.8) **Einstein versus Debye***

In both the Einstein model and the Debye model the high-temperature heat capacity is of the form

$$C = 3Nk_B(1 - \kappa/T^2 + \ldots).$$

▷ For the Einstein model calculate κ in terms of the Einstein temperature.

▷ For the Debye model calculate κ in terms of the Debye temperature.

From your results give an approximate ratio $T_{Einstein}/T_{Debye}$. Compare your result to the values for silver given in Fig. 2.4. (The ratio you calculate should be close to the ratio stated in the caption of the figure. It is not exactly the same though. Why might it not be?)

(2.9) **Einstein and Debye*&**

There are certain materials, usually made of several different types of atoms, which have some contribution to the heat capacity which is Debye-like, coming from acoustic waves with velocity v, and also have some contribution to the heat capacity, which is Einstein-like, coming from localized vibrational excitations of fixed frequency ω. (We will understand this situation better in chapter 10.) Suppose the high temperature heat capacities of the two contributions are equal. Derive the heat capacity of such a system as a function of temperature. Be very careful to keep track of the total number of degrees of freedom in the system.

(2.10) **Lindemann Criterion for Melting&**

Fredrick Lindemann, the First Viscount Cherwell of Oxford, proposed that a solid should melt when the root mean square of the vibrational displacement of the atoms is roughly .07 of the distance between neighboring atoms (the exact number depending on the particular microscopic structure of the material). Given that the distance between neighboring atoms in diamond is roughly 1.5 Angstroms, the atomic weight is 12, and the Einstein temperature is 1320K, estimate the melting temperature. The actual melting point is about 3800K.

Electrons in Metals: Drude Theory

3

Even in ancient times it was understood that certain substances (now known as metals) were somehow different from other materials in the world.[1] The defining characteristic of a metal is that it conducts electricity. At some level the reason for this conduction boils down to the fact that electrons are mobile in these materials. In later chapters we will be concerned with the question of why electrons are mobile in some materials but not in others, being that all materials have electrons in them! For now, we take as given that there are mobile electrons and we would like to understand their properties.

J.J. Thomson's 1896 discovery of the electron ("corpuscles of charge" that could be pulled out of metal) raised the question of how these charge carriers might move within the metal. In 1900 Paul Drude[2] realized that he could apply Boltzmann's kinetic theory of gases to understanding electron motion within metals. This theory was remarkably successful, providing a first understanding of metallic conduction.[3]

Having studied the kinetic theory of gases in previous courses, Drude theory should be very easy to understand. We will make three assumptions about the motion of electrons

(1) Electrons have a scattering[4] time τ. The probability of scattering within a time interval dt is dt/τ.

(2) Once a scattering event occurs, we assume the electron returns to momentum $\mathbf{p} = 0$.

(3) In between scattering events, the electrons, which are charge $-e$ particles, respond to the externally applied electric field $\mathbf{E}$ and magnetic field $\mathbf{B}$.

The first two of these assumptions are exactly those made in the kinetic theory of gases.[5] The third assumption is just a logical generalization to account for the fact that, unlike gas molecules, electrons are charged and must therefore respond to electromagnetic fields.

[1] Human mastery of metals such as copper (around 8000 BC), bronze (around 3300 BC), and iron (around 1200 BC), completely changed agriculture, weaponry, and pretty much every other aspect of life.

[2] Pronounced roughly "Drood-a".

[3] Sadly, neither Boltzmann nor Drude lived to see how much influence this theory really had—in unrelated tragic events, both of them committed suicide in 1906. Boltzmann's famous student, Ehrenfest, also committed suicide some years later. Why so many highly successful statistical physicists took their own lives is a bit of a mystery.

[4] In the kinetic theory of gas, one can estimate the scattering time based on the velocity, density, and scattering cross-section of the molecules of the gas. In Drude theory, estimates of τ are far more difficult for several reasons. First, the electrons interact via long range Coulomb interaction, so it is hard to define a cross-section. Secondly, there are many things in a solid that an electron can hit besides other electrons. As such, we will simply treat τ as a phenomenological parameter.

[5] Ideally we would do a better job with our representation of the scattering of particles. Every collision should consider two particles having initial momenta $\mathbf{p}_1^{initial}$ and $\mathbf{p}_2^{initial}$ and then scattering to final momenta $\mathbf{p}_1^{final}$ and $\mathbf{p}_2^{final}$ so as to conserve both energy and momentum. Unfortunately, keeping track of things so carefully makes the problem extremely difficult to solve. Assumption 1 is not so crazy as an approximation being that there really is a typical time between scattering events in a gas. Assumption 2 is a bit more questionable, but *on average* the final momentum after a scattering event is indeed zero (if you average momentum as a vector). However, obviously it is not correct that every particle has zero kinetic energy after a scattering event. This is a defect of the approach.

We consider an electron with momentum $\mathbf{p}$ at time t and ask what momentum it will have at time $t+dt$. There are two terms in the answer. There is a probability dt/τ that it will scatter to momentum zero. If it does not scatter to momentum zero (with probability $1-dt/\tau$) it simply accelerates as dictated by its usual equations of motion $d\mathbf{p}/dt = \mathbf{F}$. Putting the two terms together we have

$$\langle \mathbf{p}(t+dt)\rangle = \left(1 - \frac{dt}{\tau}\right)(\mathbf{p}(t) + \mathbf{F}dt) + \mathbf{0}\, dt/\tau$$

or keeping terms only to linear order in dt then rearranging,[6]

$$\frac{d\mathbf{p}}{dt} = \mathbf{F} - \frac{\mathbf{p}}{\tau} \tag{3.1}$$

where here the force $\mathbf{F}$ on the electron is just the Lorentz force

$$\mathbf{F} = -e(\mathbf{E} + \mathbf{v} \times \mathbf{B}).$$

One can think of the scattering term $-\mathbf{p}/\tau$ as just a drag force on the electron. Note that in the absence of any externally applied field the solution to this differential equation is just an exponentially decaying momentum

$$\mathbf{p}(t) = \mathbf{p}_{initial}\; e^{-t/\tau}$$

which is what we should expect for particles that lose momentum by scattering.

[6] Here we really mean the thermal average $\langle \mathbf{p}\rangle$ when we write $\mathbf{p}$. Since our scattering is probabilistic, we should view all quantities (such as the momentum) as being an expectation over these random events. A more detailed theory would keep track of the entire distribution of momenta rather than just the average momentum. Keeping track of distributions in this way leads one to the Boltzmann Transport Equation, which we will not discuss.

3.1 Electrons in Fields

3.1.1 Electrons in an Electric Field

Let us start by considering the case where the electric field is non-zero but the magnetic field is zero. Our equation of motion is then

$$\frac{d\mathbf{p}}{dt} = -e\mathbf{E} - \frac{\mathbf{p}}{\tau}.$$

In steady state, $d\mathbf{p}/dt = 0$ so we have

$$m\mathbf{v} = \mathbf{p} = -e\tau\mathbf{E}$$

with m the mass of the electron and $\mathbf{v}$ its velocity.

Now, if there is a density n of electrons in the metal each with charge $-e$, and they are all moving at velocity $\mathbf{v}$, then the electrical current is given by

$$\mathbf{j} = -en\mathbf{v} = \frac{e^2\tau n}{m}\mathbf{E}$$

or in other words, the conductivity of the metal, defined via $\mathbf{j} = \sigma\mathbf{E}$ is given by[7]

$$\sigma = \frac{e^2\tau n}{m}. \tag{3.2}$$

By measuring the conductivity of the metal (assuming we know both the charge and mass of the electron) we can determine the product of the electron density and scattering time of the electron.

[7] A related quantity is the *mobility*, defined by $\mathbf{v} = \mu\mathbf{E}$, which is given in Drude theory by $\mu = |e|\tau/m$. We will discuss mobility further in Section 17.1.1.

3.1.2 Electrons in Electric and Magnetic Fields

Let us continue on to see what other predictions come from Drude theory. Consider the transport equation (Eq. 3.1) for a system in both an electric and a magnetic field. We now have

$$\frac{d\mathbf{p}}{dt} = -e(\mathbf{E} + \mathbf{v} \times \mathbf{B}) - \mathbf{p}/\tau.$$

Again setting this to zero in steady state, and using $\mathbf{p} = m\mathbf{v}$ and $\mathbf{j} = -ne\mathbf{v}$, we obtain an equation for the steady state current

$$0 = -e\mathbf{E} + \frac{\mathbf{j} \times \mathbf{B}}{n} + \frac{m}{ne\tau}\mathbf{j}$$

or

$$\mathbf{E} = \left(\frac{1}{ne}\mathbf{j} \times \mathbf{B} + \frac{m}{ne^2\tau}\mathbf{j}\right).$$

We now define the 3 by 3 resistivity matrix $\underset{\sim}{\rho}$ which relates the current vector to the electric field vector

$$\mathbf{E} = \underset{\sim}{\rho}\mathbf{j}$$

such that the components of this matrix are given by

$$\rho_{xx} = \rho_{yy} = \rho_{zz} = \frac{m}{ne^2\tau}$$

and if we imagine $\mathbf{B}$ oriented in the $\hat{z}$ direction, then

$$\rho_{xy} = -\rho_{yx} = \frac{B}{ne}$$

and all other components of $\underset{\sim}{\rho}$ are zero. This off-diagonal term in the resistivity is known as the *Hall resistivity*, named after Edwin Hall who discovered in 1879 that when a magnetic field is applied perpendicular to a current flow, a voltage can be measured perpendicular to both current and magnetic field (see Fig. 3.1). If you are adventurous you might consider a further generalization of Drude theory to finite frequency conductivity, where it gives some interesting (and frequently accurate) predictions (see Exercise 3.1.e).

The Hall coefficient R_H is defined as

$$R_H = \frac{\rho_{yx}}{|B|}$$

which in the Drude theory is given by

$$R_H = \frac{-1}{ne}.$$

This then allows us to measure the density of electrons in a metal.

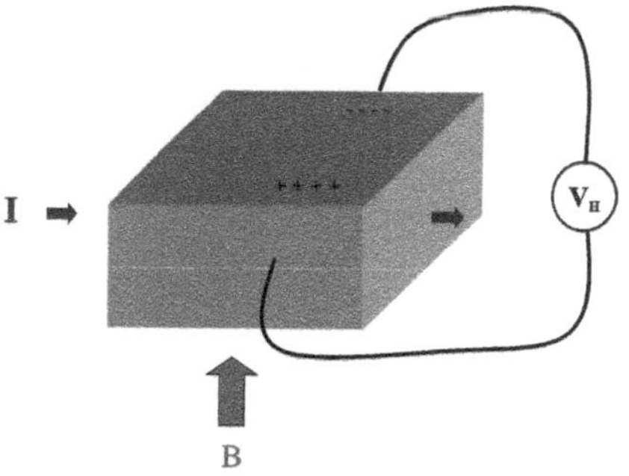

Fig. 3.1 Edwin Hall's 1879 experiment. The voltage measured perpendicular to both the magnetic field and the current is known as the Hall voltage which is proportional to B and inversely proportional to the electron density (at least in Drude theory).

Aside: One can also consider turning this experiment on its head. If you know the density of electrons in your sample you can use a Hall measurement to determine the magnetic field. This is known as a Hall sensor. Since it is hard to measure small voltages, Hall sensors typically use materials, such as semiconductors, where the density of electrons is low so R_H and hence the resulting voltage is large.

Let us then calculate $n = -1/(eR_H)$ for various metals and divide it by the density of atoms n_{atomic} (see Table 3.1). This should give us the number of free electrons per atom. Later on we will see that it is frequently not so hard to estimate the number of electrons in a system. A short description is that electrons bound in the core shells of the atoms are never free to travel throughout the crystal, whereas the electrons in the outer shell may be free (we will discuss in Chapter 16 when these electrons are free and when they are not). The number of electrons in the outermost shell is known as the *valence* of the atom.

Table 3.1 Comparison of the valence of various atoms to the valence predicted from the measured Hall coefficient.

Material	$\frac{1}{-e\,R_H\,n_{atomic}}$	Valence
Li	.8	1
Na	1.2	1
K	1.1	1
Cu	1.5	1
Be	-0.2*	2
Mg	-0.4	2
Ca	1.5	2

Here n_{atomic} is the density of atoms in the metal and R_H is the measured Hall coefficient. In Drude theory, the middle column should give the number of electrons per atom, i.e., the valence. For monovalent atoms, the agreement is fairly good. But for divalent atoms, the sign can even come out wrong! The * next to Be indicates that its Hall coefficient is anisotropic. Depending on which angle you run the current you can get either sign of the Hall coefficient!

We see from Table 3.1 that for many metals this Drude theory analysis seems to make sense—the "valence" of lithium, sodium, and potassium (Li, Na, and K) are all one, which agrees roughly with the measured number of electrons per atom. The effective valence of copper (Cu) is also one, so it is not surprising either. However, something has clearly gone seriously wrong for Be and Mg. In this case, the sign of the Hall coefficient has come out incorrect. From this result, one might conclude that the charge carrier for beryllium and magnesium (Be and Mg) have the opposite charge from that of the electron! We will see in Section 17.1.1 that this is indeed true and is a result of the so-called band structure of these materials. However, for many metals, simple Drude theory gives quite reasonable results. We will see in Chapter 17 that Drude theory is particularly good for describing semiconductors.

If we believe the Hall effect measurement of the density of electrons in metals, using Eq. 3.2 we can then extract a scattering time from the expression for the conductivity. The Drude scattering time comes out to be in the range of $\tau \approx 10^{-14}$ seconds for most metals near room temperature.

3.2 Thermal Transport

Drude was brave enough to attempt to further calculate the thermal conductivity κ due to mobile electrons[8] using Boltzmann's kinetic theory. Without rehashing the derivation, this result should look familiar to you from your previous encounters with the kinetic theory of gas:[9]

$$\kappa = \frac{1}{3} n c_v \langle v \rangle \lambda \tag{3.3}$$

where c_v is the heat capacity per particle, $\langle v \rangle$ is the average thermal velocity and $\lambda = \langle v \rangle \tau$ is the scattering length. For a conventional (monatomic) gas the heat capacity per particle is

$$c_v = \frac{3}{2} k_B$$

[8] In any experiment there will also be some amount of thermal conductivity from structural vibrations of the material as well—so-called phonon thermal conductivity (we will meet phonons in Chapter 9). However, for most metals, the thermal conductivity is mainly due to electron motion and not from vibrations.

[9] The thermal conductivity κ is defined by $\mathbf{j}_q = -\kappa \nabla T$ where $\mathbf{j}_q$ is the heat current density. The rough intuition for Eq. 3.3 is that a density n of electrons each carries an amount of heat $c_v T$ at a velocity $\langle v \rangle$ for a distance λ before scattering.

and

$$\langle v \rangle = \sqrt{\frac{8k_BT}{\pi m}} \, . \tag{3.4}$$

Assuming this all holds true for electrons, we obtain

$$\kappa = \frac{4}{\pi} \frac{n\tau k_B^2 T}{m} \, .$$

While this quantity still has the unknown parameter τ in it, it is the same quantity that occurs in the electrical conductivity (Eq. 3.2). Thus we may look at the ratio of thermal conductivity to electrical conductivity, known as the Lorenz number[10,11]

$$L = \frac{\kappa}{T\sigma} = \frac{4}{\pi} \left(\frac{k_B}{e}\right)^2 \approx 0.94 \times 10^{-8} \text{ WattOhm/K}^2 \, .$$

A slightly different prediction is obtained by realizing that we have used $\langle v \rangle^2$ in our calculation, whereas perhaps we might have instead used $\langle v^2 \rangle$ which would have then given us[12]

$$L = \frac{\kappa}{T\sigma} = \frac{3}{2} \left(\frac{k_B}{e}\right)^2 \approx 1.11 \times 10^{-8} \text{ WattOhm/K}^2 \, .$$

This result was viewed as a huge success, being that it was known for almost half a century that almost all metals have roughly the same value of this ratio—a fact known as the Wiedemann–Franz law. In fact the value predicted for this ratio is fairly close to that measured experimentally (see Table 3.2). The result appears to be off by about a factor of 2, but still that is very good, considering that before Drude no one had any idea why this ratio should be a constant at all!

In retrospect we now realize that this calculation is completely incorrect (despite its successful result). The reason we know there is a problem is because we do not actually measure a specific heat of $c_v = \frac{3}{2}k_B$ per electron in metals (for certain systems where the density of electrons is very low, we do in fact measure this much specific heat, but not in metals). In fact, in most metals we measure only a vibrational (Debye) specific heat, plus a very small term linear in T at low temperatures (see Fig. 2.5). So why does this calculation give such a good result? It turns out (and we will see in Chapter 4) that we have made two mistakes that roughly cancel each other. We have used a specific heat that is way too large, but we have also used a velocity that is way too small. We will see later that both of these mistakes are due to Fermi statistics of the electron (which we have so far ignored) and the Pauli exclusion principle.

We can see the problem much more clearly in some other quantities. The so-called *Peltier effect* is the fact that running electrical current through a material also transports heat. The Peltier coefficient Π is defined by

$$\mathbf{j}^q = \Pi \, \mathbf{j}$$

where $\mathbf{j}^q$ is the heat current density, and $\mathbf{j}$ is the electrical current density.

[10] This is named after Ludvig Lorenz, not Hendrik Lorentz who is famous for the Lorentz force and Lorentz contraction. However, just to confuse matters, the two of them worked on similar topics and there is even a Lorentz–Lorenz equation

[11] The dimensions here might look a bit funny, but κ, the thermal conductivity, is measured in Watt/(K-m) and σ is measured in 1/(Ohm-m). To see that WattOhm/K^2 is the same as $(k_B/e)^2$ note that k_B is J/K and e is Coulomb (C). So we need to show that $(\text{J/C})^2$ is WattOhm $(\text{J/C})^2 = (\text{J/sec})(\text{J/C})(1/(\text{C/sec}) =$ WattVolt/Amp = WattOhm.

[12] In kinetic theory $c_v T = \frac{1}{2} m \langle v^2 \rangle$.

Table 3.2 Lorenz numbers $\kappa/(T\sigma)$ for various metals in units of 10^{-8} WattOhm/K^2

Material	L
Lithium (Li)	2.22
Sodium (Na)	2.12
Copper (Cu)	2.20
Iron (Fe)	2.61
Bismuth (Bi)	3.53
Magnesium (Mg)	2.14

The prediction of Drude theory is that the Lorentz number should be on the order of 1×10^{-8} WattOhm/K^2.

Aside: The Peltier effect is used for thermoelectric refrigeration devices. Running electricity through a thermoelectric material forces heat to be transported through that material. You can thus transport heat away from one object and towards another. A good thermoelectric device has a high Peltier coefficient, but must also have a low resistivity, because running a current through a material with resistivity R will result in power $I^2 R$ being dissipated, thus heating it up.

In kinetic theory the thermal current is

$$\mathbf{j}^q = \frac{1}{3}(c_v T) n \mathbf{v} \tag{3.5}$$

where $c_v T$ is the heat carried by one particle (with $c_v = 3k_B/2$ the heat capacity per particle) and n is the density of particles (and 1/3 is a geometric factor that is approximate anyway). Similarly, the electrical current is

$$\mathbf{j} = -en\mathbf{v}.$$

Thus the Peltier coefficient is

$$\Pi = \frac{-c_v T}{3e} = \frac{-k_B T}{2e} \tag{3.6}$$

so the ratio (known as thermopower, or Seebeck coefficient) $S = \Pi/T$ is given by

$$S = \frac{\Pi}{T} = \frac{-k_B}{2e} = -0.43 \times 10^{-4}\text{V/K} \tag{3.7}$$

in Drude theory. For most metals the actual value of this ratio is roughly 100 times smaller! (See Table 3.3.) This is a reflection of the fact that we have used $c_v = 3k_B/2$, whereas the actual specific heat per particle is much much lower (which we will understand in the next chapter when we consider Fermi statistics more carefully). Further (analogous to the Hall coefficient), for certain metals the sign of the Seebeck coefficient is predicted incorrectly as well.

Table 3.3 Seebeck coefficients of various metals at room temperature, in units of 10^{-6} V/K

Material	S
Sodium (Na)	-5
Potassium (K)	-12.5
Copper (Cu)	1.8
Beryllium (Be)	1.5
Aluminum (Al)	-1.8

Note that the magnitude of the Seebeck coefficient is roughly one hundredth of the value predicted by Drude theory in Eqn. 3.7. For Cu and Be, the sign comes out wrong as well!

Chapter Summary

- Drude theory is based on the kinetic theory of gases.
- Assumes some scattering time τ, resulting in a conductivity $\sigma = ne^2\tau/m$.
- Hall coefficient measures density of electrons.
- Successes of Drude theory:
  - Wiedemann–Franz ratio $\kappa/(\sigma T)$ comes out close to right for most materials
  - many other transport properties predicted correctly (for example, conductivity at finite frequency)
  - Hall coefficient measurement of the density seems reasonable for many metals.

- Failures of Drude theory:
  - Hall coefficient frequently is measured to have the wrong sign, indicating a charge carrier with charge opposite to that of the electron
  - there is no $3k_B/2$ heat capacity per particle measured for electrons in metals. This then makes the Peltier and Seebeck coefficients come out wrong by a factor of 100.

The latter of the two shortcomings will be addressed in the next chapter, whereas the former of the two will be addressed in Chapter 17, where we discuss band theory.

Despite the shortcomings of Drude theory, it nonetheless was the only theory of metallic conductivity for a quarter of a century (until the Sommerfeld theory improved it), and it remains quite useful today—particularly for semiconductors and other systems with low densities of electrons (see Chapter 17).

References

Many books cover Drude theory at some level:

- Ashcroft and Mermin, chapter 1
- Burns, chapter 9 part A
- Singleton, sections 1.1–1.4
- Hook and Hall, section 3.3, sort-of

Hook and Hall aim mainly at free electron (Sommerfeld) theory (our next chapter), but they end up doing Drude theory anyway (they don't use the word "Drude").

Exercises

(3.1) **Drude Theory of Transport in Metals**

(a)‡ Assume a scattering time τ and use Drude theory to derive an expression for the conductivity of a metal.

(b) Define the resistivity matrix $\underset{\sim}{\rho}$ as $\mathbf{E} = \underset{\sim}{\rho}\mathbf{j}$. Use Drude theory to derive an expression for the matrix $\underset{\sim}{\rho}$ for a metal in a magnetic field. (You may assume $\mathbf{B}$ parallel to the $\hat{z}$ axis. The under-tilde means that the quantity $\underset{\sim}{\rho}$ is a matrix.) Invert this matrix to obtain an expression for the conductivity matrix $\underset{\sim}{\sigma}$.

(c) Define the Hall coefficient.

▷ Estimate the magnitude of the Hall voltage for a specimen of sodium in the form of a rod of rectangular cross-section 5mm by 5mm carrying a current of 1A down its long axis in a magnetic field of 1T perpendicular to the long axis. The density of sodium atoms is roughly 1 gram/cm^3, and sodium has atomic mass of roughly 23. You may assume that there is one free electron per sodium atom (sodium has *valence* 1).

▷ What practical difficulties would there be in measuring the Hall voltage and resistivity of such a specimen. How might these difficulties be addressed).

(d) What properties of metals does Drude theory not explain well?

(e)* Consider now an applied AC field $\mathbf{E} \sim e^{i\omega t}$ which induces an AC current $\mathbf{j} \sim e^{i\omega t}$. Modify the above calculation (in the presence of a magnetic field) to obtain an expression for the complex AC conductivity matrix $\underline{\sigma}(\omega)$. For simplicity in this case you may assume that the metal is very clean, meaning that $\tau \to \infty$, and you may assume that $\mathbf{E} \perp \mathbf{B}$. You might again find it convenient to assume $\mathbf{B}$ parallel to the $\hat{z}$ axis. (This exercise might look hard, but if you think about it for a bit, it isn't really much harder than what you did above!)

▷ At what frequency is there a divergence in the conductivity? What does this divergence mean? (When τ is finite, the divergence is cut off.)

▷ Explain how could one use this divergence (known as the cyclotron resonance) to measure the mass of the electron. (In fact, in real metals, the measured mass of the electron is generally not equal to the well-known value $m_e = 9.1095 \times 10^{-31}$ kg. This is a result of *band structure* in metals, which we will explain in Part VI.)

(3.2) **Scattering Times**

The following table gives electrical resistivities ρ, densities n, and atomic weights w for the metals silver and lithium:

	ρ (Ωm)	n (g/cm^3)	w
Ag	1.59×10^{-8}	10.5	107.8
Li	9.28×10^{-8}	0.53	6.94

▷ Given that both Ag and Li are monovalent (i.e., have one free electron per atom), calculate the Drude scattering times for electrons in these two metals.

In the kinetic theory of gas, one can estimate the scattering time using the equation

$$\tau = \frac{1}{n\langle v\rangle\sigma}$$

where n is the gas density, $\langle v\rangle$ is the average velocity (see Eq. 3.4), and σ is the cross-section of the gas molecule—which is roughly πd^2 with d the molecule diameter. For a nitrogen molecule at room temperature, we can use $d = .37$nm.

▷ Calculate the scattering time for nitrogen gas at room temperature and pressure and compare your result to the Drude scattering times for electrons in Ag and Li metals.

(3.3) **Ionic Conduction and Two Carrier Types**

In certain materials, particularly at higher temperature, positive ions can move throughout the sample in response to applied electric fields, resulting in what is known as ionic conduction. Since this conduction is typically poor, it is mainly observable in materials where there are no free electrons that would transport current. However, occasionally it can occur that a material has both electrical conduction and ionic conduction of roughly the same magnitude—such materials are known as mixed ion–electron conductors.

Suppose free electrons have density n_e and scattering time τ_e (and have the usual electron mass m_e and charge $-e$). Suppose that the free ions have density n_i, scattering time τ_i, mass m_i and charge $+e$. Using Drude theory,

(a) Calculate the electrical resistivity.

(b) Calculate the thermal conductivity.

(c)* Calculate the Hall resistivity.

(3.4) **Plasma Oscillation*&**

Consider a slab of metal of thickness d in the $\hat{x}$ direction (and arbitrary area perpendicular to this). If the electron density in the metal is displaced in the $+\hat{x}$ direction, charge builds up on the boundary of the slab, and an electric field results in the slab (like in a plate capacitor). The electrons in the metal respond to the electric field and are pushed back to their original position. This restoring force (like a Hooke's law spring) results in oscillations of electron density, known as a plasma oscillation.

(a)* Assume the metal is very clean. Use the finite frequency Drude conductivity in zero magnetic field (see Exercise 3.1.e with B set to zero) and calculate the plasma frequency of the metal.

(b)** Consider the case where the scattering time τ is not infinite. What happens to the plasma frequency? How do you interpret this?

(c)** Set the scattering time to ∞ again, but let the magnetic field be nonzero. What happens to the plasma frequency now?

More Electrons in Metals: Sommerfeld (Free Electron) Theory

4

In 1925 Pauli discovered the exclusion principle, that no two electrons may be in the exact same state. In 1926, Fermi and Dirac separately derived what we now call Fermi–Dirac statistics.[1] Upon learning about these developments, Sommerfeld[2] realized that Drude's theory of metals could easily be generalized to incorporate Fermi statistics, which is what we shall presently do.

4.1 Basic Fermi–Dirac Statistics

Given a system of free[3] electrons with chemical potential[4] μ the probability of an eigenstate of energy E being occupied[5] is given by the Fermi factor (See Fig. 4.1)

$$n_F(\beta(E-\mu)) = \frac{1}{e^{\beta(E-\mu)}+1}. \tag{4.1}$$

At low temperature the Fermi function becomes a step function (states below the chemical potential are filled, those above the chemical potential are empty), whereas at higher temperatures the step function becomes more smeared out.

We will consider the electrons to be in a box of size $V = L^3$ and, as with our discussion in Section 2.2.1, it is easiest to imagine that

[1] Fermi–Dirac statistics were actually derived first by Pascual Jordan in 1925. Unfortunately, the referee of the manuscript, Max Born, misplaced it and it never got published. Many people believe that were it not for the fact that Jordan later joined the Nazi party, he might have won the Nobel Prize along with Born and Walther Bothe.

[2] Sommerfeld never won a Nobel Prize, although he was nominated for it 81 times—more than any other physicist. He was also a research advisor for more Nobel laureates than anyone else in history, including Heisenberg, Pauli, Debye, Bethe, Pauling, and Rabi.

[3] Here "free" means that they do not interact with each other, with the background crystal lattice, with impurities, or with anything else for that matter.

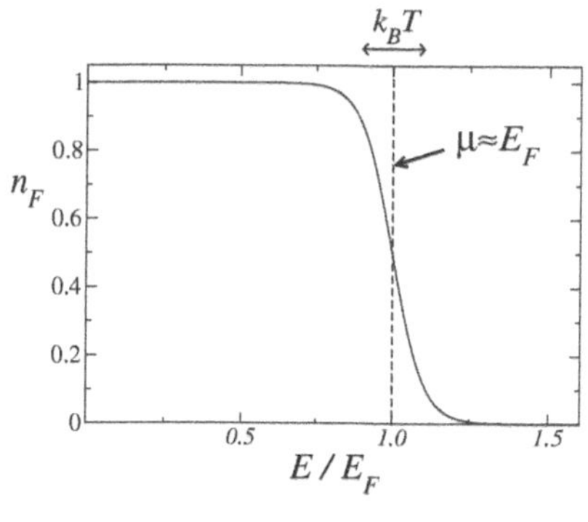

Fig. 4.1 The Fermi distribution for $k_BT \ll E_F$. The dashed line marks the chemical potential μ, which is approximately E_F. At $T = 0$ the distribution is a step, but for finite T it gets smeared over a range of energies of width a few times k_BT.

[4] In case you did not properly learn about chemical potential in your statistical physics course, it can be *defined* via Eq. 4.1, by saying that μ is whatever constant needs to be inserted into this equation to make it true. It can also be defined as an appropriate thermodynamical derivative such as $\mu = \partial U/\partial N|_{V,S}$ with U the total energy and N the number of particles or $\mu = \partial G/\partial N|_{T,P}$, with G the Gibbs potential. However, such a definition can be tricky if one worries about the discreteness of the particle number—since N must be an integer, the derivative may not be well defined. As a result the definition in terms of Eq. 4.1 is frequently best (i.e., we are treating μ as a Lagrange multiplier).

[5] When we say that there are a particular set of N orbitals occupied by electrons, we really mean that the overall wavefunction of the system is an antisymmetric function which can be expressed as a Slater determinant of N single electron wavefunctions. We will never need to actually write out such Slater determinant wavefunctions except in Section 23.3, which is somewhat more advanced material.

the box has periodic boundary conditions.[6] The plane wavefunctions are of the form $e^{i\mathbf{k}\cdot\mathbf{r}}$ where due to the boundary conditions $\mathbf{k}$ must take value $(2\pi/L)(n_1, n_2, n_3)$ with n_i integers. These plane waves have corresponding energies

$$\epsilon(\mathbf{k}) = \frac{\hbar^2|\mathbf{k}|^2}{2m} \tag{4.2}$$

with m the electron mass. Thus the total number of electrons in the system is given by

$$N = 2\sum_{\mathbf{k}} n_F(\beta(\epsilon(\mathbf{k}) - \mu)) = 2\frac{V}{(2\pi)^3}\int \mathbf{dk}\; n_F(\beta(\epsilon(\mathbf{k}) - \mu)) \tag{4.3}$$

where the prefactor of 2 accounts for the two possible spin states for each possible wavevector $\mathbf{k}$. In fact, in a metal, N will usually be given to us, and this equation will define the chemical potential as a function of temperature.

We now define a useful concept:

Definition 4.1 *The* ***Fermi energy****, E_F is the chemical potential at temperature $T = 0$.*

This is also sometimes called the *Fermi level*. The states that are filled at $T = 0$ are sometimes called the *Fermi sea*. Frequently one also defines a *Fermi temperature* $T_F = E_F/k_B$, and also the *Fermi wavevector* k_F defined via

$$E_F = \frac{\hbar^2 k_F^2}{2m} \tag{4.4}$$

and correspondingly a *Fermi momentum* $p_F = \hbar k_F$ and a *Fermi velocity*[7]

$$v_F = \hbar k_F/m. \tag{4.5}$$

[6] As mentioned in Section 2.2.1, any properties of the bulk of the solid should be independent of the type of boundary conditions we choose. If you have doubts, you can try repeating all the calculations using hard wall boundary conditions, and you will find all the same results (It is more messy, but not too much harder!).

[7] Yes, Fermi got his name attached to many things. To help spread the credit around I've called this section "Basic Fermi–Dirac Statistics" instead of just "Basic Fermi Statistics".

Aside: Frequently people think of the Fermi energy as the energy of the most energetic occupied electron state in system. While this is correct in the case where you are filling a continuum of states, it can also lead you to errors in cases where the energy eigenstates are discrete (see the related footnote 4 of this chapter), or more specifically when there is a gap between the most energetic occupied electron state in the system, and the least energetic unoccupied electron state. More correctly the Fermi energy, i.e., the chemical potential at $T = 0$, will be half-way between the most energetic occupied electron state, and the least energetic unoccupied electron state (see Exercise 4.6).

Let us now calculate the Fermi energy in a (three-dimensional) metal with N electrons in it. At $T = 0$ the Fermi function (Eq. 4.1) becomes a step function (which we write as Θ. I.e., $\Theta(x) = 1$ for $x \geqslant 0$ and $\Theta(x) = 0$ for $x < 0$), so that Eq. 4.3 becomes

$$N = 2\frac{V}{(2\pi)^3}\int \mathbf{dk}\; \Theta(E_F - \epsilon(\mathbf{k})) = 2\frac{V}{(2\pi)^3}\int^{|k|<k_F} \mathbf{dk}.$$

The final integral here is just an integral over a ball of radius k_F. Thus the integral gives us the volume of this ball ($4\pi/3$ times the cube of the radius) yielding

$$N = 2\frac{V}{(2\pi)^3}\left(\frac{4}{3}\pi k_F^3\right). \tag{4.6}$$

In other words, at $T = 0$ the electrons simply fill a ball in k-space of radius k_F. The surface of this ball, a sphere (the "Fermi sphere") of radius k_F is known as the *Fermi surface*—a term more generally defined as the surface dividing filled from unfilled states at zero temperature.

Using the fact that the density is defined as $n = N/V$ we can rearrange Eq. 4.6 to give

$$k_F = (3\pi^2 n)^{1/3}$$

and correspondingly

$$E_F = \frac{\hbar^2(3\pi^2 n)^{2/3}}{2m}. \tag{4.7}$$

Since we know roughly how many free electrons there are in a metal (say, one per atom for monovalent metals such as sodium or copper), we can estimate the Fermi energy, which, say for copper, turns out to be on the order of 7 eV, corresponding to a Fermi temperature of about 80,000 K(!). This amazingly high energy scale is a result of Fermi statistics and the very high density of electrons in metals. It is crucial to remember that for all metals, $T_F \gg T$ for any temperature anywhere near room temperature. In fact metals melt (and even vaporize!) at temperatures far far below their Fermi temperatures.

Similarly, one can calculate the Fermi velocity, which, for a typical metal such as copper, may be as large as 1% the speed of light! Again, this enormous velocity stems from the Pauli exclusion principle—all the lower momentum states are simply filled, so if the density of electrons is very high, the velocities will be very high as well.

With a Fermi energy that is so large, and therefore a Fermi sea that is very deep, any (not insanely large) temperature can only make excitations of electrons that are already very close to the Fermi surface (i.e., they can jump from just below the Fermi surface to just above with only a small energy increase). The electrons deep within the Fermi sea, near $\mathbf{k} = \mathbf{0}$, cannot be moved by any reasonably low-energy perturbation simply because there are no available unfilled states for them to move into unless they absorb a very large amount of energy.

4.2 Electronic Heat Capacity

We now turn to examine the heat capacity of electrons in a metal. Analogous to Eq. 4.3, the total energy of our system of electrons is given now by

$$\begin{aligned} E_{total} &= \frac{2V}{(2\pi)^3}\int \mathbf{dk}\, \epsilon(\mathbf{k})\, n_F(\beta(\epsilon(\mathbf{k})-\mu)) \\ &= \frac{2V}{(2\pi)^3}\int_0^\infty 4\pi k^2 dk\, \epsilon(\mathbf{k})\, n_F(\beta(\epsilon(\mathbf{k})-\mu)) \end{aligned}$$

where the chemical potential is defined as above by

$$N = \frac{2V}{(2\pi)^3} \int \mathbf{dk}\, n_F(\beta(\epsilon(\mathbf{k}) - \mu)) = \frac{2V}{(2\pi)^3} \int_0^\infty 4\pi k^2 dk\; n_F(\beta(\epsilon(\mathbf{k}) - \mu)).$$

(In both equations we have changed to spherical coordinates to obtain a one-dimensional integral and a factor of $4\pi k^2$ out front.)

It is convenient to replace k in this equation by the energy ϵ by using Eq. 4.2 or equivalently

$$k = \sqrt{\frac{2\epsilon m}{\hbar^2}}$$

so that

$$dk = \sqrt{\frac{m}{2\epsilon\hbar^2}}\; d\epsilon.$$

We can then rewrite these expressions as

$$E_{total} = V \int_0^\infty d\epsilon\; \epsilon\; g(\epsilon)\; n_F(\beta(\epsilon - \mu)) \tag{4.8}$$

$$N = V \int_0^\infty d\epsilon\; g(\epsilon)\; n_F(\beta(\epsilon - \mu)) \tag{4.9}$$

where

$$\begin{aligned} g(\epsilon)d\epsilon &= \frac{2}{(2\pi)^3} 4\pi k^2 dk = \frac{2}{(2\pi)^3} 4\pi \left(\frac{2\epsilon m}{\hbar^2}\right) \sqrt{\frac{m}{2\epsilon\hbar^2}}\; d\epsilon \\ &= \frac{(2m)^{3/2}}{2\pi^2\hbar^3} \epsilon^{1/2} d\epsilon \end{aligned} \tag{4.10}$$

is the *density of states per unit volume*. The definition[8] of this quantity is such that $g(\epsilon)d\epsilon$ is the total number of eigenstates (including both spin states) with energies between ϵ and $\epsilon + d\epsilon$.

[8] Compare the physical meaning of this definition to that of the density of states for sound waves given in Eq. 2.5.

From Eq. 4.7 we can simply derive $(2m)^{3/2}/\hbar^3 = 3\pi^2 n/E_F^{3/2}$, thus we can simplify the density of states expression to

$$g(\epsilon) = \frac{3n}{2E_F}\left(\frac{\epsilon}{E_F}\right)^{1/2} \tag{4.11}$$

which is a fair bit simpler. Note that the density of states has dimensions of a density (an inverse volume) divided by an energy. It is clear that this is the dimensions it must have, given Eq. 4.9 for example.

Note that the expression Eq. 4.9 should be thought of as defining the chemical potential given the number of electrons in the system and the temperature. Once the chemical potential is fixed, then Eq. 4.8 gives us the total kinetic energy of the system. Differentiating that quantity would give us the heat capacity. Unfortunately there is no way to do this analytically in all generality. However, we can use to our advantage that $T \ll T_F$ for any reasonable temperature, so that the Fermi factors n_F are close to a step function. Such an expansion was first used by Sommerfeld, but it is algebraically rather complicated[9] (see Ashcroft and Mermin Chapter 2 to see how it is done in detail). However, it is

[9] Such a calculation requires, among other things, the evaluation of some very nasty integrals which turn out to be related to the Riemann zeta function (see Section 2.3).

not hard to make an estimate of what such a calculation must give—which we shall now do.

When $T = 0$ the Fermi function is a step function and the chemical potential is (by definition) the Fermi energy. For small T, the step function is smeared out as we see in Fig. 4.1. Note, however, that in this smearing, the number of states that are removed from below the chemical potential is almost exactly the same as the number of states that are added above the chemical potential.[10] Thus, for small T, one does not have to move the chemical potential much from the Fermi energy in order to keep the number of particles fixed in Eq. 4.9. We conclude that $\mu \approx E_F$ for any low temperature. (In more detail we find that $\mu(T) = E_F + \mathcal{O}(T/T_F)^2$, see Ashcroft and Mermin Chapter 2.)

Thus we can focus on Eq. 4.8 with the assumption that $\mu = E_F$. At $T = 0$ let us call the kinetic energy[11] of the system $E(T = 0)$. At finite temperature, instead of a step function in Eq. 4.8 the step is smeared out as in Fig. 4.1. We see in the figure that only electrons within an energy range of roughly $k_B T$ of the Fermi surface can be excited—in general they are excited above the Fermi surface by an energy of about $k_B T$. Thus we can approximately write

$$E(T) = E(T = 0) + (\tilde{\gamma}/2)[Vg(E_F)(k_B T)](k_B T) + \ldots.$$

Here $Vg(E_F)$ is the density of states near the Fermi surface (recall g is the density of states per unit volume), so the number of particles close enough to the Fermi surface to be excited is $Vg(E_F)(k_B T)$, and the final factor of $(k_B T)$ is roughly the amount of energy that each one gets excited by. Here $\tilde{\gamma}$ is some constant which we cannot get right by such an approximate argument (but it can be derived more carefully, and it turns out that $\tilde{\gamma} = \pi^2/3$, see Ashcroft and Mermin).

We can then derive the heat capacity

$$C = \partial E/\partial T = \tilde{\gamma} k_B g(E_F) k_B T V$$

which then using Eq. 4.11 we can rewrite as

$$C = \tilde{\gamma}\left(\frac{3Nk_B}{2}\right)\left(\frac{T}{T_F}\right). \qquad (4.12)$$

The first term in brackets is just the classical result for the heat capacity of a gas, but the final factor T/T_F is tiny (0.01 or smaller!). This is the above promised linear T term in the heat capacity of electrons (see Fig. 2.5), which is far smaller than one would get for a classical gas.

This Sommerfeld prediction for the electronic (linear T) contribution to the heat capacity of a metal is typically not too far from being correct (see Table 4.1). A few metals, however, have specific heats that deviate from this prediction by a factor of 10 or more. Note that there are other measurements that indicate that these errors are associated with the electron mass being somehow changed in the metal. We will discover the reason for these deviations later when we study band theory (mainly in Chapter 17).

[10] Since the Fermi function has a precise symmetry around μ given by $n_F(\beta(E-\mu)) = 1 - n_F(\beta(\mu - E))$, this equivalence of states removed from below the chemical potential and states inserted above would be an exact statement if the density of states in Eq. 4.9 were independent of energy.

[11] In fact $E(T = 0) = (3/5)NE_F$, which is not too hard to show. See Exercise 4.1.

Table 4.1 Low-temperature heat capacity coefficient for some metals. All of these metals have heat capacities of the form $C = \gamma T + \alpha T^3$ at low temperature. This table gives the measured experimental (exp) value and the Sommerfeld theoretical (th) predictions for the coefficient γ in units of 10^{-4} J/(mol-K).

Material	γ_{exp}	γ_{th}
Lithium (Li)	18	7.4
Sodium (Na)	15	11
Potassium (K)	20	17
Copper (Cu)	7	5.0
Silver (Ag)	7	6.4
Beryllium (Be)	2	2.5
Bismuth (Bi)	1	5.0
Manganese (Mn)	170	5.2

The theoretical value is obtained by setting the electron density equal to the atomic density times the valence (number of free electrons per atom), then calculating the Fermi temperature from the density and using Eq. 4.12. Note that Mn has multiple possible valence states. In the theoretical calculation we assume valence of one which gives the largest possible predicted value of γ_{th}.

Realizing now that the heat capacity of the electron gas is reduced from that of the classical gas by a factor of $T/T_F \lesssim 0.01$, we can return to re-examine some of the above Drude calculations of thermal transport. We had found (see Eqs. 3.5–3.7) that Drude theory predicts a thermopower $S = \Pi/T = -c_v/(3e)$ that is too large by a factor of 100. Now it is clear that the reason for this error was that we used in this calculation (see Eq. 3.6) the heat capacity per electron for a classical gas, which is too large by roughly $T_F/T \approx 100$. If we repeat the calculation using the proper heat capacity, we will now get a prediction for thermopower which is reasonably close in magnitude to what is actually measured in experiment for many metals (see Table 3.3). Note that we still have not answered the question of why the sign of the Seebeck coefficient (and the Hall coefficient) sometimes comes out wrong![12]

[12] In fact a fully quantitative theory of the Seebeck coefficient turns out to be quite difficult, and we will not attempt such a thing in this book.

We also used the heat capacity per particle in the Drude calculation of the thermal conductivity $\kappa = \frac{1}{3}nc_v\langle v\rangle\lambda$. In this case, the c_v that Drude used was too large by a factor of T_F/T, but on the other hand the value of $\langle v\rangle^2$ that he used was too small by roughly the same factor (classically, one uses $mv^2/2 \sim k_BT$ whereas for the Sommerfeld model, one should use the Fermi velocity $mv_F^2/2 \sim k_BT_F$). Thus Drude's prediction for thermal conductivity came out roughly correct (and thus the Wiedemann–Franz law correctly holds).

4.3 Magnetic Spin Susceptibility (Pauli Paramagnetism)[13]

[13] Part VII of this book is entirely devoted to the subject of magnetism, so it might seem to be a bit out of place to discuss magnetism now. However since the calculation is an important result that hinges only on free electrons and Fermi statistics, it seems appropriate that it is discussed here. Most students will already be familiar with the necessary definitions of quantities such as magnetization and susceptibility so these should not cause confusion. However, for those who disagree with this strategy or are completely confused by this section it is OK to skip over it and return after reading a bit of Part VII.

Another property we can examine about the free electron gas is its response to an externally applied magnetic field. There are several ways that the electrons can respond to the magnetic field. First, the electrons' motion can be curved due to the Lorentz force. We have discussed this previously, and we will return to discuss how it results in an (orbital) magnetic moment in Section 19.5.[14] Secondly, the electron spins can flip over due to the applied magnetic field, which also changes the magnetic moment of the electron gas—this is the effect we will focus on here.

[14] For a free electron gas, the contribution to the magnetic susceptibility from the orbital motion of the electron is known as Landau diamagnetism and takes the value $\chi_{Landau} = -(1/3)\chi_{Pauli}$ (this effect is named after the famous Russian Nobel laureate Lev Landau[18]). We will discuss diamagnetism more in Chapter 19 . Unfortunately, calculating this diamagnetism is relatively tricky (see Blundell's book on magnetism, Section 7.6 for example).

Roughly, the Hamiltonian (neglecting the Lorentz force of the magnetic field, see Section 19.3 for more detail) becomes[15]

$$\mathcal{H} = \frac{\mathbf{p}^2}{2m} + g\mu_B \mathbf{B}\cdot\boldsymbol{\sigma}$$

[15] The sign of the last term, the so-called Zeeman coupling, may be a bit confusing. Recall that because the electron charge is negative, the electron dipole moment is actually opposite the direction of the electron spin (the current is rotating opposite the direction that the electron is spinning). Thus spins are lower energy when they are antialigned with the magnetic field! This is an annoyance caused by Benjamin Franklin, who declared that the charge left on a glass rod when rubbed with silk is positive.

where $g = 2$ is the g-factor of the electron,[16] $\mathbf{B}$ is the magnetic field,[17] and $\boldsymbol{\sigma}$ is the spin of the electron which takes eigenvalues $\pm 1/2$. Here I have defined (and will use elsewhere) the useful Bohr magneton

$$\mu_B = e\hbar/2m_e \approx .67k_B(\text{Kelvin/Telsa}).$$

Thus in the magnetic field the energy of an electron with spin up or down (with up meaning it points the same way as the applied field, and $B = |\mathbf{B}|$) are given by

$$\begin{aligned}\epsilon(\mathbf{k},\uparrow) &= \frac{\hbar^2|\mathbf{k}|^2}{2m} + \mu_B B \\ \epsilon(\mathbf{k},\downarrow) &= \frac{\hbar^2|\mathbf{k}|^2}{2m} - \mu_B B.\end{aligned}$$

The spin magnetization of the system (moment per unit volume) in the direction of the applied magnetic field will then be

$$M = -\frac{1}{V}\frac{dE}{dB} = -([\#\text{ up spins}] - [\#\text{ down spins}])\,\mu_B/V. \qquad (4.13)$$

So when the magnetic field is applied, it is lower energy for the spins to be pointing down, so more of them will point down. Thus a magnetization develops in the same direction as the applied magnetic field. This is known as *Pauli paramagnetism*. Here *paramagnetism* means that the magnetization is in the direction of the applied magnetic field. *Pauli paramagnetism* refers in particular to the spin magnetization of the free electron gas (we discuss paramagnetism in more detail in Chapter 19).

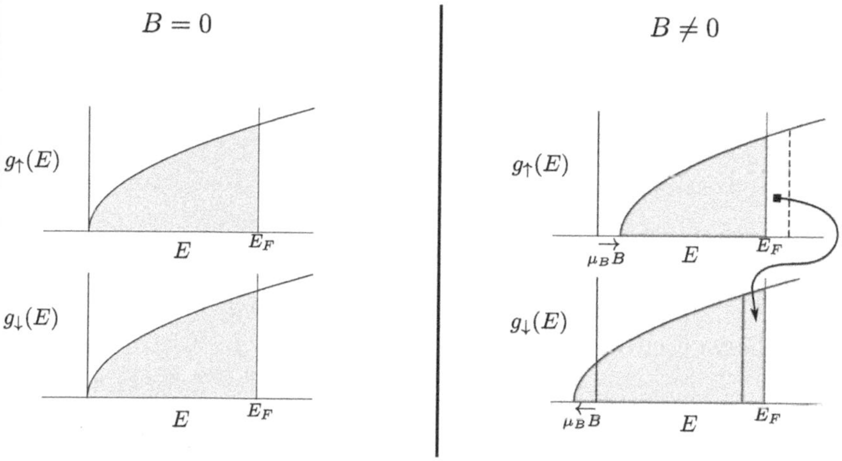

Fig. 4.2 Filling of electronic states up to the Fermi energy. **Left:** Before the magnetic field is applied the density of states for spin-up and spin-down are the same $g_\uparrow(E) = g_\downarrow(E) = g(E)/2$. Note that these functions are proportional to $E^{1/2}$ (see Eq. 4.11) hence the shape of the curve, and the shaded region indicates the states that are filled. **Right:** When the magnetic field is applied, the states with up and down spin are shifted in energy by $+\mu_B B$ and $-\mu_B B$ respectively as shown. Hence up spins pushed above the Fermi energy can lower their energies by flipping over to become down spins. The number of spins that flip (the area of the approximately rectangular sliver) is roughly $g_\uparrow(E_F)\mu_B B$.

Let us now calculate the Pauli paramagnetism of the free electron gas at $T = 0$. With zero magnetic field applied, both the spin-up and spin-down states are filled up to the Fermi energy (i.e, they fill a Fermi sphere with radius the Fermi wavevector). Near the Fermi level the density of states per unit volume for spin-up electrons is $g(E_F)/2$ and similarly the density of states per unit volume for spin-down electrons is $g(E_F)/2$. When B is applied, the spin-ups will be more costly by an energy $\mu_B B$. Thus, assuming that the chemical potential does not change, we will have $(g(E_F)/2)\mu_B B$ fewer spin-up electrons per unit volume. Similarly, the spin-downs will be less costly by the same amount, so we will have $(g(E_F)/2)\mu_B B$ more spin-downs per unit volume. Note that the total number of electrons in the system did not change, so our assumption that the chemical potential did not change is correct (recall that chemical potential is always adjusted so it gives the right total number of electrons in the system). This process is depicted in Fig. 4.2.

[16] It is a yet another constant source of grief that the letter "g" is used both for density of states and for g-factor of the electron. To avoid confusion we immediately set the g-factor to 2 and henceforth in this chapter g is reserved for density of states. Similar grief is that we now have to write $\mathcal{H}$ for Hamiltonian because $H = B/\mu_0$ is frequently used for the magnetic field with μ_0 the permeability of free space.

[17] One should be careful to use the magnetic field seen by the actual electrons—this may be different from the magnetic field applied to the sample if the sample itself develops a magnetization.

[18] Landau kept a ranking of how smart various physicists were—ranked on a logarithmic scale. Einstein was on top with a ranking of 0.5. Some of the great founders of quantum mechanics, Bose, Wigner,, Schroedinger, Heisenberg, Bohr, and Dirac received a ranking of 1. Landau modestly ranked himself a 2.5 but after winning the Nobel Prize raised himself to 2. He said that anyone ranked below 4 was not worth talking to. (Two different versions of this list have Newton either ranked 0, above Einstein, or 1, below.)

[19]See also the very closely related derivation given in Section 23.1.2.

Table 4.2 Experimentally measured magnetic susceptbilities χ_{exp} of various metals compared to Pauli's theoretical prediction χ_{Pauli}. In both cases the susceptibility is dimensionless and is listed here in units of 10^{-6} (e.g., Li has $\chi_{exp} = 3.4 \times 10^{-6}$)

Material	χ_{exp}	χ_{Pauli}
Lithium (Li)	3.4	10
Sodium (Na)	6.2	8.3
Potassium (K)	5.7	6.7
Copper (Cu)	-9.6	12
Beryllium (Be)	-23	17
Aluminum (Al)	21	16

The theoretical calculation uses Eqs. 4.14, 4.11, and 4.7 and assumes the bare mass of the electron for m.

Using Eq. 4.13, given that we have moved $g(E_F)\mu_B B/2$ up spins to down spins, the magnetization (magnetic moment per unit volume) is given by

$$M = g(E_F)\mu_B^2 B$$

and hence the magnetic susceptibility $\chi = \lim_{H\to 0} \partial M/\partial H$ is given (at $T = 0$) by[19]

$$\chi_{Pauli} = \frac{dM}{dH} = \mu_0 \frac{dM}{dB} = \mu_0\, \mu_B^2\, g(E_F) \tag{4.14}$$

with μ_0 the permeability of free space. In fact, as shown in Table 4.2, this result is not far from correct for simple metals such as Li, Na, or K. Although for some other metals we see that, once again, we get the overall sign wrong! We will return to discuss magnetic properties in Part VII.

4.4 Why Drude Theory Works So Well

In retrospect we can understand a bit more about why Drude theory was so successful. We now realize that because of Fermi statistics, treating electrons as a classical gas is incorrect—resulting in a huge overestimation of the heat capacity per particle, and in a huge underestimation of the typical velocity of particles. As described at the end of Section 4.2, these two errors can sometimes cancel giving reasonable results nonetheless.

However, we can also ask why it is that Drude was successful in calculation of transport properties such as the conductivity and the Hall coefficient. In these calculations neither the velocity of the particle nor the specific heat enter. But still, the idea that a single particle will accelerate freely for some amount of time, then will scatter back to zero momentum seems like it must be wrong, since the state at zero momentum is always fully occupied. The transport equation (Eq. 3.1) that we solve

$$\frac{d\mathbf{p}}{dt} = \mathbf{F} - \frac{\mathbf{p}}{\tau} \tag{4.15}$$

in the Drude theory describes the motion of each particle. However, we can just as well use the same equation to describe the motion of the center of mass of the entire Fermi sea! On the top of Fig. 4.3 we have a picture of a Fermi sphere of radius $\mathbf{k}_F$. The typical electron has a very large velocity on the order of the Fermi velocity v_F, but the average of all of the (vector) velocities is zero. When an electric field is applied in the bottom of Fig. 4.3 every electron in the system accelerates together in the $\hat{x}$ direction, and the center of the Fermi sea shifts. (The electric field in the figure is in the $-\hat{x}$ direction, so that the force is in the $+\hat{x}$ direction since the charge on the electron is $-e$). The shifted Fermi sea has some non-zero average velocity, known as the *drift velocity* $\mathbf{v}_{drift}$. Since the kinetic energy of the shifted Fermi sea is higher than the energy of the Fermi sea with zero average velocity, the electrons will try to scatter back (with scattering rate $1/\tau$) to lower kinetic energy and shift the Fermi sea back to its original configuration with zero drift velocity. We can then view the Drude transport equation (Eq. 4.15) as describing the motion of the average velocity (momentum) of the entire Fermi sea.

One can think about how this scattering actually occurs in the Sommerfeld model. Here, most electrons have nowhere to scatter to, since all of the available $\mathbf{k}$ states with lower energy (lower $|\mathbf{k}|$) are already filled. However, the few electrons near the Fermi surface in the thin crescent between the shifted and unshifted Fermi sea scatter into the thin unfilled crescent on the other side of the unfilled Fermi sea to lower their energies (see Fig. 4.3). Although these scattering processes happen only to a very few of the electrons, the scattering events are extremely violent in that the change in momentum is exceedingly large (scattering all the way across the Fermi sea[20]).

[20] Actually, it may be that many small scattering events walking around the edge of these crescents make up this one effective scattering event

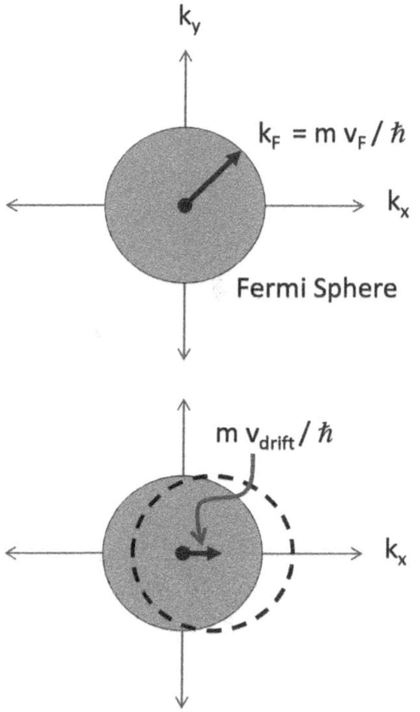

Fig. 4.3 Drift velocity and fermi velocity. The drift wavevector is the displacement of the entire Fermi sphere (which is generally very very small) whereas the Fermi wavevector is the radius of the Fermi sphere, which can be very large. Drude theory makes sense if you think of it as a transport equation for the center of mass of the entire Fermi sphere—i.e., it describes the drift velocity. Scattering of electrons only occurs between the thin crescents that are the difference between the shifted and unshifted Fermi spheres.

4.5 Shortcomings of the Free Electron Model

Although the Sommerfeld (free electron) model explains quite a bit about metals, it remains incomplete. Here are some items that are not well explained within Sommerfeld theory:

- Having discovered now that the typical velocity of electrons v_F is extremely large, and being able to measure the scattering time τ, we obtain a scattering length $\lambda = v_F \tau$ that may be 100 Ångstroms at room temperature, and might even reach 1mm at low temperature. One might wonder, if there are atoms, and hence charged atomic nuclei, every few Ångstroms in a metal, why do the electrons not scatter from these atoms? (We will discuss this in Chapter 15—the resolution is a result of Bloch's theorem.)
- Many of our results depend on the number of electrons in a metal. In order to calculate this number we have always used the chemical valence of the atom (for example, we assume one free electron per Li atom). However, in fact, except for hydrogen, there are actually many electrons per atom. Why do core electrons not "count" for calculating the Fermi energy or velocity? What about insulators where there are no electrons free?
- We have still not resolved the question of why the Hall effect sometimes comes out with the incorrect sign, as if the charge carrier were positive rather than negative as we expect for electrons of charge $-e$.
- In optical spectra of metals there are frequently many features (higher absorption at some frequencies, lower absorption at other frequencies). These features give metals their characteristic colors (for example, they make gold yellowish). The Sommerfeld model does not explain these features at all.

- The measured specific heat of electrons is much more correct than in Drude theory, but for some metals is still off by factors as large as 10. Measurements of the mass of the electron in a metal also sometimes give answers that differ from the actual mass of the electron by similar factors.
- Magnetism: Some metals, such as iron, are magnetic even without any applied external magnetic field. We will discuss magnetism in Part VII.
- Electron interaction: We have treated the electrons as if they are non-interacting fermions. In fact, the typical energy of interaction for electrons, $e^2/(4\pi\epsilon_0 r)$ with r the typical distance between electrons, is huge, roughly the same scale as the Fermi energy. Yet we have ignored the Coulomb interaction between electrons completely. Understanding why this works is an extremely hard problem that was only understood starting in the late 1950s—again due to the brilliance of Lev Landau (see margin note 18 in this chapter about Landau). The theory that explains this is frequently known as "Landau Fermi Liquid Theory", but we will not study it in this book. We will, however, study electron–electron interactions a bit more seriously in Chapter 23.

With the exception of the final two points (magnetism and electron interaction) all of these issues will be resolved once we study electronic *band structure* in Chapters 11, 15, and particularly 17. In short, we are not taking seriously the periodic structure of atoms in materials.

Chapter Summary

- Sommerfeld theory treats properly the fact that electrons are fermions.
- High density of electrons results in extremely high Fermi energy and Fermi velocity. Thermal and electric excitations are small redistributions of electrons around the Fermi surface.
- Compared to Drude theory, Sommerfeld theory obtains electron velocity ~ 100 times larger, but heat capacity per electron ~ 100 times smaller. This leaves the Wiedemann–Franz ratio roughly unchanged from Drude, but fixes problems in predications of thermal properties. Drude transport equations make sense if one considers velocities to be drift velocities, not individual electron velocities.
- Specific heat and (Pauli) paramagnetic susceptibility can be calculated explicitly (know these derivations!) in good agreement with experiment (at least for some simple metals).
- Despite successes, there are still some serious shortcomings of Sommerfeld theory.

References

For free electron (Sommerfeld) theory, good references are:

- Ashcroft and Mermin, chapters 2–3
- Singleton, sections 1.5–1.6
- Rosenberg, sections 7.1–7.9
- Ibach and Luth, sections 6–6.5
- Kittel, chapter 6
- Burns, chapter 9B (excluding 9.14 and 9.16)
- Hook and Hall, chapter 3 (blends Drude and Sommerfeld)

Exercises

(4.1) **Fermi Surface in the Free Electron (Sommerfeld) Theory of Metals**

(a)‡ Explain what is meant by the Fermi energy, Fermi temperature and the Fermi surface of a metal.

(b)‡ Obtain an expression for the Fermi wavevector and the Fermi energy for a gas of electrons (in 3D).

▷ Show that the density of states at the Fermi surface, dN/dE_F can be written as $3N/2E_F$.

(c) Estimate the value of E_F for sodium [The density of sodium atoms is roughly 1 gram/cm^3, and sodium has atomic mass of roughly 23. You may assume that there is one free electron per sodium atom (sodium has *valence* one)]

(d) Now consider a two-dimensional Fermi gas. Obtain an expression for the density of states at the Fermi surface.

(4.2) **Velocities in the Free Electron Theory**

(a) Assuming that the free electron theory is applicable: show that the speed v_F of an electron at the Fermi surface of a metal is $v_F = \frac{\hbar}{m}(3\pi^2 n)^{1/3}$ where n is the density of electrons.

(b) Show that the mean drift speed v_d of an electron in an applied electric field E is $v_d = |\sigma E/(ne)|$, where σ is the electrical conductivity, and show that σ is given in terms of the mean free path λ of the electrons by $\sigma = ne^2\lambda/(mv_F)$.

(c) Assuming that the free electron theory is applicable to copper:

(i) calculate the values of both v_d and v_F for copper at 300K in an electric field of 1 V m^{-1} and comment on their relative magnitudes.
(ii) estimate λ for copper at 300K and comment upon its value compared to the mean spacing between the copper atoms.

You will need the following information: copper is monovalent, meaning there is one free electron per atom. The density of atoms in copper is $n = 8.45 \times 10^{28}$ m^{-3}. The conductivity of copper is $\sigma = 5.9 \times 10^7 \Omega^{-1}$m^{-1} at 300K.

(4.3) **Physical Properties of the Free Electron Gas**

In both (a) and (b) you may always assume that the temperature is much less than the Fermi temperature.

(a)‡ Give a simple but approximate derivation of the Fermi gas prediction for heat capacity of the conduction electron in metals.

(b)‡ Give a simple (not approximate) derivation of the Fermi gas prediction for magnetic susceptibility of the conduction electron in metals. Here susceptibility is $\chi = dM/dH = \mu_0 dM/dB$ at small H and is meant to consider the magnetization of the electron spins only.

(c) How are the results of (a) and (b) different from that of a classical gas of electrons?

▷ What other properties of metals may be different from the classical prediction?

(d) The experimental specific heat of potassium metal at low temperatures has the form:

$$C = \gamma T + \alpha T^3$$

where $\gamma = 2.08\,\text{mJ}\,\text{mol}^{-1}\,\text{K}^{-2}$ and $\alpha = 2.6\,\text{mJ}\,\text{mol}^{-1}\,\text{K}^{-4}$.

▷ Explain the origin of each of the two terms in this expression.

▷ Make an estimate of the Fermi energy for potassium metal.

(4.4) **Another Review of Free Electron Theory**

▷ What is the *free electron model* of a metal.

▷ Define *Fermi energy* and *Fermi temperature*.

▷ Why do metals held at room temperature feel cold to the touch even though their Fermi temperatures are much higher than room temperature?

(a) A d-dimensional sample with volume L^d contains N electrons and can be described as a free electron model. Show that the Fermi energy is given by

$$E_F = \frac{\hbar^2}{2mL^2}(N a_d)^{2/d}$$

Find the numerical values of a_d for $d = 1, 2$, and 3.

(b) Show also that the density of states at the Fermi energy is given by

$$g(E_F) = \frac{Nd}{2L^d E_F}$$

▷ Assuming the free electron model is applicable, estimate the Fermi energy and Fermi temperature of a one-dimensional organic conductor which has unit cell of length 0.8 nm, where each unit cell contributes one mobile electron.

(c) Consider relativistic electrons where $E = c|\mathbf{p}|$. Calculate the Fermi energy as a function of the density for electrons in $d = 1, 2, 3$ and calculate the density of states at the Fermi energy in each case.

(4.5) **Chemical Potential of 2D Electrons**

Show that for free electron gas in two dimensions, the chemical potential μ is independent of the temperature so long as $k_B T \ll \mu$. Hint: first examine the density of states in two dimensions.

(4.6) **Chemical Potential at $T = 0$**

Consider a system of N non-interacting electrons. At $T = 0$ the N lowest-energy eigenstates will be filled and all the higher energy eigenstates will be empty. Show that at $T = 0$ the energy of the chemical potential is precisely half way between the highest energy filled eigenstate and the lowest-energy unfilled eigenstate.

(4.7) **More Thermodynamics of Free Electrons**

(a) Show that the kinetic energy of a free electron gas in three dimensions is $E = \frac{3}{5} E_F N$.

(b) Calculate the pressure $P = -\partial E/\partial V$, and then the bulk modulus $B = -V \partial P/\partial V$.

(c) Given that the density of atoms in sodium is $2.53 \times 10^{22}\text{cm}^{-3}$ and that of potassium is $1.33 \times 10^{22}\text{cm}^{-3}$, and given that both of these metals are monovalent (i.e., have one free electron per atom), calculate the bulk modulus associated with the electrons in these materials. Compare your results to the measured values of 6.3 GPa and 3.1 GPa respectively.

(4.8) **Heat Capacity of a Free Electron Gas***

In Exercise 4.3.a we approximated the heat capacity of a free electron gas

(a*) Calculate an exact expression for the heat capacity of a 2d metal at low temperature.

(b**) Calculate an exact expression for the heat capacity of a 3d metal at low temperature.

The following integral may be useful for these calculations:

$$\int_{-\infty}^{\infty} dx \frac{x^2 e^x}{(e^x + 1)^2} = \frac{\pi^2}{3} = \zeta(2)/2$$

Note that for the 3d case you have to worry about the fact that the chemical potential will shift as a function of temperature. Why does this not happen (at least for low T) in the 2d case?

Part II

Structure of Materials

The Periodic Table

5

In Chapter 2 we found that the Debye model gave a reasonably good description of the heat capacity of solids. However, we also found a number of shortcomings of the theory. These shortcomings basically stemmed from not taking seriously the fact that solids are actually made up of individual atoms assembled in a periodic structure.

Similarly, in Chapter 4 we found that the Sommerfeld model described quite a bit about metals, but had a number of shortcomings as well—many of these were also due to not realizing that the solids are made up of individual atoms assembled in periodic structures.

As such, a large amount of this book will be devoted to understanding the effects of these individual atoms and their periodic arrangement on the electrons and on the vibrations of the solid. In other words, it is time to think microscopically about the structure of materials. To do this we start with a review of some basic atomic physics and basic chemistry.

This part of the book should provide everything you will need to know about basic chemistry crammed into a nutshell.[1] If you have had a good chemistry course, much of this material may sound familiar. But hopefully, looking at the same chemistry from the point of view of a physicist will be somewhat enlightening nonetheless.

[1]... and anything that can be put into a nutshell, probably should be.

5.1 Chemistry, Atoms, and the Schroedinger Equation

When we think about the physics of a single atom, or if we ask about why two atoms stick together, we are in some sense trying to describe the solution to a many-particle Schroedinger[2] equation describing the many electrons and many nuclei in a solid. We can at least write down the equation

$$H\Psi = E\Psi$$

where Ψ is the wavefunction describing the positions and spin states of all the electrons and nuclei in the system. The terms in the Hamiltonian

[2]Erwin Schroedinger was a Fellow at Magdalen College Oxford from 1933 to 1938, but he was made to feel not very welcome there because he had a rather "unusual" personal life—he lived with both his wife, Anny, and with his mistress, Hilde, who, although married to another man, bore Schroedinger's child, Ruth. After Oxford, Schroedinger was coaxed to live in Ireland with the understanding that this unusual arrangement would be fully tolerated. Surprisingly, all of the parties involved seemed fairly content until 1946 after Schroedinger fathered two more children with two different Irish women, whereupon Hilde decided to take Ruth back to Austria to live with her lawful husband. Anny, entirely unperturbed by this development and having her own lovers as well, remained Erwin's close companion until his death.

include a kinetic term (with inputs of the electron and nuclear masses) as well as a Coulomb interaction term between all the electrons and nuclei.[3] While this type of description of chemistry is certainly true, it is also mostly useless.[4] No one ever even tries to solve the Schroedinger equation for more than a few particles at a time. Trying to solve it for dozens of electrons in a large atom, much less 10^{23} electrons in a real solid, is completely absurd. One must try to extract useful information about the behavior of atoms from simplified models in order to obtain a qualitative understanding. (This is a great example of what I was ranting about in Chapter 1—reductionism does not work: saying that the Schroedinger equation is the whole solution is indeed misguided.) More sophisticated techniques try to turn these qualitative understandings into quantitative predictions.[5]

What we will try to do here (and in the next chapter) is to try to understand a whole lot of chemistry from the point of view of a physicist. We will construct some simple toy models to understand atoms and why they stick together; but at the end of the day, we cannot trust our simplified models too much and we really should learn more chemistry if we want to answer real chemistry questions, like whether yttrium will form a carbonate salt.

5.2 Structure of the Periodic Table

We start with some of the fundamentals of electrons in an isolated atom. Recall from basic quantum mechanics that an electron in an atomic orbital can be labeled by four quantum numbers, $|n, l, l_z, \sigma_z\rangle$, where

$$\begin{aligned} n &= 1, 2, \ldots \\ l &= 0, 1, \ldots, n-1 \\ l_z &= -l, \ldots, l \\ \sigma_z &= -1/2 \text{ or } +1/2. \end{aligned}$$

Here n is the principal quantum number, l is the angular momentum, l_z is its z-component and σ_z is the z-component of spin.[6] Recall that the angular momentum shells with $l = 0, 1, 2, 3$ are sometimes known as s,p,d,f, ... respectively in atomic language.[7] These shells can accommodate $2, 6, 10, 14, \ldots$ electrons respectively including both spin states.

When we consider multiple electrons in one atom, we need to decide which orbitals are filled and which ones are empty. The first rule is known as the Aufbau principle[8] and the second is sometimes called Madelung's rule.

> **Aufbau Principle** (paraphrased): Shells should be filled starting with the lowest available energy state. An entire shell is filled before another shell is started.[9]

> **Madelung's Rule**: The energy ordering is from lowest value of $n + l$ to the largest; and when two shells have the same value of $n + l$, fill the one with the smaller n first.[10]

[3] To have a fully functioning "Theory of Everything" as far as all of chemistry, biology, and most of everything that matters to us (besides the sun and atomic energy) is concerned, one needs only Coulomb interaction plus the kinetic term in the Hamiltonian, plus a bit of spin–orbit (relativistic effects).

[4] If you want to annoy your chemist friends (or enemies) try repeatedly telling them that they are actually studying the Schroedinger equation.

[5] As emphasized in Chapter 1 even the world's largest computers cannot solve the Schroedinger equation for a system of more than a few electrons. Nobel Prizes (in chemistry) were awarded to Walter Kohn and John Pople for developing computational methods that can obtain highly accurate approximations. These approaches have formed much of the basis of modern quantum chemistry. Despite the enormous sucuss of these computer approaches, our simple models are still crucial for developing understanding. To quote the Nobel laureate Eugene Wigner "It is nice to know that the computer understands the problem, but I would like to understand it too".

[6] You probably discussed these quantum numbers in reference to the eigenstates of a hydrogen atom. The orbitals of any atom can be labeled similarly.

[7] A mnemonic for this order is "Some Poor Dumb Fool". Another one (if you want to go to higher orbitals) is "Smart Physicist Don't Find Giraffes Hiding".

[8] Aufbau means "construction" or "building up" in German.

[9] It is important to realize that a given orbital is different in different atoms. For example, the 2s orbital in a nitrogen atom is different from the 2s orbital in an iron atom. The reason for this is that the charge of the nucleus is different *and also* that one must account for the interaction of an electron in an orbital with all of the other electrons in that atom.

[10] Depending on your country of origin, Madelung's rule might instead be known as Klechkovsky's rule.

This ordering rule means that shells should be filled in the order[11]

$$1s, 2s, 2p, 3s, 3p, 4s, 3d, 4p, 5s, 4d, 5p, 6s, 4f, \ldots.$$

A simple mnemonic for this order can be constructed by drawing the diagram shown in Fig. 5.1. So for example, let us consider an isolated nitrogen atom which has atomic number 7 (i.e., 7 electrons). Nitrogen (N) has a filled 1s shell (containing two electrons, one spin-up, one spin-down), has a filled 2s shell (containing two electrons, one spin-up, one spin-down), and has three remaining electrons in the 2p shell. In atomic notation we would write this as $1s^2 2s^2 2p^3$.

To take a more complicated example, consider the atom praseodymium (Pr) which is a rare earth element with atomic number 59. Following Madelung's rule, we obtain an atomic[12] configuration

$$1s^2 2s^2 2p^6 3s^2 3p^6 4s^2 3d^{10} 4p^6 5s^2 4d^{10} 5p^6 6s^2 4f^3.$$

Note that the "exponents" properly add up to 59.

There are a few atoms that violate this ordering (Madelung's) rule. One example is copper which typically fills the 3d shell by "borrowing" an electron from the (putatively lower energy) 4s shell. Also, when an atom is part of a molecule or is in a solid, the ordering may change a little. However, the general trend given by this rule is rather robust.

This shell filling sequence is, in fact, the rule which defines the overall structure of the periodic table with each "block" of the periodic table representing the filling of some particular shell (see the periodic table given in Fig. 5.2). For example, the first line of the periodic table has the elements H and He, which have atomic fillings $1s^x$ with $x = 1, 2$ respectively (and the 1s shell holds at most 2 electrons). The left of the second line of the table contains Li and Be which have atomic fillings $1s^2 2s^x$ with $x = 1, 2$ respectively. The right of the second line of the table shows B, N, C, O, F, Ne which have atomic fillings $1s^2 2s^2 2p^x$ with $x = 1 \ldots 6$ and recall that the 2p shell can hold at most 6 electrons. One can continue and reconstruct the entire periodic table this way!

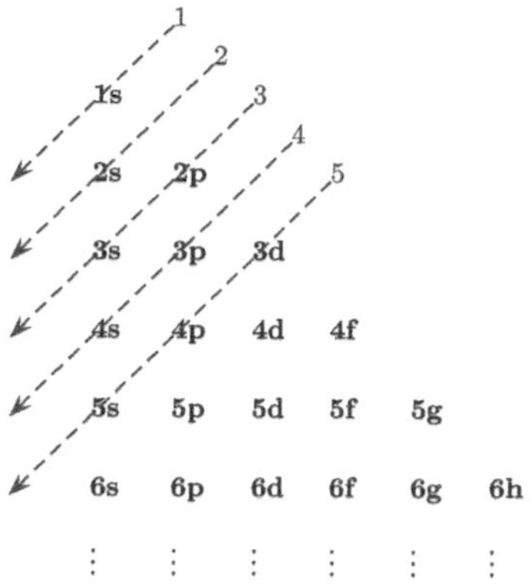

Fig. 5.1 Ordering of filling orbitals in atoms (Madelung's rule).

[12]This tediously long atomic configuration can be abbreviated as $[Xe]6s^2 4f^3$ where [Xe] represents the atomic configuration of xenon, which, being a noble gas, is made of entirely filled shells.

5.3 Periodic Trends

The periodic table, proposed in 1869 by Dmitri Mendeleev,[13] is structured so that elements with similar chemical properties lie in the same

[13]One of the scandals of Nobel history is that Mendeleev was deprived of the prize several times by the strong opposition of the Swedish chemist Svantes Arrhenius, who held a great deal of influence over the Nobel committee. Mendeleev had critiqued one of Arrhenius' theories, and Arrhenius was the type to hold a grudge.

[11]You may find it surprising that shells are filled in this order, being that for a simple hydrogen atom orbital energies increase with n and are independent of l (neglecting fine structure). However, in any atom other than hydrogen, we must also consider interaction of each electron with all of the other electrons. Treating this effect in detail is quite complex, so it is probably best to consider this ordering (Madelung's) rule to be simply empirical. Nonetheless, various approximation methods have been able to give some insight. Typical approximation schemes replace the Coulomb potential of the nucleus with some screened potential which represents the charge both of the nucleus and of all the other electrons (See for example, R. Ladder, *Phys. Rev.* **99**, 510, 1955). Note in particular that once we change the potential from the Coulomb $1/r$ form, we immediately break the energy degeneracy between different l states for a given n. While it is fairly easy to argue that the lower l values should fill first given the same n (see for example L. D. Landau and E. M. Lifshitz, *Quantum Mechanics*, Pergamon, 1974), there is no simple argument that derives the complete ordering. (See also the book by Pauling for more detail.)

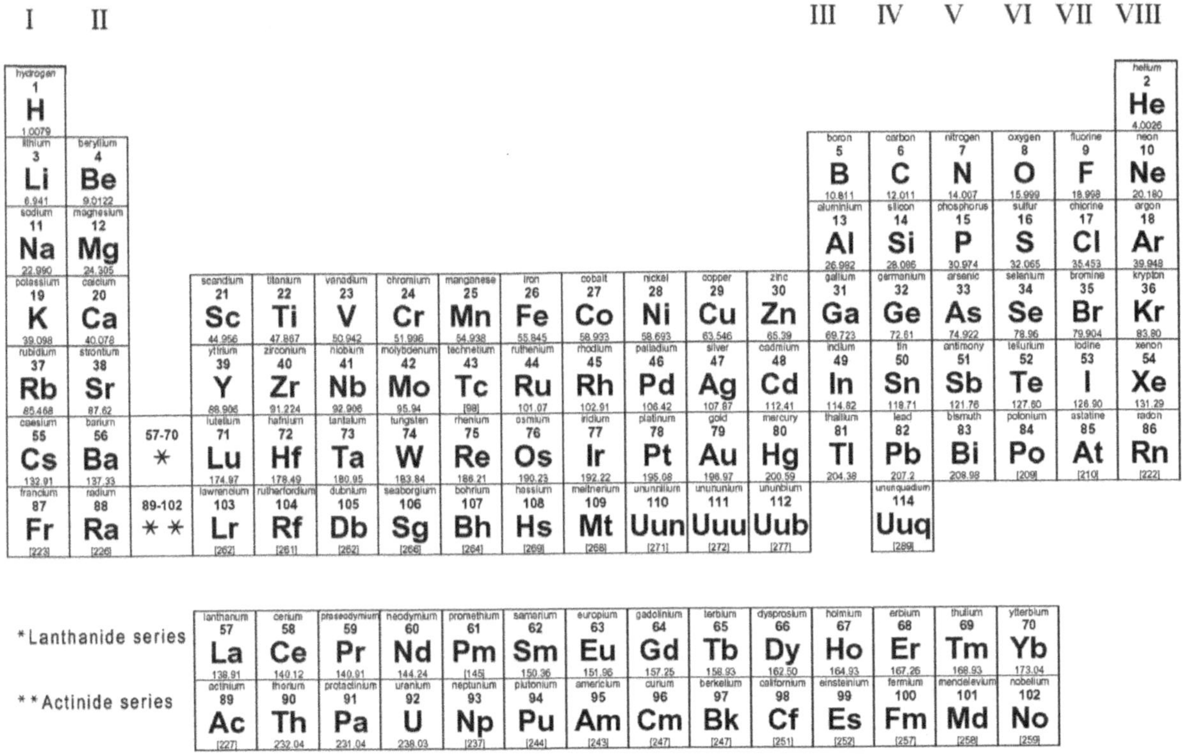
I
II
III
IV
V
VI
VII
VIII
hydrogen 1 H 1.0079
helium 2 He 4.0026
lithium 3 Li 6.941
beryllium 4 Be 9.0122
boron 5 B 10.811
carbon 6 C 12.011
nitrogen 7 N 14.007
oxygen 8 O 15.999
fluorine 9 F 18.998
neon 10 Ne 20.180
sodium 11 Na 22.990
magnesium 12 Mg 24.305
aluminium 13 Al 26.982
silicon 14 Si 28.086
phosphorus 15 P 30.974
sulfur 16 S 32.065
chlorine 17 Cl 35.453
argon 18 Ar 39.948
potassium 19 K 39.098
calcium 20 Ca 40.078
scandium 21 Sc 44.956
titanium 22 Ti 47.867
vanadium 23 V 50.942
chromium 24 Cr 51.996
manganese 25 Mn 54.938
iron 26 Fe 55.845
cobalt 27 Co 58.933
nickel 28 Ni 58.693
copper 29 Cu 63.546
zinc 30 Zn 65.39
gallium 31 Ga 69.723
germanium 32 Ge 72.61
arsenic 33 As 74.922
selenium 34 Se 78.96
bromine 35 Br 79.904
krypton 36 Kr 83.80
rubidium 37 Rb 85.468
strontium 38 Sr 87.62
yttrium 39 Y 88.906
zirconium 40 Zr 91.224
niobium 41 Nb 92.906
molybdenum 42 Mo 95.94
technetium 43 Tc [98]
ruthenium 44 Ru 101.07
rhodium 45 Rh 102.91
palladium 46 Pd 106.42
silver 47 Ag 107.87
cadmium 48 Cd 112.41
indium 49 In 114.82
tin 50 Sn 118.71
antimony 51 Sb 121.76
tellurium 52 Te 127.60
iodine 53 I 126.90
xenon 54 Xe 131.29
caesium 55 Cs 132.91
barium 56 Ba 137.33
57-70 *
lutetium 71 Lu 174.97
hafnium 72 Hf 178.49
tantalum 73 Ta 180.95
tungsten 74 W 183.84
rhenium 75 Re 186.21
osmium 76 Os 190.23
iridium 77 Ir 192.22
platinum 78 Pt 195.08
gold 79 Au 196.97
mercury 80 Hg 200.59
thallium 81 Tl 204.38
lead 82 Pb 207.2
bismuth 83 Bi 208.98
polonium 84 Po [209]
astatine 85 At [210]
radon 86 Rn [222]
francium 87 Fr [223]
radium 88 Ra [226]
89-102 **
lawrencium 103 Lr [262]
rutherfordium 104 Rf [261]
dubnium 105 Db [262]
seaborgium 106 Sg [266]
bohrium 107 Bh [264]
hassium 108 Hs [269]
meitnerium 109 Mt [268]
ununnilium 110 Uun [271]
unununium 111 Uuu [272]
ununbium 112 Uub [277]
ununquadium 114 Uuq [289]
*Lanthanide series
lanthanum 57 La 138.91
cerium 58 Ce 140.12
praseodymium 59 Pr 140.91
neodymium 60 Nd 144.24
promethium 61 Pm [145]
samarium 62 Sm 150.36
europium 63 Eu 151.96
gadolinium 64 Gd 157.25
terbium 65 Tb 158.93
dysprosium 66 Dy 162.50
holmium 67 Ho 164.93
erbium 68 Er 167.26
thulium 69 Tm 168.93
ytterbium 70 Yb 173.04
**Actinide series
actinium 89 Ac [227]
thorium 90 Th 232.04
protactinium 91 Pa 231.04
uranium 92 U 238.03
neptunium 93 Np [237]
plutonium 94 Pu [244]
americium 95 Am [243]
curium 96 Cm [247]
berkelium 97 Bk [247]
californium 98 Cf [251]
einsteinium 99 Es [252]
fermium 100 Fm [257]
mendelevium 101 Md [258]
nobelium 102 No [259]

Fig. 5.2 The periodic table of the elements. Note that the structure of the periodic table reflects Madelung's rule which dictates the order in which shells are filled. The organization of the table is such that each column has similar chemical properties. Each element is listed with its atomic number and its atomic weight.

column. For example, the chemistry of carbon, silicon, and germanium are quite similar to each other and these three elements are all in column IV. The reason for the chemical similarities between certain elements stems from the details of the fillings of atomic orbital shells, and hence is a corollary of Madelung's rule. We will see that to a large extent chemistry is determined by the electrons in the outermost shell of an atom. So for example, the fact that the chemistries of C, Si, and Ge are similar is due to the fact that (via Madelung's rule) each has only two electrons in a partially filled p-shell.[14]

[14]Carbon has has two electrons in a 2p shell; silicon (Si) has two electrons in a 3p shell; and germanium (Ge) has two electrons in a 4p shell.

The periodic table describes many of the chemical trends of the atoms. For example, going from left to right of the periodic table, the atomic radius always tends to decrease. Going from left to right, the energy required to remove an electron from an atom, the so-called *ionization energy,* also increases. Similarly, the energy gained by adding an electron to an atom, the so-called *electron affinity,* also increases from left to right (we will study ionization energy and electron affinity in more detail in Section 6.1, see Fig. 6.1). In some sense it appears that electrons are more tightly bound to the nucleus of atoms on the right of the periodic table than they are on the left of the table. To understand why this is true we must remember that electrons in atoms interact not only with the nucleus, but also with the other electrons.

5.3.1 Effective Nuclear Charge

To a large extent the interaction of electrons with each other can be understood in terms of how one electron will shield the other from the nucleus, yielding a reduced "effective nuclear charge". As an example, let us consider the case of a sodium (Na) atom, with 11 electrons, which has filled 1s, 2s, 2p shells, and then a single electron in the 3s shell. While the actual nucleus has charge +11, to the one electron in the 3s shell, the nucleus appears as if it has a charge of only +1 (= +11 from the nucleus −10 electron charges in inner shells). As shown in Fig. 5.3, the electrons in the inner orbitals are at a radius much less than that of the 3s electron, so to that one 3s electron the other electrons look like part of the nucleus. As a result, the sodium atom is very much like an electron in a 3s shell bound to a nucleus of small effective charge +1. Hence this is rather weak binding of the last electron, the atomic radius is large, this last electron is easily ionized, and there is not much binding energy if one were to add yet another electron to the atom.

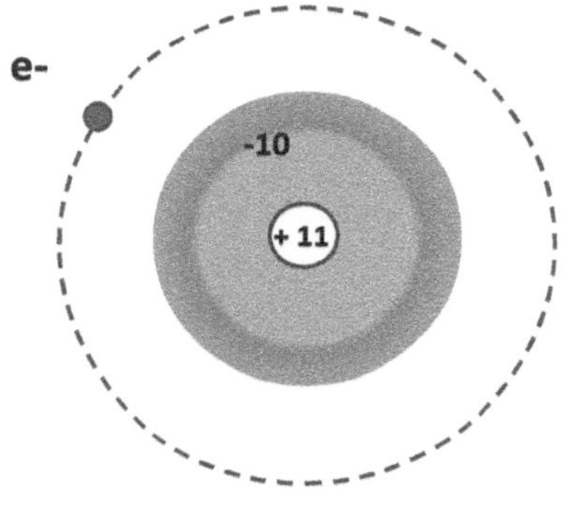

Fig. 5.3 For a sodium atom, since there is only one electron in the outermost shell (3s shell in this case), the effective nuclear charge seen by that one electron is +1.

On the other hand, let us consider the case of fluorine, which has 9 electrons. In this case, there are two electrons in the inner 1s shell and seven electrons in the outermost $n = 2$ (2s and 2p) shells. The 2s and 2p shells are at about the same radius as each other, so for simplicity we draw them as just a single shell. As shown in Fig. 5.4 the two electrons in the inner shell appear as if they are part of the nucleus. Very crudely we might then think that each electron in the outer shell now sees an effective nuclear charge of +7 (= +9 from the nucleus −2 from the electrons in the inner shells), so we would expect a very strong binding of the electrons to the nucleus. Indeed, the binding energy of the outer electrons is very large for fluorine. However, it is a bit of an overestimate to say that the effective nuclear charge is +7. The seven electrons in the outermost shell also shield each other from the nucleus to reduce the effective nuclear charge somewhat. The general rule is that an electron at a smaller radius from the nucleus screens the nuclear charge for electrons which are at a larger radius.[15] So if two electrons are at roughly the same radius (e.g., two electrons both in an $n = 2$ shell) then each screens each other by about half a charge (50% of the charge density is "inside" and 50% is outside). One might then more accurately estimate that the effective nuclear charge for fluorine is +4 (= +9 from the nucleus −2 for the two electrons in the inner shell, then each electron in the outer most shell sees 6 other electrons which are half inside its radius giving another −3). The truth lies somewhere between the two estimates of +4 and +7. (Of course the whole idea of effective

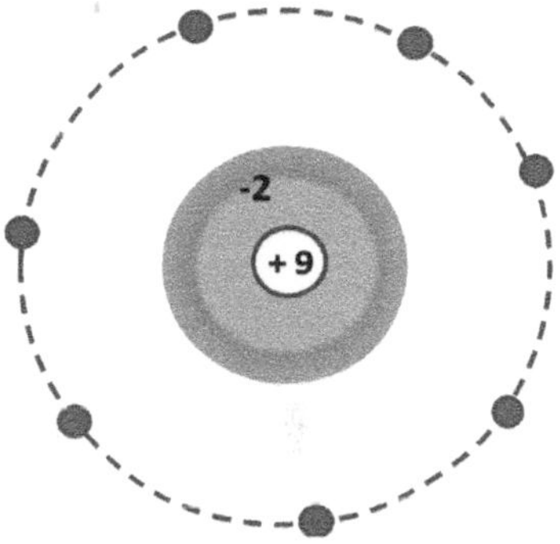

Fig. 5.4 For a fluorine atom, since there are many electrons in the outermost shell (2s and p in this case), the effective nuclear charge is quite large. Depending on how we treat the screening of electrons in the same shell we can get effective nuclear charges ranging from about +4 to +7.

[15] Recall Newton's shell theorem (Gauss's law): if there is a spherically symmetric charge distribution, from any point *outside* of the charge distribution, you can consider all of the charge to be at the center of the sphere.

nuclear charge is just an approximation anyway. Really we want to solve the Schroedinger equation for all the electrons at the same time!)

The general trend that we can deduce from this picture is that the binding energy of the outermost electrons increases as we go from left to right in the rows of the periodic table—i.e., as we put more electrons in the outermost shells the effective charge of the nucleus increases, since electrons at the same radius screen each other from the nucleus ineffectively. The stronger binding due to the larger effective nuclear charge accounts for many of the key features of the periodic table: The radius of atoms drops going left to right in the periodic table, the energy required to ionize an atom increases going left to right, and the energy gained by adding an electron (the electron affinity) also increases going left to right (not including the noble gases, which are filled shells). This general periodic trend is sometimes summarized by chemists by saying that atomic shells "'want" to be filled, or by saying that a filled shell is particularly stable. In fact, what they mean by this is simply that with increasing number of electrons in the outer shells, those electrons become bound increasingly tightly.

Chapter Summary

- Filling of atomic shells gives the structure of the periodic table (Aufbau principle and Madelung's rule).
- Elements in the same column of the periodic table (with a few exceptions) have the same number of electrons in the outermost shells, and hence have similar chemical properties.
- The idea of screening and effective nuclear charge explains a number of the periodic trends going across the rows of the periodic table.

References

Any chemistry textbook should be a useful reference. Some (not introductory references) that I like are:

- Pauling, chapters 1–3
- Murrel *et al.*, chapters 1–5

Exercises

(5.1) **Madelung's Rule**

▷ Use Madelung's rule to deduce the atomic shell filling configuration of the element tungsten (symbol W) which has atomic number 74.

▷ Element 118 has recently been discovered, and is expected to be a noble gas, i.e., is in group VIII. (No real chemistry tests have been performed on the element yet, as the nucleus decays very quickly.) Assuming that Madelung's rule continues to hold, what should the atomic number be for the next noble gas after this one?

(5.2) Effective Nuclear Charge and Ionization Energy

(a) Let us approximate an electron in the nth shell (i.e., principal quantum number n) of an atom as being like an electron in the nth shell of a hydrogen atom with an effective nuclear charge Z. Use your knowledge of the hydrogen atom to calculate the ionization energy of this electron (i.e., the energy required to pull the electron away from the atom) as a function of Z and n.

(b) Consider the two approximations discussed in the text for estimating the effective nuclear charge:

- (Approximation a)

$$Z = Z_{nuc} - N_{inside}$$

- (Approximation b)

$$Z = Z_{nuc} - N_{inside} - (N_{same} - 1)/2$$

where Z_{nuc} is the actual nuclear charge (or atomic number), N_{inside} is the number of electrons in shells inside of n (i.e., electrons with principal quantum numbers $n' < n$), and N_{same} is the total number of electrons in the nth principal shell (including the electron we are trying to remove from the atom, hence the -1).

▷ Explain the reasoning behind these two approximations.

▷ Use these approximations to calculate the ionization energies for the atoms with atomic number 1 through 21. Make a plot of your results and compare them to the actual ionization energies (you will have to look these up on a table).

Your results should be qualitatively decent. If you try this for higher atomic numbers, the simple approximations begin to break down. Why is this?

(5.3) Exceptions to Madelung's Rule

Although Madelung's rule for the filling of electronic shells holds extremely well, there are a number of exceptions to the rule. Here are a few of them:

Cu = [Ar] $4s^1 3d^{10}$
Pd = [Kr] $5s^0 4d^{10}$
Ag = [Kr] $5s^1 4d^{10}$
Au = [Xe] $6s^1 4f^{14} 5d^{10}$

▷ What should the electron configurations be if these elements followed Madelung's rule and the Aufbau principle?

▷ Explain how the statement "3d is inside of 4s" might help justify this exception in copper.

(5.4) Mendeleev's Nobel Prize

Imagine writing a letter to the Nobel committee nominating Mendeleev, the creator of the periodic table, for a Nobel Prize. Explain why the periodic table is so important. Remember that the periodic table (1869) was devised many years before the structure of the hydrogen atom was understood. (If you do not already have some background in chemistry, you may want to read the next chapter before attempting this exercise.)

(5.5) Ordering of Energy Levels*&

As mentioned in footnote 11, the fact that the energy levels of a hydrogen atom are independent of angular momentum L is very special to the $1/r$ Coulomb interaction. Using your knowledge of the hydrogen atom, recall the form of the wavefunction for the 2s and 2p orbitals.

(a) Consider adding a weak perturbation to the Coulomb attraction $V = V_{\text{Coulomb}} + \delta V$ where $\delta V = \epsilon e^{-\alpha r}$ is of the so-called Yukawa form. Using first order perturbation theory, calculate the change in the 2s and 2p energies due to this perturbation, and show they are not equal.

(b) Instead add a weak interaction $\delta V = \epsilon / r$. Show that the 2s and 2p orbitals remain degenerate.

What Holds Solids Together: Chemical Bonding

6

Having discussed some features of the periodic table, it is now worth asking ourselves why atoms stick together to form solids.

From a chemist's point of view one frequently thinks about different *types* of chemical bonds depending on the types of atoms involved, and in particular, depending on the atom's position on the periodic table (and on the atom's *electronegativity*—which is its tendency to attract electrons). In this chapter we will discuss ionic bonds, covalent bonds, van der Waals (fluctuating dipole, or molecular) bonds, metallic bonds, and hydrogen bonds.[1] Of course, they are all different aspects of the Schroedinger equation, and any given material may exhibit aspects of several of these types of bonding. Nonetheless, qualitatively it is quite useful to discuss these different types of bonds to give us intuition about how chemical bonding can occur. A brief description of the many types of bonding is given in the summary Table 6.1 at the end of this chapter. Note that the table should be considered just as rules-of-thumb, as many materials have properties intermediate between the categories listed.

[1] Insert obligatory pun about 007.

6.1 Ionic Bonds

The general idea of an ionic bond is that for certain compounds (for example, binary compounds, such as NaCl, made of one element in group I and one element in group VII), it is energetically favorable for an electron to be physically transferred from one atom to the other, leaving two oppositely charged ions which then attract each other. One writes a chemical "reaction" of the form

$$\mathrm{Na} + \mathrm{Cl} \rightarrow \mathrm{Na}^+ + \mathrm{Cl}^- \rightarrow \mathrm{NaCl}.$$

To find out if such a reaction happens, one must look at the energetics associated with the transfer of the electron.

In order to examine energetics more carefully, it is not *too* hard to imagine solving the Schroedinger equation for a single atom and determining the energy of the neutral atom, of the positive ion, and of the negative ion. Alternatively, if solving the Schroedinger equation proves

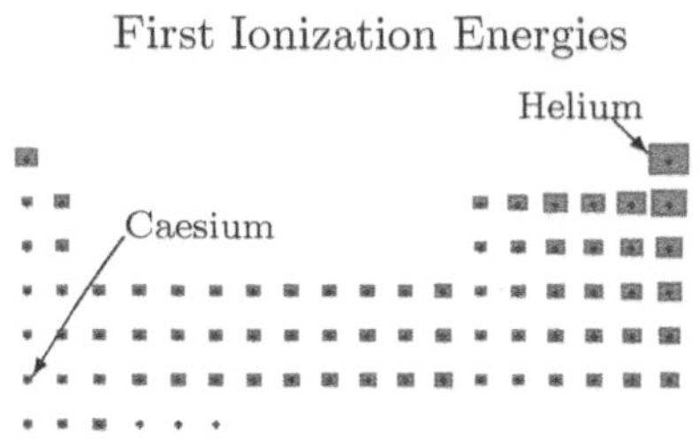

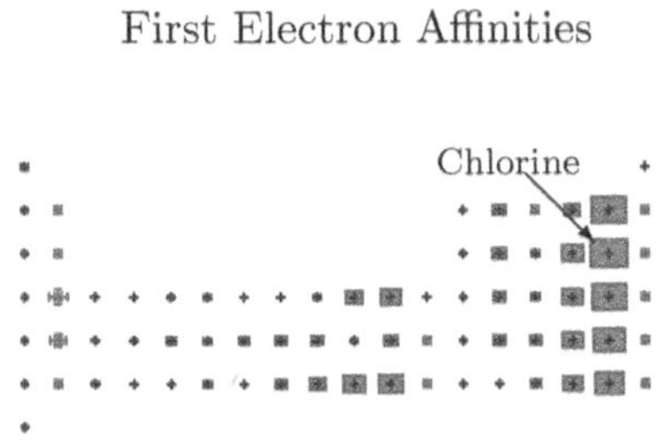

Fig. 6.1 Pictorial periodic tables of first ionization energies (top) and first electron affinities (bottom). The word "first" here means that we are measuring the energy to lose or gain a first electron starting with a neutral atom. The linear size of each box represents the magnitude of the energies (scales on the two plots differ). For reference the largest ionization energy is helium, at roughly 24.58 eV per atom, the lowest is caesium at 3.89 eV. The largest electron affinity is chlorine which gains 3.62 eV when binding to an additional electron. The few elements marked with lighter shaded crosses (including Ca and Sr) have *negative* electron affinities.

too hard, there are many ways that the energy levels of atoms (or ions) can be measured with some sort of spectroscopy. We define:

$$\begin{aligned}\text{Ionization Energy} &= \text{Energy required to remove one electron} \\ &\quad\ \text{from a neutral atom to create a positive ion,} \\ \text{Electron Affinity} &= \text{Energy gain for creating a negative ion} \\ &\quad\ \text{from a neutral atom by adding an electron.}\end{aligned}$$

To be precise, in both cases we are comparing the energy of having an electron either at position infinity, or on the atom. Further, if we are removing or adding only a single electron, then these are called *first* ionization energies and *first* electron affinities respectively (one can similarly define energies for removing or adding two electrons which would be called *second*). Finally we note that chemists typically work with systems at fixed (room) temperature and (atmospheric) pressure, in which case they are likely to be more concerned with Gibbs free energies, rather than pure energies. We will always assume that one is using the appropriate free energy for the experiment in question (and we will be sloppy and always call an energy E).

The periodic trend, which we justified in some detail in Section 5.3, is that the ionization energy is smallest on the left (groups I and II) of the periodic table and largest on the right (groups VII and VIII). To a lesser extent the ionization energy also tends to decrease towards the bottom of the periodic table (see Fig. 6.1). Similarly, electron affinity is also largest on the right and top of the periodic table, not including the group VIII noble gases which roughly do not attract electrons measurably at all (see Fig. 6.1).

The total energy change from transferring an electron from atom A to atom B is

$$\Delta E_{A+B\to A^+ + B^-} = (\text{Ionization Energy})_A - (\text{Electron Affinity})_B.$$

Note carefully the sign. The ionization energy is a positive energy that must be put in, the electron affinity is an energy that comes out.

However this ΔE is the energy to transfer an electron between two atoms very far apart. In addition, there is also[2]

$$\text{Cohesive Energy} = \text{Energy gain from } A^+ + B^- \to AB.$$

This cohesive energy is mostly a classical effect of the Coulomb interaction between the ions as one lets the ions come close together.[3] The

[2]The term "Cohesive Energy" can be ambiguous since sometimes people use it to mean the energy to put ions together into a compound, and other times they mean it to be the energy to put neutral atoms together! Here we mean the former. Furthermore, one should be cautioned that cohesive energy per atom of putting together two atoms to make a diatomic molecule (say, an NaCl molecule) would be very different from the cohesive energy per atom of putting many atoms together to make a bulk solid (say, a bulk NaCl solid).

total energy gain for forming a molecule from the two individual atoms is thus given by

$$\begin{aligned}\Delta E_{A+B\to AB} &= (\text{Ionization Energy})_A - (\text{Electron Affinity})_B \\ &\quad -(\text{Cohesive Energy of } AB).\end{aligned}$$

One obtains an ionic bond if the total ΔE for this process is less than zero.

In order to determine whether an electron is likely to be transferred between one atom and another, it is convenient to use the so-called *electronegativity*, which roughly describes how much an atom "wants" electrons, or how much an atom attracts electrons to itself. While there are various definitions of electronegativity that are used, a simple and useful definition is known as the Mulliken Electronegativity[4,5]

$$(\text{Mulliken) Electronegativity} = \frac{\text{Electron Affinity} + \text{Ionization Energy}}{2}.$$

According to the periodic trends we discussed in Section 5.3, the electronegativity is extremely large for elements in the upper right of the periodic table (not including the noble gases).

In bonding, the electron is always transferred from the atom of lower electronegativity to higher electronegativity. The greater the difference in electronegativities between two atoms the more completely the electron is transferred from one atom to the other. If the difference in electronegativities is small, then the electron is only partially transferred from one atom to the other. We will see in the next section that one can also have covalent bonding even between two identical atoms where there is no difference in electronegativities, and therefore no net transfer of electrons. Before leaving the topic of ionic bonds, it is worth discussing some of the typical physics of ionic solids. First of all, the materials are typically hard and have high melting temperatures, as the Coulomb interaction between oppositely charged ions is strong. However, since water is extremely polar, it can dissolve an ionic solid. This happens (see Fig 6.2) by arranging the water molecules such that the negative side of the molecule is close to the positive ions and the positive side of the molecule is close to the negative ions. Further, in an ionic solid the charges are bound strongly to the ions so these materials are electrically insulating (we will discuss in much more depth in Chapter 16 what makes a material an insulator).

[3] One can write a simple classical equation for a total cohesive energy for a solid

$$E_{cohesive} = -\sum_{i<j} \frac{Q_i Q_j}{4\pi\epsilon_0 |\mathbf{r}_i - \mathbf{r}_j|}$$

where Q_i is the charge on the ith ion, and $\mathbf{r}_i$ is its position. This sum is sometimes known as the *Madelung Energy.* It might look like one could make the cohesive energy infinitely large by letting two ions come to the same position! However, when atoms approach each other within roughly an atomic radius the approximation of the atom as a point charge breaks down. One thus needs a more quantum-mechanical treatment to determine, *ab initio*, how close two oppositely charged ions will come to each other.

[4] This electronegativity can be thought of as approximately the negative of the chemical potential via

$$\begin{aligned}\frac{1}{2}(E_{\text{affinity}} + E_{\text{ion}}) &= \frac{1}{2}([E_N - E_{N+1}] + [E_{N-1} - E_N]) = \frac{E_{N-1} - E_{N+1}}{2} \\ &\approx -\frac{\partial E}{\partial N} \approx -\mu.\end{aligned}$$

See however footnote 4 from Section 4.1 on defining chemical potential for systems with discrete energy levels and discrete number of electrons.

[5] Both Robert Mulliken and Linus Pauling won Nobel Prizes in chemistry for their work understanding chemical bonding including the concept of electronegativity. Pauling won a second Nobel Prize, in peace, for his work towards banning nuclear weapons testing. (Only four people have ever won two Nobels: Marie Curie, Linus Pauling, John Bardeen, and Fredrick Sanger. We should all know these names!) Pauling was criticized later in his life for promoting high doses of vitamin C to prevent cancer and other ailments, sometimes apparently despite scientific evidence to the contrary.

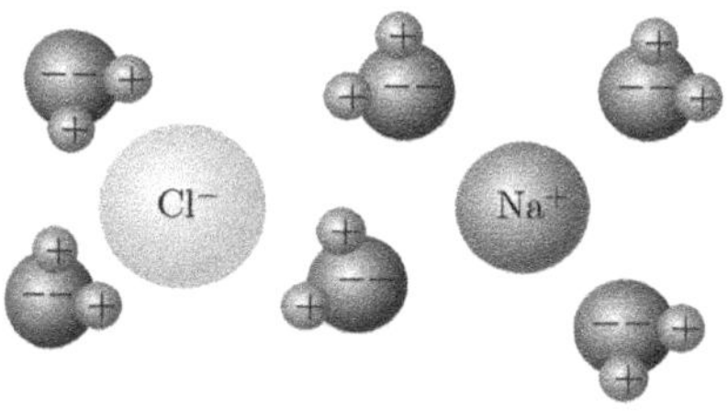

Fig. 6.2 Salt, NaCl, dissolved in water. Ionic compounds typically dissolve easily in water since the polar water molecules can screen the highly charged, but otherwise stable, ions.

6.2 Covalent Bond

A covalent bond is a bond where electrons are shared roughly equally between two atoms. There are several pictures that can be used to describe the covalent bond.

6.2.1 Particle in a Box Picture

Let us model a hydrogen atom as a box of size L for an electron (for simplicity, let us think about a one-dimensional system). The energy of a single electron in a box is (this should look familiar!)

$$E = \frac{\hbar^2 \pi^2}{2mL^2} \; .$$

Now suppose two such atoms come close together. An electron that is shared between the two atoms can now be delocalized over the positions of both atoms, thus it is in a box of size $2L$ and has lower energy

$$E = \frac{\hbar^2 \pi^2}{2m(2L)^2} \; .$$

This reduction in energy that occurs by delocalizing the electron is the driving force for forming the chemical bond. The new ground-state orbital is known as a *bonding* orbital.

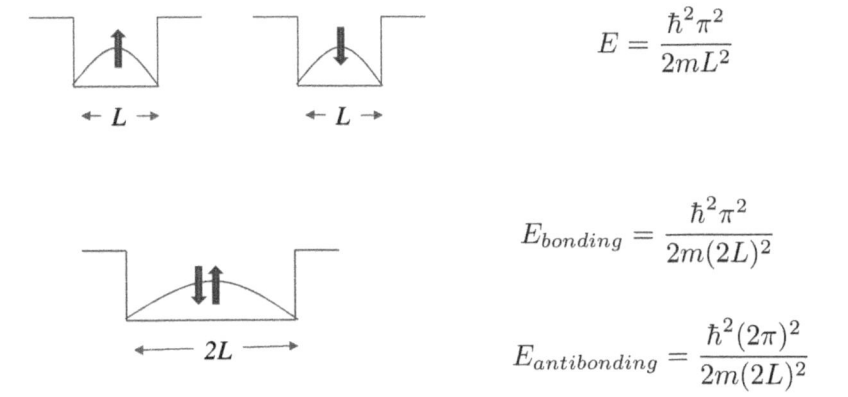

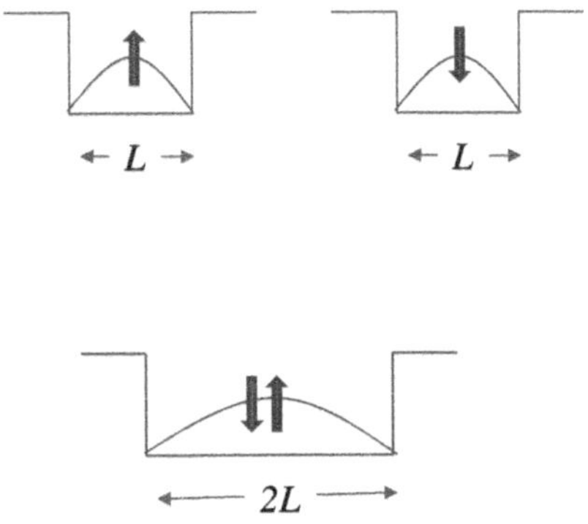

$$E = \frac{\hbar^2 \pi^2}{2mL^2}$$

$$E_{bonding} = \frac{\hbar^2 \pi^2}{2m(2L)^2}$$

$$E_{antibonding} = \frac{\hbar^2 (2\pi)^2}{2m(2L)^2}$$

Fig. 6.3 Particle in a box picture of covalent bonding. Two separated hydrogen atoms are like two different boxes each with one electron in the lowest eigenstate. When the two boxes are pushed together, one obtains a larger box—thereby lowering the energy of the lowest eigenstate—which is known as the *bonding* orbital. The two electrons can take opposite spin states and can thereby both fit in the bonding orbital. The first excited state is known as the *antibonding* orbital

If each atom starts with a single electron (i.e., it is a hydrogen atom) then when the two atoms come together to form a lower energy (bonding) orbital, then *both* electrons can go into this same ground-state orbital since they can take opposite spin states. This bonding process is depicted in the molecular orbital diagram shown in the top of Fig. 6.4. Of course the reduction in energy of the two electrons must compete against the Coulomb repulsion between the two nuclei, and the Coulomb repulsion of the two electrons with each other, which is a much more complicated calculation.

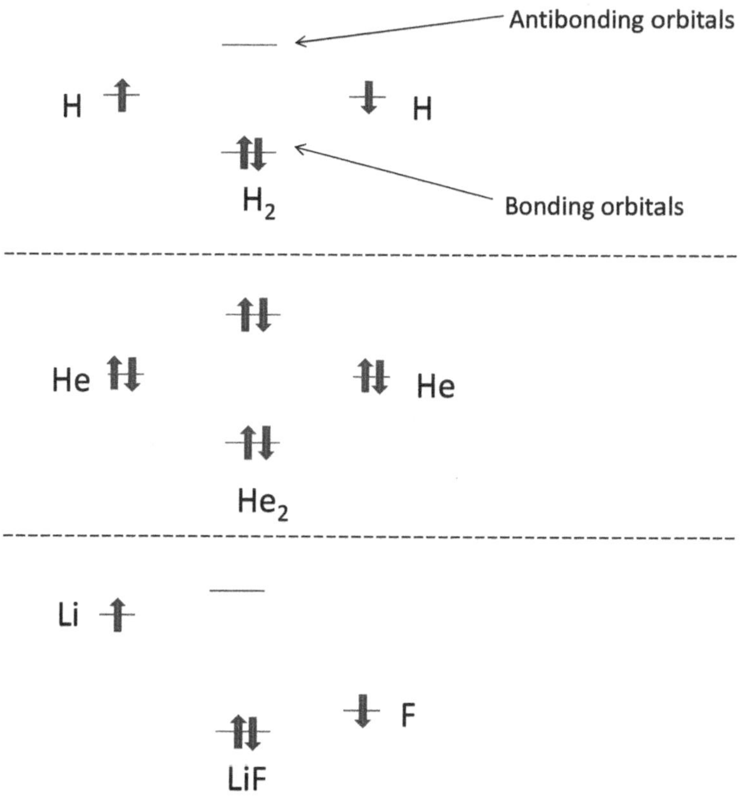

Fig. 6.4 Molecular orbital picture of bonding. In this type of picture, on the far left and far right are the orbital energies of the individual atoms well separated from each other (energy is the vertical axis). In the middle of the diagram are the orbital energies when the atoms come together to form a molecule. **Top:** Two hydrogen atoms (one having a spin-up electron and one having spin-down) come together to form a H_2 molecule. In the particle-in-a-box picture, the lowest-energy eigenstate is reduced in energy when the atoms come together and both electrons go into this bonding orbital. **Middle:** In the case of helium, since there are two electrons per atom, the bonding orbitals are filled, and the antibonding orbitals must be filled as well. The total energy is not reduced by the two helium atoms coming together (thus helium does not form He_2). **Bottom:** In the case of LiF, the energies of the lithium and the fluorine orbitals are different. As a result, the bonding orbital is mostly composed of the orbital on the F atom—meaning that the bonding electrons are mostly transferred from Li to F—forming a more ionic bond. See Exercise 6.3.

Now suppose we had started with two helium atoms, where each atom has two electrons, then when the two atoms come together there is not enough room in the single ground-state wavefunction. In this case, two of the four electrons must occupy the first excited orbital—which in this case turns out to be exactly the same electronic energy as the original ground-state orbital of the original atoms. Since no energy is gained by these electrons when the two atoms come together these excited orbitals are known as *antibonding* orbitals. (In fact it requires energy to push the two atoms together if one includes Coulomb repulsions between the nuclei.) This is depicted in the molecular orbital diagram in the middle of Fig. 6.4.

6.2.2 Molecular Orbital or Tight Binding Theory

In this section we make slightly more quantitative some of the idea of the previous section. Let us write a Hamiltonian for two hydrogen atoms. Since the nuclei are heavy compared to the electrons, we will fix the nuclear positions and solve the Schroedinger equation for the electrons as a function of the distance between the nuclei. This fixing of the position of nuclei is known as a "Born–Oppenheimer" approximation.[6,7] We hope to calculate the eigenenergies of the system as a function of the distance between the positively charged nuclei.

For simplicity, let us consider a single electron and two identical positive nuclei. We write the Hamiltonian as

$$H = K + V_1 + V_2$$

with

$$K = \frac{\mathbf{p}^2}{2m}$$

being the kinetic energy of the electron and

$$V_i = \frac{-e^2}{4\pi\epsilon_0|\mathbf{r} - \mathbf{R}_i|}$$

is the Coulomb interaction energy between the electron at position $\mathbf{r}$ and the nucleus at position $\mathbf{R}_i$.

Generally this type of Schroedinger equation is hard to solve exactly (in fact it *can* be solved exactly in this case, but it is not particularly enlightening to do so). Instead, we will attempt a variational solution. Let us write a trial wavefunction as

$$|\psi\rangle = \phi_1|1\rangle + \phi_2|2\rangle \tag{6.1}$$

where ϕ_i are complex coefficients, and the kets $|1\rangle$ and $|2\rangle$ are known as "atomic" orbitals or "tight binding" orbitals.[8] The form of Eq. 6.1 is frequently known as a "linear combination of atomic orbitals" or LCAO.[9] The orbitals which we use here can be taken as the ground-state solution of the Schroedinger equation when there is only one nucleus present:

$$\begin{aligned} (K + V_1)|1\rangle &= \epsilon_0|1\rangle \\ (K + V_2)|2\rangle &= \epsilon_0|2\rangle \end{aligned} \tag{6.2}$$

where ϵ_0 is the ground-state energy of the single atom.[10] I.e., $|1\rangle$ is a ground-state orbital for an electron bound to nucleus 1 and $|2\rangle$ is a ground-state orbital for an electron bound to nucleus 2.

We will now make a rough approximation that $|1\rangle$ and $|2\rangle$ are orthogonal so we can choose a normalization such that

$$\langle i|j\rangle = \delta_{ij} \quad . \tag{6.3}$$

When the two nuclei get very close together, this orthogonality is clearly no longer even close to correct. In Exercise 6.5 we repeat this calculation more correctly where we do not assume orthonormality.[11] But fortunately most of what we learn does not depend too much on whether the orbitals are orthogonal or not, so for simplicity we will assume orthonormal orbitals.

[6] Max Born (also the same guy from Born–von Karman boundary conditions) was one of the founders of quantum physics (see also margin note 1 from Chapter 4) winning a Nobel Prize in 1954. His daughter, and biographer, Irene, married into the Newton-John family, and had a daughter named Olivia, who became a pop icon and film star in the 1970s. Her most famous role was in the movie of *Grease* playing opposite John Travolta. When I was a kid, she was every teenage guy's dreamgirl (her, or Farrah Fawcett).

[7] J. Robert Oppenheimer later became the head scientific manager of the American atomic bomb project during the second world war. After this giant scientific and military triumph, he pushed for control of nuclear weapons leading to his being accused of being a communist sympathizer during the "Red" scares of the 1950s and he ended up having his security clearance revoked.

[8] The term "tight binding" is from the idea that an atomic orbital is tightly bound to its nucleus.

[9] The LCAO approach can be improved systematically by using more orbitals and more variational coefficients—which then can be optimized with the help of a computer. This general idea formed the basis of the quantum chemistry work of John Pople. See margin note 5 of the prior chapter.

[10] Here ϵ_0 is *not* a dielectric constant or the permittivity of free space, but rather the energy of an electron in an orbital (at some point we just run out of new symbols to use for new quantities!).

[11] Alternatively we could have orthonormality at the price of using orbitals which are not solutions to the Schroedinger equation for a single nucleus.

An effective Schroedinger equation can be written down for our variational wavefunction which (unsuprisingly) takes the form of an eigenvalue problem[12]

$$\sum_j H_{ij}\phi_j = E\phi_i \tag{6.4}$$

where

$$H_{ij} = \langle i|H|j\rangle$$

is a two-by-two matrix in this case (the equation generalizes in the obvious way to the case where there are more than two orbitals).

Recalling our definition of $|1\rangle$ as being the ground-state energy of $K + V_1$, we can write[13]

$$\begin{aligned}
H_{11} &= \langle 1|H|1\rangle = \langle 1|K+V_1|1\rangle + \langle 1|V_2|1\rangle = \epsilon_0 + V_{cross} & (6.5)\\
H_{22} &= \langle 2|H|2\rangle = \langle 2|K+V_2|2\rangle + \langle 2|V_1|2\rangle = \epsilon_0 + V_{cross} & (6.6)\\
H_{12} &= \langle 1|H|2\rangle = \langle 1|K+V_2|2\rangle + \langle 1|V_1|2\rangle = 0 \;-\; t & (6.7)\\
H_{21} &= \langle 2|H|1\rangle = \langle 2|K+V_2|1\rangle + \langle 2|V_1|1\rangle = 0 \;-\; t^* \,. & (6.8)
\end{aligned}$$

In the first two lines

$$V_{cross} = \langle 1|V_2|1\rangle = \langle 2|V_1|2\rangle$$

is the Coulomb potential felt by orbital $|1\rangle$ due to nucleus 2, or equivalently the Coulomb potential felt by orbital $|2\rangle$ due to nucleus 1. In the second two lines (Eqs. 6.7 and 6.8) we have also defined the so-called *hopping* term[14,15]

$$t = -\langle 1|V_2|2\rangle = -\langle 1|V_1|2\rangle.$$

The reason for the name "hopping" will become clear in a moment. Note that in the second two lines (Eqs. 6.7 and 6.8) the first term vanishes because of orthogonality of $|1\rangle$ and $|2\rangle$ (invoking also Eqs. 6.2).

At this point our Schroedinger equation is reduced to a two-by-two matrix equation of the form

$$\begin{pmatrix} \epsilon_0 + V_{cross} & -t \\ -t^* & \epsilon_0 + V_{cross} \end{pmatrix} \begin{pmatrix} \phi_1 \\ \phi_2 \end{pmatrix} = E \begin{pmatrix} \phi_1 \\ \phi_2 \end{pmatrix}. \tag{6.9}$$

The interpretation of this equation is roughly that orbitals $|1\rangle$ and $|2\rangle$ both have energies ϵ_0 which is shifted by V_{cross} due to the presence of the other nucleus. In addition the electron can "hop" from one orbital to the other by the off-diagonal t term. To understand this interpretation more fully, we realize that in the *time-dependent* Schroedinger equation, if the matrix were diagonal a wavefunction that started completely in orbital $|1\rangle$ would stay on that orbital for all time. However, with the off-diagonal term, the time-dependent wavefunction can oscillate between the two orbitals.

Diagonalizing this two-by-two matrix we obtain eigenenergies

$$E_\pm = \epsilon_0 + V_{cross} \pm |t|$$

[12] To derive this eigenvalue equation we start with an expression for the energy

$$E = \frac{\langle \psi|H|\psi\rangle}{\langle \psi|\psi\rangle}$$

then with ψ written in the variational form of Eq. 6.1, we minimize the energy by setting $\partial E/\partial \phi_i = \partial E/\partial \phi_i^* = 0$. See Exercise 6.2.

[13] In atomic physics courses, the quantities V_{cross} and t are often called a *direct* and *exchange* terms and are sometimes denoted $\mathcal{J}$ and $\mathcal{K}$. We avoid this terminology because the same words are almost always used to describe 2-electron interactions in condensed matter.

[14] The minus sign is a convention for the definition of t. Note that we can (and often will) choose t to be real and positive by multiplying the wavefunction $|1\rangle$ by an appropriate complex phase (a so-called "gauge transformation"). Note also that t has dimensions of energy.

[15] The second equality here can be obtained by rewriting

$$H_{12} = \langle 1|K+V_1|2\rangle + \langle 1|V_2|2\rangle$$

and allowing $K + V_1$ to act to the left.

where the lower energy orbital is the bonding orbital whereas the higher energy orbital is the antibonding. The corresponding wavefunctions are then[16]

[16] The bonding wavefunction corresponds to $\phi_1 = 1/\sqrt{2}$ and $\phi_2 = \pm 1/\sqrt{2}$ whereas the antibonding wavefunction corresponds to $\phi_1 = 1/\sqrt{2}$ and $\phi_2 = \mp 1/\sqrt{2}$. See Eq. 6.1.

$$|\psi_{bonding}\rangle = \frac{1}{\sqrt{2}}(\,|1\rangle \pm |2\rangle\,) \quad (6.10)$$

$$|\psi_{antibonding}\rangle = \frac{1}{\sqrt{2}}(\,|1\rangle \mp |2\rangle\,). \quad (6.11)$$

I.e., these are the symmetric and antisymmetric superpositions of orbitals. The signs $\pm$ and $\mp$ depend on the sign of t, where the lower energy one is always called the bonding orbital and the higher energy one is called antibonding. To be precise $t > 0$ makes $(|1\rangle + |2\rangle)/\sqrt{2}$ the lower energy bonding orbital. Roughly one can think of these two wavefunctions as being the lowest two "particle-in-a-box" orbitals— the lowest-energy wavefunction does not change sign as a function of position, whereas the first excited state changes sign once, i.e., it has a single node (for the case of $t > 0$ the analogy is precise).

It is worth briefly considering what happens if the two nuclei being bonded together are not identical. In this case the energy ϵ_0 for an electron to sit on orbital 1 would be different from that of orbital 2 (see bottom of Fig. 6.4). The matrix equation, Eq. 6.9, would no longer have equal entries along the diagonal, and the magnitude of ϕ_1 and ϕ_2 would no longer be equal in the ground state as they are in Eq. 6.10. Instead, the lower-energy orbital would be more greatly filled in the ground state. As the energies of the two orbitals become increasingly different, the electron is more completely transferred entirely onto the lower-energy orbital, essentially becoming an ionic bond.

Aside: In Section 23.3, we will consider a more general tight binding model with more than one electron in the system and with Coulomb interactions between electrons as well. That calculation is more complicated, but shows very similar results. That calculation is also much more advanced, but might be fun to read for the adventurous.

Note again that V_{cross} is the energy that the electron on orbital 1 feels from nucleus 2. However, we have not included the fact that the two nuclei also interact, and to a first approximation, this Coulomb repulsion between the two nuclei will cancel[17] the attractive energy between the nucleus and the electron on the opposite orbital. Thus, including this energy we will obtain

[17] If you think of a positively charged nucleus and a negatively charged electron surrounding the nucleus, from far outside of that electron's orbital radius the atom looks neutral. Thus a second nucleus will neither be attracted nor repelled from the atom so long as it remains outside of the electron cloud of the atom.

$$\tilde{E}_\pm \approx \epsilon_0 \pm |t|.$$

As the nuclei get closer together, the hopping term $|t|$ increases, giving an energy level diagram as shown in Fig. 6.5. This picture is obviously unrealistic, as it suggests that two atoms should bind together at zero distance between the nuclei. The problem here is that our assumptions and approximations begin to break down as the nuclei get closer together (for example, our orbitals are no longer orthogonal, V_{cross} does not exactly cancel the Coulomb energy between nuclei, etc.).

A more realistic energy level diagram for the bonding and antibonding states is given in Fig. 6.6. Note that the energy diverges as the nuclei get pushed together (this is from the Coulomb repulsion between nuclei). As such there is a minimum energy of the system when the nuclei are at some non-zero distance apart from each other, which then becomes the ground-state distance of the nuclei in the resulting molecule.

Aside: In Fig. 6.6 there is a minimum of the bonding energy when the nuclei are some particular distance apart. This optimal distance will be the distance of the bond between two atoms. However, at finite temperature, the distance will fluctuate around this minimum (think of a particle in a potential well at finite temperature). Since the potential well is steeper on one side than on the other, at finite temperature, the "particle" in this well will be able to fluctuate to larger distances a bit more than it is able to fluctuate to smaller distances. As a result, the average bond distance will increase at finite temperature. This thermal expansion will be explored further in Chapter 8.

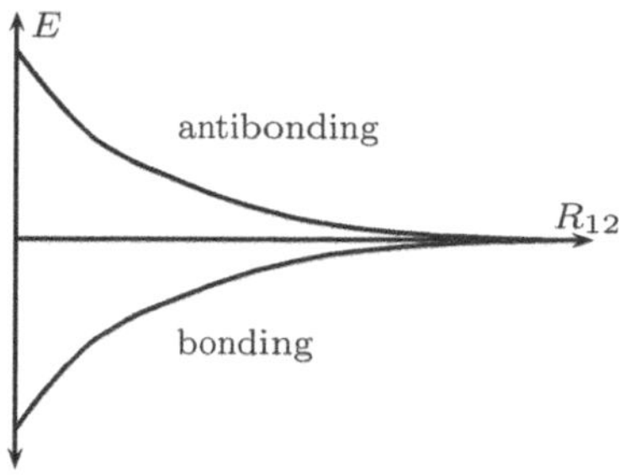

Fig. 6.5 Model tight binding energy levels as a function of distance between the nuclei of the atoms.

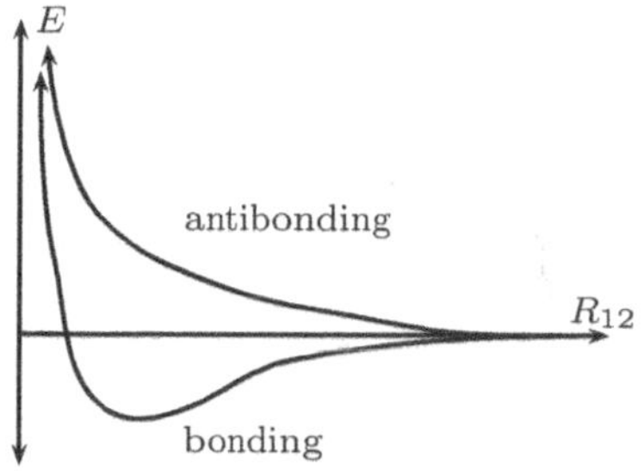

Fig. 6.6 More realistic energy levels as a function of distance between the nuclei of the atoms.

Covalently bonded materials tend to be strong and tend to be electrical semiconductors[18] or insulators (roughly since electrons are tied up in the local bonds—we will give a more detailed description of what makes a material an insulator in Chapter 16). The directionality of the orbitals makes these materials retain their shape well (non-ductile) so they are brittle. They do not dissolve in polar solvents such as water in the same way that ionic materials do.

6.3 Van der Waals, Fluctuating Dipole Forces, or Molecular Bonding

When two atoms (or two molecules) are very far apart from each other, there remains an attraction between them due to what is known as van der Waals[19] forces, sometimes known as fluctuating dipole forces, or molecular bonding. In short, both atoms have a dipole moment, which may be zero on average, but can fluctuate "momentarily" due to quantum mechanics.[20] If the first atom obtains a momentary dipole moment, the second atom can polarize—also obtaining a dipole moment to lower its energy. As a result, the two atoms (momentarily dipoles) will attract each other.[21]

This type of bonding between atoms is very typical of inert atoms (such as noble gases: He, Ne, Kr, Ar, Xe) whose electrons do not participate in covalent bonds or ionic bonds. It is also typical of bonding between inert[22] molecules such as nitrogen molecules N_2 where there is no possibility for the electrons in this molecule to form covalent or ionic

[18] We have not defined the word "semiconductor" yet, but we will return to it later in depth (for example, in Chapter 17).

[19] J. D. van der Waals was awarded the Nobel Prize in physics in 1910 for his work on the structure of liquids and gases. You may remember the van der Waals equation of state from your thermodynamics courses. There is a crater named after him on the far side of the moon.

[20] This is a slightly imprecise, but useful, interpretation of quantum-mechanical uncertainty as being temporal fluctuation.

[21] Some people also call the force between two permanent dipoles a van der Waals force — and also the force between one permanent and one induced (or momentary) dipole. We may or may not agree that these should be called van der Waals forces, but no matter what we call it, we all must agree that there is a force!

[22] Whereas the noble gases are inert because they have filled atomic orbital shells, the nitrogen molecule is inert essentially because it has a filled shell of *molecular* orbitals—all of the bonding orbitals are filled, and there is a large energy gap to any antibonding orbitals.

bonds between molecules. This bonding is weak compared to covalent or ionic bonds, but it is also long ranged in comparison since the electrons do not need to hop between atoms.

To be more quantitative, let us consider an electron orbiting a nucleus (say, a proton). If the electron is at a fixed position, there is a dipole moment $\mathbf{p} = e\mathbf{d}$, where $\mathbf{d}$ is the vector from the electron to the proton. With the electron "orbiting" (i.e, in an unperturbed eigenstate), the average dipole moment is zero. However, if an electric field is applied to the atom, the atom will develop a polarization (i.e., it will be more likely for the electron to be found on one side of the nucleus than on the other). We write

$$\mathbf{p} = \chi \mathbf{E}$$

where χ is known as the polarizability (also known as electric susceptibility). This polarizability can be calculated for, say, a hydrogen atom explicitly.[23] At any rate, it is some positive quantity.

[23] This is a good exercise in quantum mechanics. See for example, E. Merzbacher, *Quantum Mechanics*, Wiley 1961. See also Exercise 6.6.

Now, let us suppose we have two such atoms, separated by a distance r in the $\hat{x}$ direction (see Fig. 6.7). Suppose one atom momentarily has a dipole moment $\mathbf{p}_1$ and for definiteness, suppose this dipole moment is in the $\hat{z}$ direction. Then the second atom will feel an electric field

$$E = \frac{p_1}{4\pi\epsilon_0 r^3}$$

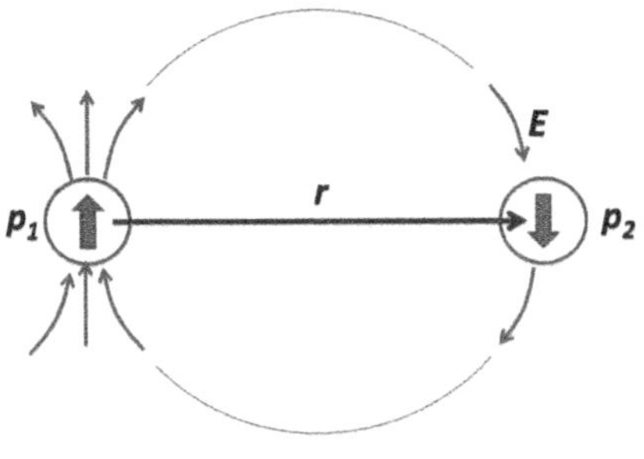

Fig. 6.7 An atom with a polarization p_1 induces a polarization p_2 in a second atom.

in the negative $\hat{z}$ direction. The second atom then, due to its polarizability, develops a dipole moment $p_2 = \chi E$ which in turn is attracted to the first atom. The potential energy between these two dipoles is[24]

$$U = \frac{-|p_1||p_2|}{4\pi\epsilon_0 r^3} = \frac{-p_1\chi E}{(4\pi\epsilon_0)r^3} = \frac{-|p_1|^2\chi}{(4\pi\epsilon_0 r^3)^2}. \qquad (6.12)$$

[24] Had we chosen the polarization to not be perpendicular to the separation between atoms we would have an additional factor of $(1+3\cos^2\theta)$ with θ the angle between $\mathbf{p}_1$ and $\mathbf{r}$.

Therefore there is a force $-dU/dr$ which is attractive and proportional to $1/r^7$.

You can check that independent of the direction of the original dipole moment, the force is always attractive and proportional to $1/r^7$. Although there will be a (non-negative) prefactor which depends on the angle between the dipole moment $\mathbf{p}_1$ and $\mathbf{r}$ the direction between the two atoms.

Note that this argument appears to depend on the fact that the dipole moment $\mathbf{p}_1$ of the first atom is non-zero, whereas on average the atom's dipole moment is actually zero. However in Eq. 6.12 in fact what enters is $|\mathbf{p}_1|^2$ which has a non-zero expectation value (this is precisely the calculation that $\langle x \rangle$ for an electron in a hydrogen atom is zero, but $\langle x^2 \rangle$ is non-zero).

While these fluctuating dipolar forces are generally weak, they are the only forces that occur when electrons cannot be shared or transferred between atoms—either in the case where the electrons are not chemically active or when the atoms are far apart. Although individually weak, when considering the van der Waals forces of many atoms put together, the total forces can be quite strong. A well-known example of a van der Waals force is the force that allows lizards such as geckos to climb up even very smooth walls (such as glass windows). They have hair on their feet that makes very close contact with the atoms of the wall, and they then stick to the wall mostly due to van der Waals forces!

6.4 Metallic Bonding

It is sometimes hard to distinguish metallic bonding from covalent bonding. Roughly, however, one defines a metallic bond to be the bonding that occurs in a metal. These bonds are similar to covalent bonds in the sense that electrons are shared between atoms, but in this case the electrons become delocalized throughout the crystal (we will discuss how this occurs in Section 11.2). We should think of the delocalized free electrons as providing the glue that holds together the positive ions that they have left behind.

Since the electrons are completely delocalized, the bonds in metals tend not to be directional. Metals are thus often ductile and malleable. Since the electrons are free, metals are good conductors of electricity as well as of heat.

6.5 Hydrogen Bonds

The hydrogen atom is extremely special due to its very small size. As a result, the bonds formed with hydrogen atoms are qualitatively different from other bonds. When the hydrogen atom forms a covalent or ionic bond with a larger atom, being small, the hydrogen nucleus (a proton) simply sits on the surface of its partner. This then makes the molecule (hydrogen and its partner) into a dipole. These dipoles can then attract charges, or other dipoles, as usual.

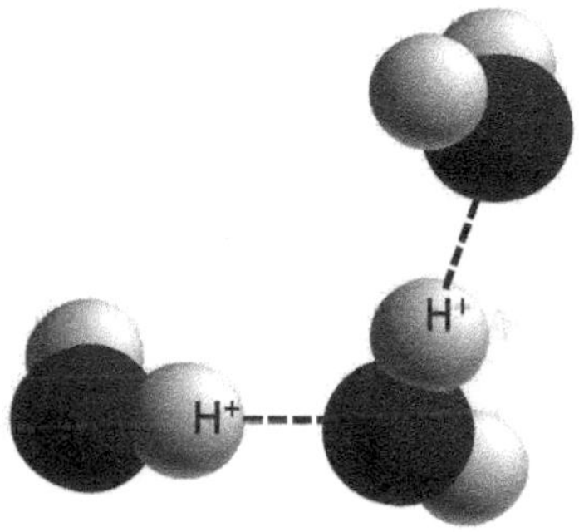

Fig. 6.8 Hydrogen bonds in H_2O. A hydrogen on one water molecule is attracted to an oxygen on another molecule forming a weak hydrogen bond. These bonds are strong enough to form ice below the 273.15 K.

What is special about hydrogen is that when it forms a bond, and its electron is attracted away from the proton onto (or partially onto) its partner, the unbonded side of the proton left behind is a naked positive charge—unscreened by any electrons in core orbitals. As a result, this positive charge is particularly effective in being attracted to other clouds of electrons.

A very good example of the hydrogen bond is water, H_2O. Each oxygen atom is bound to two hydrogens (however because of the atomic orbital structure, these atoms are not collinear). The hydrogens, with their positive charge remain attracted to oxygens of other water molecules. In ice, these attractions are strong enough to form a weak but stable bond between water molecules, thus forming a crystal. Sometimes one can think of the hydrogen atom as forming "half" a bond with two oxygen atoms, thus holding the two oxygen atoms together.

Hydrogen bonding is also extremely important in biological molecules where, for example, hydrogen bonds hold together strands of DNA.

Chapter Summary (Table)

Type of Bonding	Description	Typical of which compounds	Typical Properties
Ionic	Electron is transferred from one atom to another, and the resulting ions attract each other.	Binary compounds made of constituents with very different electronegativity: e.g., group I–VII compounds such as NaCl.	• Hard, very brittle • High melting temperature • Electrical insulator • Water soluble
Covalent	Electron is shared between two atoms forming a bond. Energy lowered by delocalization of wavefunction.	Compounds made of constituents with similar electronegativities (e.g., III–V compounds such as GaAs), or solids made of one element only such as diamond (C).	• Very hard (brittle) • High melting temperature • Electrical insulators or semiconductors
Metallic	Electrons are delocalized throughout the solid forming a glue between positive ions.	Metals. Left and middle of periodic table.	• Ductile, malleable (due to non-directional nature of bond). Can be hardened by adding certain impurities. • Lower melting temperature • Good electrical and thermal conductors
Molecular (van der Waals, fluctuating dipole)	No transfer of electrons. Temporary or permanent dipole moments on constituents align to cause attraction. Bonding strength increases with size of molecule or polarity of constituent.	Noble gas solids, solids made of non-polar (or slightly polar) molecules binding to each other (wax).	• Soft, weak • Low melting temperature • Electrical insulators
Hydrogen	Involves hydrogen ion bound to one atom but still attracted to another. Special case because H is so small.	Important in organic and biological materials. Holds together ice.	• Weak bond (stronger than vdW though) • Important for maintaining shape of DNA and proteins

Table 6.1 Types of bonds. This table should be thought of as providing rough rules. Many materials show characteristics intermediate between two (or more!) classes. Chemists often subdivide each of these classes even further.

References on Chemical Bonding

- Rosenberg, sections 1.11–1.19
- Ibach and Luth, chapter 1
- Hook and Hall, section 1.6
- Kittel, chapter 3 up to elastic strain
- Dove, chapter 5
- Ashcroft and Mermin, chapters 19–20
- Burns, sections 6.2–6.6 and also chapters 7 and 8
- Pauling, chapters 1–3
- Murrel *et al.*, chapters 1–6 (This has some detail on LCAO method.)

Most of these references are far more than you probably want to know. The first four on the list are good starting points.

Exercises

(6.1) **Chemical Bonding**

(a) Qualitatively describe five different types of chemical bonds and why they occur.

▷ Describe which combinations of what types of atoms are expected to form which types of bonds (make reference to location on the periodic table).

▷ Describe some of the qualitative properties of materials that have these types of bonds.

(Yes, you can just copy the table out of the chapter summary, but the point of this exercise is to learn the information in the table!)

(b) Describe qualitatively the phenomenon of van der Waals forces. Explain why the force is attractive and proportional to $1/R^7$ where R is the distance between two atoms.

(6.2) **Covalent Bonding in Detail***

(a) *Linear Combination of Atomic Orbitals:*

In Section 6.2.2 we considered two atoms each with a single atomic orbital. We called the orbital $|1\rangle$ around nucleus 1 and $|2\rangle$ around nucleus 2. More generally we may consider any set of wavefunctions $|n\rangle$ for $n = 1, \ldots, N$. For simplicity, let us assume this basis is orthonormal $\langle n|m\rangle = \delta_{n,m}$ (More generally, one cannot assume that the basis set of orbitals is orthonormal. In Exercise 6.5 we properly consider a non-orthonormal basis.)

Let us write a trial wavefunction for our ground state as

$$|\Psi\rangle = \sum_n \phi_n |n\rangle.$$

This is known as a linear combination of atomic orbitals, LCAO, or tight binding (it is used heavily in numerical simulation of molecules).

We would like to find the lowest-energy wavefunction we can construct in this form, i.e., the best approximation to the actual ground-state wavefunction. (The more states we use in our basis, generally, the more accurate our results will be.) We claim that the ground state is given by the solution of the effective Schroedinger equation

$$\mathcal{H}\boldsymbol{\phi} = E\boldsymbol{\phi} \tag{6.13}$$

where $\boldsymbol{\phi}$ is the vector of N coefficients ϕ_n, and $\mathcal{H}$ is the N by N matrix

$$\mathcal{H}_{n,m} = \langle n|H|m\rangle$$

with H the Hamiltonian of the full system we are considering. To prove this, let us construct the energy

$$E = \frac{\langle\psi|H|\psi\rangle}{\langle\psi|\psi\rangle}$$

▷ Show that minimizing this energy with respect to each ϕ_n gives the same eigenvalue equation, Eq. 6.13. (Caution: ϕ_n is generally complex! If you are not comfortable with complex differentiation, write everything in terms of real and imaginary parts of each ϕ_n.) Similarly, the second eigenvalue of the effective Schroedinger equation will be an approximation to the first excited state of the system.

(b) *Two-orbital covalent bond*

Let us return to the case where there are only two orbitals in our basis. This pertains to a case where we have two identical nuclei and a single electron which will be shared between them to form a covalent bond. We write the full Hamiltonian as

$$H = \frac{\mathbf{p}^2}{2m} + V(\mathbf{r} - \mathbf{R_1}) + V(\mathbf{r} - \mathbf{R_2}) = K + V_1 + V_2$$

where V is the Coulomb interaction between the electron and the nucleus, R_1 is the position of the first nucleus and R_2 is the position of the second nucleus. Let ϵ be the energy of the atomic orbital around one nucleus in the absence of the other. In other words

$$\begin{aligned}(K + V_1)|1\rangle &= \epsilon|1\rangle \\ (K + V_2)|2\rangle &= \epsilon|2\rangle\end{aligned}$$

Define also the cross-energy element

$$V_{cross} = \langle 1|V_2|1\rangle = \langle 2|V_1|2\rangle$$

and the hopping matrix element

$$t = -\langle 1|V_2|2\rangle = -\langle 1|V_1|2\rangle$$

These are not typos!

▷ Why can we write V_{cross} and t equivalently using either one of the expressions given on the right-hand side?

▷ Show that the eigenvalues of our Schroedinger equation Eq. 6.13 are given by

$$E = \epsilon + V_{cross} \pm |t|$$

▷ Argue (perhaps using Gauss's law) that V_{cross} should roughly cancel the repulsion between nuclei, so that, in the lower eigenstate the total energy is indeed lower when the atoms are closer together.

▷ This approximation must fail when the atoms get sufficiently close. Why?

(6.3) **LCAO and the Ionic–Covalent Crossover**

For Exercise 6.2.b consider now the case where the atomic orbitals $|1\rangle$ and $|2\rangle$ have unequal energies $\epsilon_{0,1}$ and $\epsilon_{0,2}$. As the difference in these two energies increases show that the bonding orbital becomes more localized on the lower-energy atom. For simplicity you may use the orthogonality assumption $\langle 1|2\rangle = 0$. Explain how this calculation can be used to describe a crossover between covalent and ionic bonding.

(6.4) **Ionic Bond Energy Budget**

The ionization energy of a sodium atom is about 5.14 eV. The electron affinity of a chlorine atom is about 3.62 eV. When a single sodium atom bonds with a single chlorine atom, the bond length is roughly 0.236 nm. Assuming that the cohesive energy is purely Coulomb energy, calculate the total energy released when a sodium atom and a chlorine atom come together to form a NaCl molecule. Compare your result to the experimental value of 4.26 eV. Qualitatively account for the sign of your error.

(6.5) **LCAO Done Right***

(a)* In Exercise 6.2 we introduced the method of linear combination of atomic orbitals. In that exercise we assumed that our basis of orbitals is orthonormal. In this exercise we will relax this assumption.

Consider now many orbitals on each atom (and potentially many atoms). Let us write

$$|\psi\rangle = \sum_{i=1}^{N} \phi_i |i\rangle$$

for an arbitrary number N of orbitals. Let us write the N by N overlap matrix $\mathcal{S}$ whose elements are

$$\mathcal{S}_{i,j} = \langle i|j\rangle$$

In this case do *not* assume that $\mathcal{S}$ is diagonal.

Using a similar method as in Exercise 6.2, derive the new "Schroedinger equation"

$$\mathcal{H}\phi = E\mathcal{S}\phi \qquad (6.14)$$

with the same notation for $\mathcal{H}$ and ϕ as in Exercise 6.2. This equation is known as a "generalized eigenvalue problem" because of the $\mathcal{S}$ on the right-hand side.

(b)** Let us now return to the situation with only two atoms and only one orbital on each atom but such that $\langle 1|2\rangle = \mathcal{S}_{1,2} \neq 0$. Without loss of generality we may assume $\langle i|i\rangle = 1$ and $\mathcal{S}_{1,2}$ is real. If the atomic orbitals are s-orbitals then we may assume also that t is real and positive (why?).

Use Eq. 6.14 to derive the eigenenergies of the system.

(6.6) **Van der Waals Bonding in Detail***

(a) Here we will do a much more precise calculation of the van der Waals force between two hydrogen atoms. First, let the positions of the two nuclei be separated by a vector $\mathbf{R}$, and let the vector from nucleus 1 to electron 1 be $\mathbf{r}_1$ and let the vector from nucleus 2 to electron 2 be $\mathbf{r}_2$ as shown in the following figure.

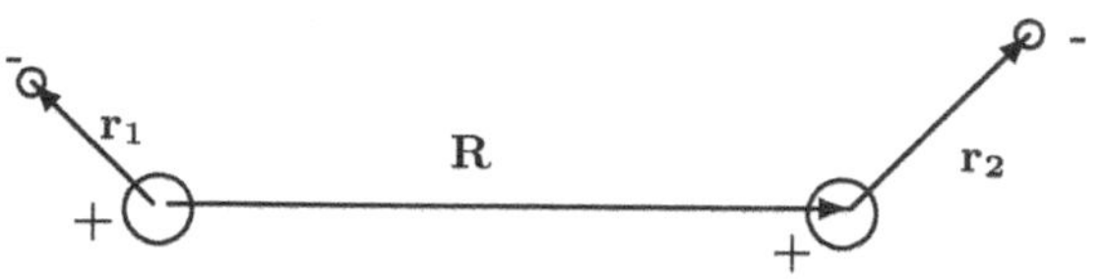

Let us now write the Hamiltonian for both atoms (assuming fixed positions of nuclei, i.e., using Born–Oppenheimer approximation) as

$$
\begin{aligned}
H &= H_0 + H_1 \\
H_0 &= \frac{\mathbf{p}_1{}^2}{2m} + \frac{\mathbf{p}_2{}^2}{2m} - \frac{e^2}{4\pi\epsilon_0|\mathbf{r}_1|} - \frac{e^2}{4\pi\epsilon_0|\mathbf{r}_2|} \\
H_1 &= \frac{e^2}{4\pi\epsilon_0|\mathbf{R}|} + \frac{e^2}{4\pi\epsilon_0|\mathbf{R}-\mathbf{r}_1+\mathbf{r}_2|} \\
&\quad - \frac{e^2}{4\pi\epsilon_0|\mathbf{R}-\mathbf{r}_1|} - \frac{e^2}{4\pi\epsilon_0|\mathbf{R}+\mathbf{r}_2|}
\end{aligned}
$$

Here H_0 is the Hamiltonian for two non-interacting hydrogen atoms, and H_1 is the interaction between the atoms.

Without loss of generality, let us assume that $\mathbf{R}$ is in the $\hat{x}$ direction. Show that for large $\mathbf{R}$ and small $\mathbf{r}_i$, the interaction Hamiltonian can be written as

$$H_1 = \frac{e^2}{4\pi\epsilon_0|\mathbf{R}|^3}(z_1z_2 + y_1y_2 - 2x_1x_2) + \mathcal{O}(1/R^4)$$

where x_i, y_i, z_i are the components of $\mathbf{r}_i$. Show that this is just the interaction between two dipoles.

(b) *Perturbation Theory:*

The eigenvalues of H_0 can be given as the eigenvalues of the two atoms separately. Recall that the eigenstates of hydrogen are written in the usual notation as $|n,l,m\rangle$ and have energies $E_n = -\text{Ry}/n^2$ with $\text{Ry} = me^4/(32\pi^2\epsilon_0^2\hbar^2) = e^2/(8\pi\epsilon_0 a_0)$ the Rydberg (here $l \geqslant 0$, $|m| \leqslant l$ and $n \geqslant l+1$). Thus the eigenstates of H_0 are written as $|n_1,l_1,m_1;n_2,l_2,m_2\rangle$ with energies $E_{n_1,n_2} = -\text{Ry}(1/n_1^2 + 1/n_2^2)$. The ground state of H_0 is $|1,0,0;1,0,0\rangle$.

▷ Perturbing H_0 with the interaction H_1, show that to first order in H_1 there is no change in the ground-state energy. Thus conclude that the leading correction to the ground-state energy is proportional to $1/R^6$ (and hence the force is proportional to $1/R^7$).

▷ Recalling second-order perturbation theory show that we have a correction to the total energy given by

$$\delta E = \sum_{\substack{n_1,n_2\\l_1,l_2\\m_1,m_2}} \frac{|\langle 1,0,0;1,0,0|\ H_1\ |n_1,l_1,m_1;n_2,l_2,m_2\rangle|^2}{E_{0,0} - E_{n_1,n_2}}$$

▷ Show that the force must be attractive.

(c)* *Bounding the binding energy:*

First, show that the numerator in this expression is zero if either $n_1 = 1$ or $n_2 = 1$. Thus the smallest E_{n_1,n_2} that appears in the denominator is $E_{2,2}$. If we replace E_{n_1,n_2} in the denominator with $E_{2,2}$ then the $|\delta E|$ we calculate will be greater than than the $|\delta E|$ in the exact calculation. On the other hand, if we replace E_{n_1,n_2} by 0, then the $|\delta E|$ will always be less than the δE of the exact calculation.

▷ Make these replacements, and perform the remaining sum by identifying a complete set. Derive the bound

$$\frac{6e^2a_0^5}{4\pi\epsilon_0 R^6} \leqslant |\delta E| \leqslant \frac{8e^2a_0^5}{4\pi\epsilon_0 R^6}$$

You will need the matrix element for a hydrogen atom

$$\langle 1,0,0|x^2|1,0,0\rangle = a_0^2$$

where $a_0 = 4\pi\epsilon_0\hbar^2/(me^2)$ is the Bohr radius. (This last identity is easy to derive if you remember that the ground-state wavefunction of a hydrogen atom is proportional to $e^{-r/2a_0}$.)

Types of Matter

7

Once we understand how it is that atoms bond together, we can examine what types of matter can be formed. This chapter will give a *very brief* and obviously crude, but obligatory, overview of some of these types of matter.

Atoms can obviously bond together the form regular crystals. A crystal is made of small units reproduced many times and built into a regular array. The macroscopic morphology of a crystal can reflect its underlying structure (See Fig. 7.1). We will spend much of the remainder of this book studying crystals.

Fig. 7.1: **Left:** Small units reproduced periodically to form a crystal. This particular figure depicts NaCl (table salt), with the larger spheres being Cl^- ions and the smaller spheres being Na^+ ions. **Right:** The macroscopic morphology of a crystal often will reflect the underlying microscopic structure. These are large crystals of salt (also known as halite). Photograph by Piotr Włodarczyk, used by kind permission.

It is also possible that atoms will bind together to form molecules, and the molecules will stick together via weak van der Waals bonds to form so-called *molecular crystals* (see Fig. 7.2).

Fig. 7.2 A molecular crystal. **Left:** 60 atoms of carbon bind together to form a large molecule known as a buckyball.[1] **Right:** the buckyballs stick together by weak van der Waals bonds to form a molecular crystal.

[1]The name "buckyball" is a nickname for Buckminsterfullerene, named after Richard Buckminster Fuller, the famed developer of the geodesic dome, which buckyballs are supposed to resemble; although the shape is actually precisely that of a soccer ball. This name was chosen by the discoverers of the buckyball, Harold Kroto, Robert Curl, and Richard Smalley, who were awarded a Nobel Prize in chemistry for their discovery despite their choice of nomenclature (probably the name "soccerballene" would have been better).

Fig. 7.3 Cartoon of a liquid. In liquids, molecules are not in an ordered configuration and are free to move around (i.e, the liquid can flow). However, the liquid molecules do attract each other and at any moment in time you can typically define neighbors.

Another form of matter is liquid. Here, atoms are attracted to each other, but not so strongly that they form permanent bonds (or the temperature is high enough to make the bonds unstable). Liquids (and gases)[2] are disordered configurations of molecules where the molecules are free to move around into new configurations (see Fig. 7.3).

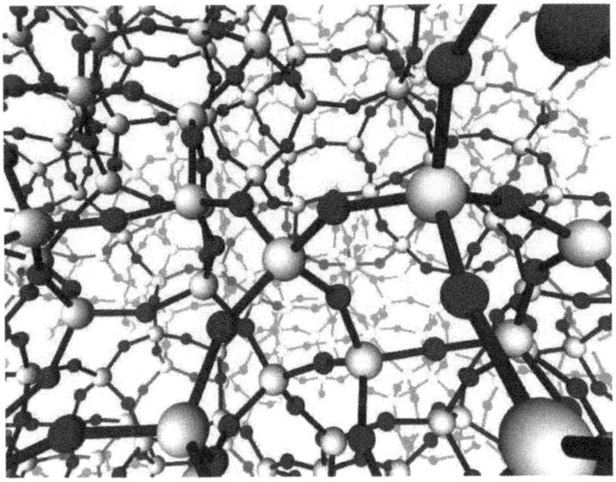

Fig. 7.4. Molecular structure of an amorphous solid: Silica (SiO_2) can either be a crystal (such as quartz) or it can be amorphous (such as window glass). In this depiction of amorphous silica, the Si atoms are the lighter shaded atoms, each having four bonds and the O atoms are the dark atoms, each having two bonds. Here the atoms are disordered, but are bonded together and cannot flow.

Somewhere midway between the idea of a crystal and the idea of a liquid is the possibility of amorphous[3] solids (including glasses). In this case the atoms are bonded into position in a disordered configuration. Unlike a liquid, the atoms cannot flow freely (see Fig. 7.4)

[3]The word "amorphous" is from Greek, meaning "without form".

Many more possibilities exist. For example, one may have so-called liquid crystals, where the system orders in some ways but remains disordered in other ways. For example, in Fig. 7.5.b the system is crys-

[2]As we should have learned in our stat-mech and thermo courses, there is no "fundamental" difference between a liquid and a gas. Generally liquids are high density and not very compressible, whereas gases are low density and very compressible. A single substance (say, water) may have a phase transition between its gas and liquid phase (boiling), but one can also go continuously from the gas to liquid phase without boiling by going to high pressure before raising the temperature and going around the critical point (going "supercritical").

talline (ordered) in one direction, but remains disordered within each plane. One can also consider cases where the molecules are always oriented the same way but are at completely random positions (known as a "nematic", see Fig. 7.5.c). There are a huge variety of possible liquid crystal phases of matter. In every case it is the interactions between the molecules ("bonding" of some type, whether it be weak or strong) that dictates the configurations.

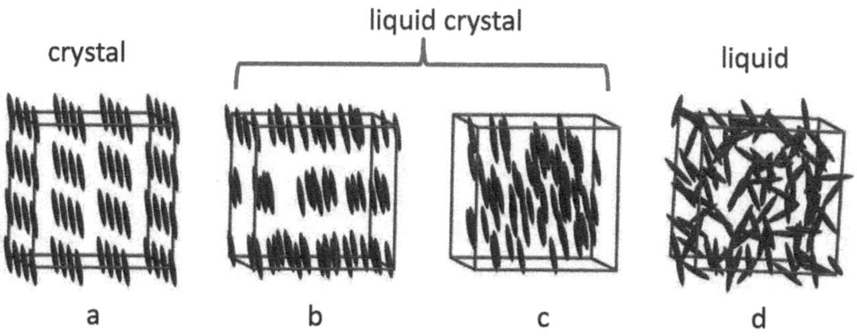

Fig. 7.5 Cartoon of liquid crystals. Liquid crystals have some of the properties of a solid and some of the properties of a liquid. **(a)** The far left is a crystal of molecules—all the molecules are positionally ordered and all are oriented in the same direction. **(b)** In the middle left picture the molecules retain their orientation, and retain some of their positional order—they group into discrete layers—thus being "crystalline" in the vertical direction. But within each layer, they are disordered and even can flow within the layer (this is known as a smectic-C phase). **(c)** In this figure, the positional order is lost, the positions of the molecules are random, but the molecules all retain their orientations (this is known as a nematic phase). **(d)** On the far right, the system is a true liquid, there is no positional order or orientational order.

One can also have so-called quasi-crystals which are ordered but non-periodic arrangements. In a quasi-crystal, such as the one shown in Fig. 7.6, component units are assembled together with a set of regular rules which appears to make a periodic structure, but in fact the pattern is non-repeating.[4] Although quasicrystals made of atoms are extremely rare in nature,[5] many man-made quasicrystalline materials are now known.

[5]The first naturally occuring quasicrystal was found in 2009. It is believed to be part of a meteorite.

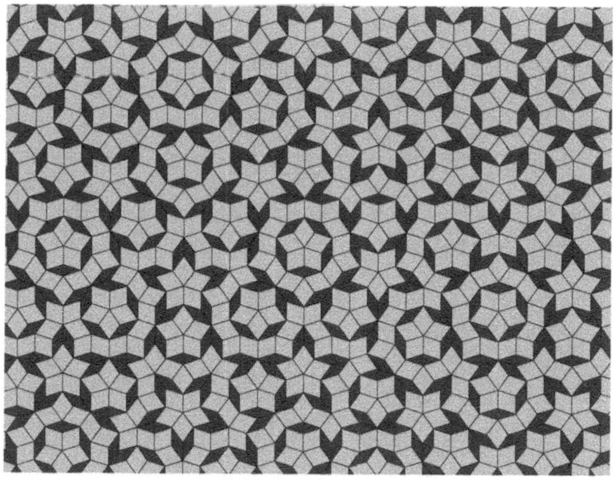

Fig. 7.6 This quasicrystal, known as Penrose tiling, can be assembled by following a simple set of rules. While the pattern looks regular it is actually non-periodic as it never repeats.

[4]The fact that chemical compounds can have regular but non-repeating structures was extremely controversial at first. After discovering this phenomenon in 1982, Dan Shechtman's claims were initially rejected by the scientific community. The great Linus Pauling was particularly critical of the idea (see margin note 5 in Chapter 6). Eventually, Shechtman was proven right and was awarded the Nobel Prize in chemistry in 2011.

One should also be aware of polymers,[6] which are long chains of atoms. Examples include DNA, collagen (see Fig. 7.7), polypropylene, etc.

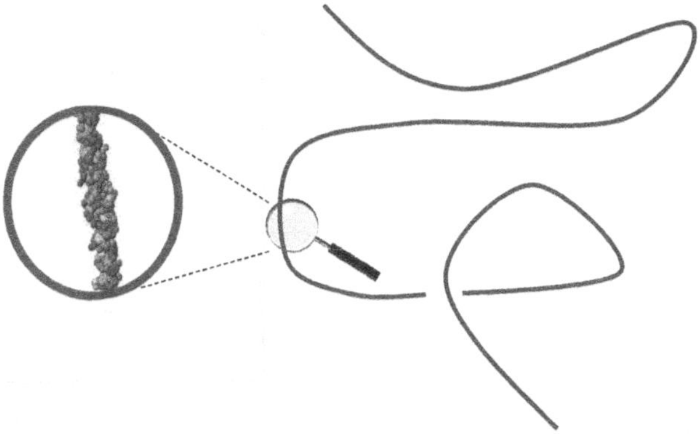

Fig. 7.7 Cartoon of a polymer. A polymer is a long chain of atoms. Shown here is the biological polymer collagen.

And there are many more types of condensed matter systems that we simply do not have time to discuss.[7] One can even engineer artificial types of order which do not occur naturally. Each one of these types of matter has its own interesting properties and if we had more time we would discuss them all in depth! Given that there are so many types of matter, it may seem odd that we are going to spend essentially the entire remainder of this book focused on simple crystalline solids. There are very good reasons for this however. First of all, the study of solids is one of the most successful branches of physics—both in terms of how completely we understand them and also in terms of what we have been able to do practically with this understanding. (For example, the entire modern semiconductor industry is a testament to how successful our understanding of solids is!) More importantly, however, the physics that we learn by studying solids forms an excellent starting point for trying to understand the many more complex forms of matter that exist.

[7] Particularly interesting are forms such as superfluids, where quantum mechanics dominates the physics. But alas, we must save discussion of this for another book.

References

- Dove, chapter 2 gives discussion of many types of matter.
- Chaikin and Lubensky gives a much broader discussion of types of matter.

[6] Here is a really cool experiment to do in your kitchen. Cornstarch is a polymer—a long chain of atoms. Take a box of cornstarch and make a mixture of roughly half cornstarch and half water (you may have to play with the proportions). The concoction should still be able to flow. If you put your hand into it, it will feel like a liquid and be gooey. But if you take a tub of this and hit it with a hammer very quickly, it will feel as hard as a brick, and it will even crack (then it turns back to goo). In fact, you can make a deep tub of this stuff and although it feels completely like a fluid, you can run across the top of it. (If you are too lazy to do this, try Googling "Ellen cornstarch" to see a YouTube video of the experiment. You might also Google "cornstarch, speaker" to see what happens when you put this mess on top of an acoustic speaker.) This mixture, sometimes known as "Oobleck" in a nod to Dr. Seuss, is an example of a "non-Newtonian" fluid—its effective viscosity depends on how fast the force is applied to the material. The reason that polymers have this property is that the long polymer strands get tangled with each other. If a force is applied slowly the strands can unentangle and flow past each other. But if the force is applied quickly they cannot unentangle fast enough and the material acts just like a solid.

Part III

Toy Models of Solids in One Dimension

One-Dimensional Model of Compressibility, Sound, and Thermal Expansion

8

In the first few chapters (Chapters 2–4) we found that our simple models of solids, and electrons in solids, were insufficient in several ways. In order to improve our understanding, we decided that we needed to take the periodic microstructure of crystals more seriously. In this part of the book we finally begin this more careful microscopic consideration. To get a qualitative understanding of the effects of the periodic lattice, it is frequently sufficient to think in terms of simple one-dimensional systems. This is our strategy for the next few chapters. Once we have introduced a number of important principles in one dimension, we will address the complications associated with higher dimensionality.

In Chapter 6 we discussed bonding between atoms. We found, particularly in the discussion of covalent bonding, that the lowest-energy configuration would have the atoms some optimal distance apart (see Fig. 6.6, for example). Given this shape of the energy as a function of distance between atoms we will be able to come to some interesting conclusions.

For simplicity, let us imagine a one-dimensional system of atoms (atoms in a single line). The potential $V(x)$ between two neighboring atoms is drawn in Fig. 8.1.

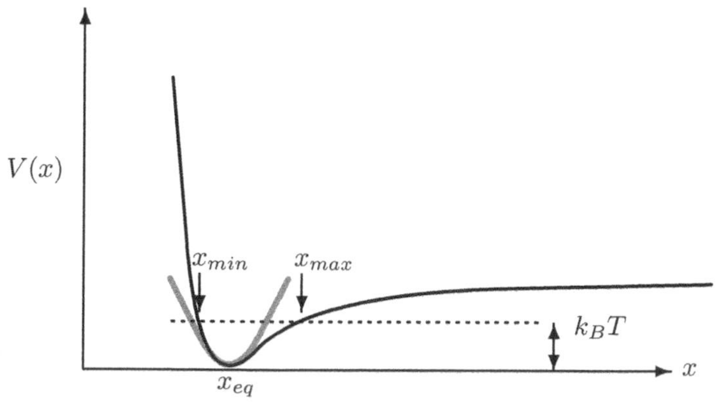

Fig. 8.1 Potential between neighboring atoms (thin black). The thick light gray curve is a quadratic approximation to the minimum (it may look crooked but in fact the thick gray curve is symmetric and the thin black curve is asymmetric). The equilibrium position is x_{eq}. At finite temperature T, the system can oscillate between x_{max} and x_{min} which are not symmetric around the minimum. Thus as T increases the *average* position moves out to larger distance and the system expands.

The classical equilibrium position is the position at the bottom of the well (marked x_{eq} in the figure). The distance between atoms at low temperature should then be x_{eq}. (Quantum mechanics can change this value and increase it a little bit. See Exercise 8.4.)

Let us now Taylor expand the potential around its minimum.

$$V(x) \approx V(x_{eq}) + \frac{\kappa}{2}(x - x_{eq})^2 - \frac{\kappa_3}{3!}(x - x_{eq})^3 + \ldots.$$

Note that there is no linear term (if there were a linear term, then the position x_{eq} would not be the minimum). If there are only small deviations from the position x_{eq} the higher terms are much much smaller than the leading quadratic term and we can throw these terms out. This is a rather crucial general principle that *any* smooth potential, close enough to its minimum, is quadratic.

Compressibility (or Elasticity)

We thus have a simple Hooke's law quadratic potential around the minimum. If we apply a force to compress the system (i.e., apply a pressure to our model one-dimensional solid) we find

$$-\kappa(\delta x_{eq}) = F$$

where the sign is so that a positive (compressive) pressure reduces the distance between atoms. This is obviously just a description of the compressibility (or elasticity) of a solid. The usual description of compressibility is

$$\beta = -\frac{1}{V}\frac{\partial V}{\partial P}$$

(one should ideally specify if this is measured at fixed T or at fixed S. Here, we are working at $T = S = 0$ for simplicity). In the one-dimensional case, we write the compressibility as[1]

$$\beta = -\frac{1}{L}\frac{\partial L}{\partial F} = \frac{1}{\kappa x_{eq}} = \frac{1}{\kappa a} \tag{8.1}$$

[1] Here β is not the inverse temperature! Unfortunately the same symbol is conventionally used for both quantities.

with L the length of the system and x_{eq} the spacing between atoms. Here we make the conventional definition that the equilibrium distance between identical atoms in a system (the so-called *lattice constant*) is written as a.

Sound

You may recall from your study of fluids that in an isotropic compressible fluid, one predicts sound waves with velocity

$$v = \sqrt{\frac{B}{\rho}} = \sqrt{\frac{1}{\rho\beta}} \tag{8.2}$$

where ρ is the mass density of the fluid, B is the bulk modulus, which is $B = 1/\beta$ with β the (adiabatic) compressibility. While in a real solid the compressibility is anisotropic and the speed of sound depends in detail on the direction of propagation, in our model one-dimensional solid this is not a problem.

We calculate that the density is m/a with m the mass of each particle and a the equilibrium spacing between particles. Thus using our previous result (Eq. 8.1 in Eq. 8.2), we predict a sound wave with velocity

$$v = \sqrt{\frac{\kappa a^2}{m}} \, . \tag{8.3}$$

Shortly (in Section 9.2) we will rederive this expression from the microscopic equations of motion for the atoms in the one-dimensional solid.

Thermal Expansion

So far we have been working at zero temperature, but it is worth thinking at least a little bit about thermal expansion. This has been mentioned previously in the **Aside** at the end of section 6.2.2 and will be fleshed out more completely in Exercises 8.2–8.4 (in fact even in the exercises the treatment of thermal expansion will be very crude, but that should still be enough to give us the general idea of the phenomenon[2]).

Let us consider again Fig. 8.1 but now at finite temperature. We can imagine the potential as a function of distance between atoms as being like a ball rolling around in a potential. At zero energy, the ball sits at the minimum of the distribution. But if we give the ball some finite temperature (i.e., some energy) it will oscillate around the minimum. At fixed energy $k_B T$ the ball rolls back and forth between the points x_{min} and x_{max} where $V(x_{min}) = V(x_{max}) = k_B T$. But away from the minimum the potential is asymmetric, so $|x_{max} - x_{eq}| > |x_{min} - x_{eq}|$ so on average the particle has a position $\langle x \rangle > x_{eq}$. This is in essence the reason for thermal expansion! We will obtain positive thermal expansion for any system where $\kappa_3 > 0$ (i.e., at small x the potential is steeper) which almost always is true for real solids.

[2]Explaining thermal expansion more correctly is, unfortunately, rather messy! See for example, Ashcroft and Mermin.

Chapter Summary

- Forces between atoms determine ground-state structure.
- These same forces, perturbing around the ground state, determine elasticity, sound velocity, and thermal expansion.
- Thermal expansion comes from the non-quadratic part of the interatomic potential.

References

Sound and Compressibility:

- Goodstein, section 3.2b
- Hook and Hall, section 2.2
- Ibach and Luth, beginning of section 4.5 (more advanced treatment)

Thermal Expansion. Most reference that discuss thermal expansion go into a large amount of depth. The following are fairly concise:

- Kittel, chapter 5, section on thermal expansion
- Hook and Hall, section 2.7.1

Exercises

(8.1) **Potentials Between Atoms**

As a model of thermal expansion, we study the distance between two nearest-neighbor atoms in an anharmonic potential that looks roughly like this

where x is the distance between the two neighboring atoms. This potential can be expanded around its minimum as

$$V(x) = \frac{\kappa}{2}(x - x_0)^2 - \frac{\kappa_3}{3!}(x - x_0)^3 + \ldots \qquad (8.4)$$

where the minimum is at position x_0 and $\kappa_3 > 0$. For small energies, we can truncate the series at the cubic term. (Note that we are defining the energy at the bottom of the well to be zero here.)

A very accurate approximate form for interatomic potentials (particularly for inert atoms such as helium or argon) is given by the so-called Lennard-Jones potential

$$V(x) = 4\epsilon\left[\left(\frac{\sigma}{x}\right)^{12} - \left(\frac{\sigma}{x}\right)^{6}\right] + \epsilon \qquad (8.5)$$

where ϵ and σ are constants that depend on the particular atoms we are considering.

▷ What is the meaning of the exponent 6 in the second term of this expression (i.e., why is the exponent necessarily chosen to be 6).

▷ By expanding Eq. 8.5 around its minimum, and comparing to Eq. 8.4, calculate the values of the coefficients x_0, κ, and κ_3 for the Lennard-Jones potential in terms of the constants ϵ and σ. We will need these results in Exercise 8.3.

(8.2) **Classical Model of Thermal Expansion**

(i) In classical statistical mechanics, we write the expectation of x as

$$\langle x \rangle_\beta = \frac{\int dx\, x\, e^{-\beta V(x)}}{\int dx\, e^{-\beta V(x)}}$$

Although one cannot generally do such integrals for arbitrary potential $V(x)$ as in Eq. 8.4, one can expand the exponentials as

$$e^{-\beta V(x)} = e^{-\frac{\beta\kappa}{2}(x-x_0)^2}\left[1 + \frac{\beta\kappa_3}{6}(x - x_0)^3 + \ldots\right]$$

and let limits of integration go to $\pm\infty$.

▷ Why is this expansion of the exponent and the extension of the limits of integration allowed?

▷ Use this expansion to derive $\langle x \rangle_\beta$ to lowest order in κ_3, and hence show that the coefficient of thermal expansion is

$$\alpha = \frac{1}{L}\frac{dL}{dT} \approx \frac{1}{x_0}\frac{d\langle x \rangle_\beta}{dT} = \frac{1}{x_0}\frac{k_B\,\kappa_3}{2\kappa^2}$$

with k_B Boltzmann's constant.

▷ In what temperature range is the above expansion valid?

▷ While this model of thermal expansion in a solid is valid if there are only two atoms, why is it invalid for the case of a many-atom chain? (Although actually it is not so bad as an approximation!)

(8.3) **Properties of Solid Argon**

For argon, the Lennard-Jones constants ϵ and σ from Eq. 8.5 are given by ϵ = 10meV and σ = .34nm. You will need to use some of the results from Exercise 8.1.

(a) *Sound*

Given that the atomic weight of argon is 39.9, estimate the sound wave velocity in solid argon. The actual value of the longitudinal velocity is about 1600 m/sec.

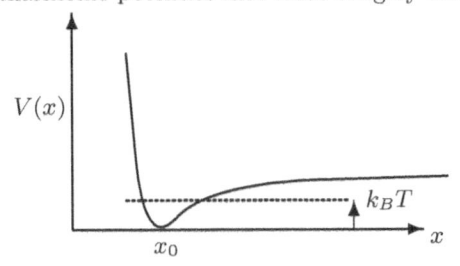

(b) *Thermal Expansion*

Using the results of Exercise 8.2, estimate the thermal expansion coefficient α of argon. Note: *You can do this part even if you couldn't completely figure out Exercise 8.2!*

The actual thermal expansion coefficient of argon is approximately $\alpha = 2\times 10^{-3}/\mathrm{K}$ at about 80K. However, at lower temperature α drops quickly. In the next exercise will use a more sophisticated quantum model to understand why this is so.

(8.4) **Quantum Model of Thermal Expansion**

(a) In quantum mechanics we write a Hamiltonian

$$H = H_0 + \delta V$$

where

$$H_0 = \frac{p^2}{2m} + \frac{\kappa}{2}(x - x_0)^2 \qquad (8.6)$$

is the Hamiltonian for the free Harmonic oscillator, and δV is the perturbation (see Eq. 8.4)

$$\delta V = -\frac{\kappa_3}{6}(x - x_0)^3 + \ldots$$

where we will throw out quartic and higher terms.

▷ What value of m should be used in Eq. 8.6?

Using perturbation theory it can be shown that, to lowest order in κ_3 the following equation holds

$$\langle n|x|n\rangle = x_0 + E_n\kappa_3/(2\kappa^2) \qquad (8.7)$$

where $|n\rangle$ is the eigenstate of the Harmonic oscillator whose energy is

$$E_n = \hbar\omega(n + \frac{1}{2}) + \mathcal{O}(\kappa_3) \qquad n \geqslant 0$$

with $\omega = \sqrt{\kappa/m}$. In (c) we will prove Eq. 8.7. For now, take it as given.

▷ Note that even when the oscillator is in its ground state, the expectation of x deviates from x_0. Physically why is this?

(b)* Use Eq. 8.7 to calculate the quantum expectation of x at any temperature. We write

$$\langle x\rangle_\beta = \frac{\sum_n \langle n|x|n\rangle e^{-\beta E_n}}{\sum_n e^{-\beta E_n}}$$

▷ Derive the coefficient of thermal expansion.

▷ Examine the high temperature limit and show that it matches that of Exercise 8.2.

▷ In what range of temperatures is our perturbation expansion valid?

▷ In light of the current quantum calculation, when is the classical calculation from Exercise 8.2 valid?

▷ Why does the thermal expansion coefficient drop at low temperature?

(c)** Prove Eq. 8.7 by using lowest-order perturbation theory.

Hint: It is easiest to perform this calculation by using raising and lowering (ladder) operators. Recall that one can define operators a and $a^\dagger$ such that $[a, a^\dagger] = 1$ and

$$\begin{aligned} a^\dagger|n\rangle_0 &= \sqrt{n+1}|n+1\rangle_0 \\ a|n\rangle_0 &= \sqrt{n}|n-1\rangle_0 \ . \end{aligned}$$

Note that these are kets and operators for the unperturbed Hamiltonian H_0. In terms of these operators, we have the operator $x - x_0$ given by

$$x - x_0 = \sqrt{\frac{\hbar}{2m\omega}}(a + a^\dagger).$$

(8.5) **Gruneisen Parameter*&**

Another measure of the non-harmonicity of the interaction between atoms is the change in frequency of oscillations as the distance between atoms is changed. We will see how this is related to thermal expansion.

In the case of the Einstein model of a solid, there is only a single frequency ω to keep track of. We define the so-called Gruneisen parameter

$$\gamma = -\left.\frac{\partial \log \omega}{\partial \log V}\right|_T .$$

Using the fact that the entropy of a quantum harmonic oscillator is a function of the quantity $x = \beta\hbar\omega$ only, derive

$$\gamma = \left.\frac{\partial \log S}{\partial \log V}\right|_T \left.\frac{\partial \log T}{\partial \log S}\right|_V = \frac{\alpha}{\beta_T c_V}$$

where in the last step you must use a thermodynamic Maxwell relation. Here c_V is the heat capacity per unit volume at constant volume, $\alpha = (\partial \log V/\partial T)_P$ is the expansion coefficient and $\beta_T = -(\partial \log V/\partial P)_T$ the isothermal compressibility.

Vibrations of a One-Dimensional Monatomic Chain

9

In Chapter 2 we considered the Boltzmann, Einstein, and Debye models of vibrations in solids. In this chapter we will consider a more detailed model of vibration in a solid, first classically, and then quantum-mechanically. We will be able to better understand what these early attempts to understand vibrations achieved and we will be able to better understand their shortcomings.

Let us consider a chain of identical atoms of mass m where the equilibrium spacing between atoms is a (we will sometimes call this quantity the *lattice constant*). Let us define the position of the nth atom to be x_n and the equilibrium position of the nth atom to be $x_n^{eq} = na$.

Once we allow motion of the atoms, we will have x_n deviating from its equilibrium position, so we define the small variable

$$\delta x_n = x_n - x_n^{eq} \quad .$$

Note that in our simple model we are allowing motion of the masses only in one dimension (i.e., we are allowing longitudinal motion of the chain, not transverse motion).

As discussed in the previous chapter, if the system is at low enough temperature we can consider the potential holding the atoms together to be quadratic. Thus, our model of a solid is a chain of masses held together with springs each having equilibrium length a as shown in Fig. 9.1. Because of the quadratic potentials, and thus the relation to simple harmonic motion, this model is frequently known as a *harmonic* chain.

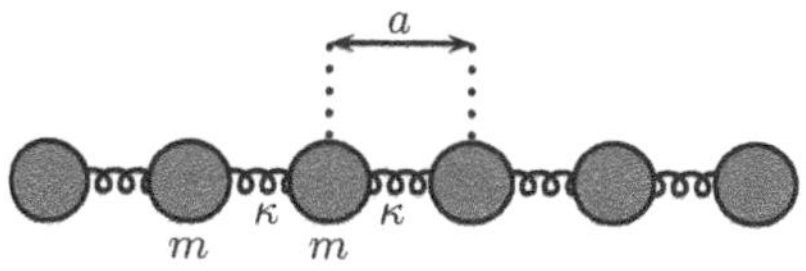

Fig. 9.1 The one-dimensional monatomic harmonic chain. Each ball has mass m and each spring has spring constant κ. The lattice constant, or spacing between successive masses at rest, is a.

With this quadratic interatomic potential, we can write the total potential energy of the chain to be

$$\begin{aligned} V_{tot} &= \sum_i V(x_{i+1} - x_i) = \sum_i \frac{\kappa}{2}(x_{i+1} - x_i - a)^2 \\ &= \sum_i \frac{\kappa}{2}(\delta x_{i+1} - \delta x_i)^2. \end{aligned}$$

The force on the nth mass on the chain is then given by

$$F_n = -\frac{\partial V_{tot}}{\partial x_n} = \kappa(\delta x_{n+1} - \delta x_n) + \kappa(\delta x_{n-1} - \delta x_n).$$

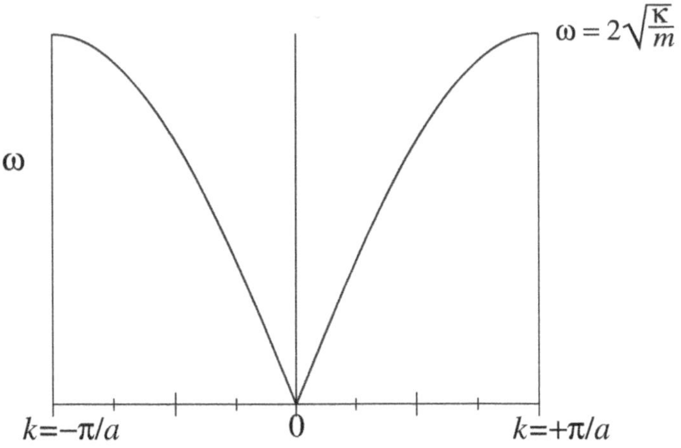

Fig. 9.2 Dispersion relation for vibrations of the one-dimensional monatomic harmonic chain. The dispersion is periodic in $k \to k + 2\pi/a$.

Thus we have Newton's equation of motion

$$m\,\ddot{\delta x}_n = F_n = \kappa(\delta x_{n+1} + \delta x_{n-1} - 2\delta x_n). \tag{9.1}$$

To remind the reader, for any coupled system, a *normal mode* is defined to be a collective oscillation where all particles move at the same frequency. We now attempt a solution to Newton's equations by using an ansatz[1] that describes the normal modes as waves

$$\delta x_n = Ae^{i\omega t - ikx_n^{eq}} = Ae^{i\omega t - ikna}$$

where A is an amplitude of oscillation, and k and ω are the wavevector and frequency of the proposed wave.

Now the reader might be confused about how it is that we are considering complex values of δx_n. Here we are using complex numbers for convenience but actually we implicitly mean to take the real part (this is analogous to what one does in circuit theory with oscillating currents!). Since we are taking the real part, it is sufficient to consider only $\omega \geqslant 0$, however, we must be careful that k can then have either sign, and these are inequivalent once we have specified that ω is positive.

Plugging our ansatz into Eq. 9.1 we obtain

$$-m\omega^2 Ae^{i\omega t - ikna} = \kappa Ae^{i\omega t}\left[e^{-ika(n+1)} + e^{-ika(n-1)} - 2e^{-ikan}\right]$$

or

$$m\omega^2 = 2\kappa[1 - \cos(ka)] = 4\kappa \sin^2(ka/2). \tag{9.2}$$

We thus obtain the result

$$\omega = 2\sqrt{\frac{\kappa}{m}}\left|\sin\left(\frac{ka}{2}\right)\right|. \tag{9.3}$$

In general a relationship between a frequency (or energy) and a wavevector (or momentum) is known as a *dispersion relation*. This particular dispersion relation is shown in Fig. 9.2

[1] In case you have not seen this word before, "ansatz" means "educated guess to be later verified". The word is from German meaning "approach" or "attempt".

9.1 First Exposure to the Reciprocal Lattice

Note that in Fig. 9.2 we have only plotted the dispersion for $-\pi/a \leqslant k \leqslant \pi/a$. The reason for this is obvious from Eq. 9.3—the dispersion relation is actually periodic in $k \to k + 2\pi/a$. In fact this is a very important general principle:

> **Principle 9.1:** A system which is periodic in real space with a periodicity a will be periodic in reciprocal space with periodicity $2\pi/a$.

In this principle we have used the word *reciprocal* space which means k-space. In other words this principle tells us that if a system looks the same when $x \to x + a$ then in k-space the dispersion will look the same when $k \to k + 2\pi/a$. We will return to this principle many times in later chapters.

The periodic unit (the "unit cell") in k-space is conventionally known as the *Brillouin zone*.[2,3] This is our first exposure to the concept of a Brillouin zone, but it will play a very central role in later chapters. The "first Brillouin zone" is a unit cell in k-space centered around the point $k = 0$. Thus in Fig. 9.2 we have shown only the first Brillouin zone, with the understanding that the dispersion is periodic for higher k. The points $k = \pm\pi/a$ are known as the *Brillouin-zone boundary* and are defined in this case as being points which are symmetric around $k = 0$ and are separated by $2\pi/a$.

It is worth pausing for a second and asking why we expect that the dispersion curve should be periodic in $k \to k + 2\pi/a$. Recall that we defined our vibration mode to be of the form

$$\delta x_n = A e^{i\omega t - ikna} \tag{9.4}$$

If we take $k \to k + 2\pi/a$ we obtain

$$\delta x_n = A e^{i\omega t - i(k+2\pi/a)na} = A e^{i\omega t - ikna} e^{-i2\pi n} = A e^{i\omega t - ikna}$$

where here we have used

$$e^{-i2\pi n} = 1$$

for any integer n. What we have found here is that shifting $k \to k + 2\pi/a$ gives us back exactly the same oscillation mode the we had before we shifted k. The two are physically exactly equivalent!

In fact, it is similarly clear that shifting k to any $k + 2\pi p/a$ with p an integer will give us back exactly the same wave also since

$$e^{-i2\pi np} = 1$$

as well. We can thus define a set of points in k-space (reciprocal space) which are all physically equivalent to the point $k = 0$. This set of points is known as the *reciprocal lattice*. The original periodic set of points

[2] Leon Brillouin was one of Sommerfeld's students. He is famous for many things, including for being the "B" in the "WKB" approximation. If you haven't learned about WKB, you really should!

[3] The pronunciation of "Brillouin" is something that gives English speakers a great deal of difficulty. If you speak French you will probably cringe at the way this name is butchered. (I did badly in French in school, so I'm probably one of the worst offenders.) According to online dictionaries it is properly pronounced somewhere between the following words: brēwan, breel-wahn, bree(y)lwa(n), and bree-l-(uh)-wahn. At any rate, the "l" and the "n" should both be very weak. I've also been told that it can be thought of as Brie, the cheese, and Rouen the town in France (which I've never known how to pronounce, so that doesn't help much).

$x_n = na$ is known as the *direct lattice* or *real-space lattice* to distinguish it from the reciprocal lattice, when necessary.

The concept of the reciprocal lattice will be extremely important later on. We can see the analogy between the direct lattice and the reciprocal lattice as follows:

$$\begin{array}{lclllllll} x_n & = & \ldots & -2a, & -a, & 0, & a, & 2a, & \ldots \\ G_n & = & \ldots & -2\left(\frac{2\pi}{a}\right), & -\frac{2\pi}{a}, & 0, & \frac{2\pi}{a}, & 2\left(\frac{2\pi}{a}\right), & \ldots \end{array}$$

Note that the defining property of the reciprocal lattice in terms of the points in the real lattice can be given as

$$e^{iG_m x_n} = 1. \tag{9.5}$$

A point G_m is a member of the reciprocal lattice if and only if Eq. 9.5 is true for all x_n in the real lattice.

Aliasing:

The fact that a wavevector k describes the same wave as the wavevector $k + G_m$ can cause a great deal of confusion. For example, we usually think of wavelength as being $2\pi/k$. But if k is equivalent to $k+G_m$, how do we know if we should choose $2\pi/k$ or $2\pi/(k + G_m)$? The resolution of this puzzle (and many related conundrums[4]) is to realize that k and $k + G_m$ are only equivalent so long as one only measures the wave at lattice points $x_n = na$ and not at arbitrary points x along the axis. Indeed, in our wave ansatz, Eq. 9.4, the wave is only defined at these lattice positions (i.e., a displacement is defined for each mass). In Fig. 9.3 it is shown how the waves corresponding to k and $k+2\pi/a$ take the same values at lattice points $x_n = na$, but disagree between lattice points. As a result, it is somewhat meaningless to ask if the wavelength is $2\pi/k$ or $2\pi/(k + 2\pi/a)$, as both describe the same vibrational wave ansatz that we have used! This phenomenon, that two waves with different wavelengths will look the same if they are sampled only at lattice points, is often known as *aliasing* of waves.[5]

[4] Another closely related question is how we should interpret phase velocity $v_{phase} = \omega/k$ if k is the same as $k+G_m$.

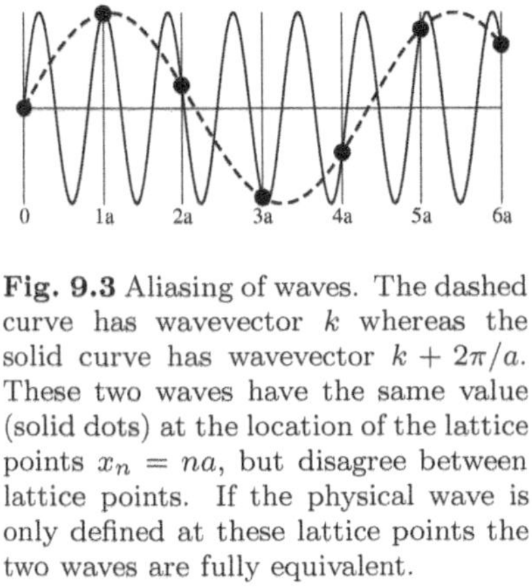

Fig. 9.3 Aliasing of waves. The dashed curve has wavevector k whereas the solid curve has wavevector $k + 2\pi/a$. These two waves have the same value (solid dots) at the location of the lattice points $x_n = na$, but disagree between lattice points. If the physical wave is only defined at these lattice points the two waves are fully equivalent.

[5] This terminology came from radio engineers, who found that one wavelength could "disguise" as another.

9.2 Properties of the Dispersion of the One-Dimensional Chain

We now return to more carefully examine the properties of the dispersion we calculated (Eq. 9.3).

Sound Waves:

Recall that a sound wave[6] is a vibration that has a long wavelength (compared to the interatomic spacing). In this long-wavelength regime, we find the dispersion we just calculated to be linear with wavevector

[6] For reference it is good to remember that humans can hear sound wavelengths roughly between 1cm and 10m. This is very long wavelength compared to interatomic spacings.

$\omega = v_{sound} k$ as expected for sound with

$$v_{sound} = a\sqrt{\frac{\kappa}{m}}$$

(to see this, just expand the sin in Eq. 9.3). Note that this sound velocity matches the velocity predicted from Eq. 8.3!

However, we note that at larger k, the dispersion is no longer linear. This is in disagreement with what Debye assumed in his calculation in Section 2.2. So clearly this is a shortcoming of the Debye theory. In reality the dispersion of normal modes of vibration is linear only at long wavelength.

At shorter wavelength (larger k) one typically defines two different velocities. The *group velocity*, the speed at which a wavepacket moves, is given by

$$v_{group} = d\omega/dk,$$

and the *phase velocity*, the speed at which the individual maxima and minima move, is given by[7]

$$v_{phase} = \omega/k.$$

These two match in the case of a linear dispersion, but otherwise are different. Note that the group velocity becomes zero at the Brillouin zone boundaries $k = \pm\pi/a$ (i.e., the dispersion is flat). As we will see many times later on, this is a general principle!

[7] The difference between group velocity and phase velocity is something that often causes confusion. If this is not already clear to you, I recommend looking on the web. There are many nice sites that give illustrations of the two.

Counting Normal Modes:

Let us now ask how many normal modes there are in our system. Naively it would appear that we can put any k such that $-\pi/a \leqslant k < \pi/a$ into Eq. 9.3 and obtain a normal mode with wavevector k and frequency $\omega(k)$. However, this is not precisely correct.

Let us assume our system has exactly N masses in a row, and for simplicity let us assume that our system has periodic boundary conditions, i.e., particle x_0 has particle x_1 to its right and particle x_{N-1} to its left. Another way to say this is to let, $x_{n+N} = x_n$, i.e., this one-dimensional system forms a big circle. In this case we must be careful that the wave ansatz Eq. 9.4 makes sense as we go all the way around the circle. We must therefore have

$$e^{i\omega t - ikna} = e^{i\omega t - ik(N+n)a}.$$

Or equivalently we must have

$$e^{ikNa} = 1.$$

This requirement restricts the possible values of k to be of the form

$$k = \frac{2\pi p}{Na} = \frac{2\pi p}{L}$$

where p is an integer and L is the total length of the system. Thus k becomes quantized rather than a continuous variable (this is exactly the same argument as we saw previously in Section 2.2.1!). This means that the k-axis in Fig. 9.3 is actually a discrete set of many many individual points; the spacing between two of these consecutive points being $2\pi/(Na) = 2\pi/L$.

Let us now count how many normal modes we have. As mentioned in Section 9.1 in our discussion of the Brillouin zone, adding $2\pi/a$ to k brings one back to exactly the same physical wave. Thus we only ever need consider k values within the first Brillouin zone (i.e., $-\pi/a \leqslant k < \pi/a$, and since π/a is the same as $-\pi/a$ we choose to count one but not the other). Thus the total number of normal modes is

$$\begin{aligned}\text{Total Number of Modes} &= \frac{\text{Range of } k}{\text{Spacing between neighboring } k} \\ &= \frac{2\pi/a}{2\pi/(Na)} = N. \end{aligned} \tag{9.6}$$

There is precisely one normal mode per mass in the system—that is, one normal mode per degree of freedom in the whole system. This is what Debye insightfully predicted in order to cut off his divergent integrals in Section 2.2.3!

9.3 Quantum Modes: Phonons

We now make a rather important leap from classical to quantum physics.

> **Quantum Correspondence:** If a classical harmonic system (i.e., any quadratic Hamiltonian) has a normal oscillation mode at frequency ω the corresponding quantum system will have eigenstates with energy
>
> $$E_n = \hbar\omega(n + \frac{1}{2}). \tag{9.7}$$

Presumably you know this well in the case of a single harmonic oscillator. The only thing different here is that our harmonic oscillator can be a collective normal mode not just the motion of a single particle. This quantum correspondence principle will be the subject of Exercises 9.1 and 9.7.

Thus at a given wavevector k, there are many possible eigenstates, the ground state being the $n = 0$ eigenstate which has only the zero-point energy $\hbar\omega(k)/2$. The lowest-energy excitation is of energy $\hbar\omega(k)$ greater than the ground state corresponding to the excited $n = 1$ eigenstate. Generally all excitations at this wavevector occur in energy units of $\hbar\omega(k)$, and the higher values of energy correspond classically to oscillations of increasing amplitude.

Each excitation of this "normal mode" by a step up the harmonic oscillator excitation ladder (increasing the quantum number n) is known as a "phonon".

Definition 9.1 *A* ***phonon*** *is a discrete quantum of vibration.*[8]

This is entirely analogous to defining a single quantum of light as a photon. As is the case with the photon, we may think of the phonon as actually being a particle, or we can think of the phonon as being a quantized wave.

If we think about the phonon as being a particle (as with the photon) then we see that we can put many phonons in the same state (ie., the quantum number n in Eq. 9.7 can be increased to any value), thus we conclude that phonons, like photons, are bosons. As with photons, at finite temperature there will be a non-zero number of phonons "occupying" a given mode (i.e., n will be on average non-zero) as described by the Bose occupation factor

$$n_B(\beta\hbar\omega) = \frac{1}{e^{\beta\hbar\omega} - 1}$$

with $\beta = 1/(k_B T)$ and ω the oscillation frequency of the mode. Thus, the energy expectation of the phonons at wavevector k is given by

$$E_k = \hbar\omega(k)\left(n_B(\beta\hbar\omega(k)) + \frac{1}{2}\right).$$

We can use this type of expression to calculate the heat capacity of our one-dimensional model[9]

$$U_{total} = \sum_k \hbar\omega(k)\left(n_B(\beta\hbar\omega(k)) + \frac{1}{2}\right)$$

where the sum over k here is over all possible normal modes, i.e, $k = 2\pi p/(Na)$ such that $-\pi/a \leqslant k < \pi/a$. Thus we really mean

$$\sum_k \rightarrow \sum_{\substack{p=-N/2 \\ k=(2\pi p)/(Na)}}^{p=(N/2)-1}$$

Since for a large system, the k points are very close together, we can convert the discrete sum into an integral (something we should be very familiar with by now from Section 2.2.1) to obtain

$$\sum_k \rightarrow \frac{Na}{2\pi}\int_{-\pi/a}^{\pi/a} dk \;.$$

Note that we can use this continuum integral to count the total number of modes in the system

$$\frac{Na}{2\pi}\int_{-\pi/a}^{\pi/a} dk = N$$

as predicted by Debye.

[8] I do not like the definition of a phonon as "a quantum of vibrational energy" which many books use. The vibration does indeed carry energy, but it carries other quantum numbers (such as crystal momentum) as well, so why specify energy only?

[9] The observant reader will note that we are calculating $C_V = dU/dT$ the heat capacity at constant volume. Why constant volume? As we saw when we studied thermal expansion, the crystal does not expand unless we include third (or higher) order terms in the interatomic potential, which are not in this model!

Using this integral form of the sum, we have the total energy given by

$$U_{total} = \frac{Na}{2\pi} \int_{-\pi/a}^{\pi/a} dk \, \hbar\omega(k) \left(n_B(\beta\hbar\omega(k)) + \frac{1}{2} \right)$$

from which we could calculate the heat capacity as dU/dT.

These two previous expressions look exactly like what Debye would have obtained from his calculation (for a one-dimensional version of his model)! The only difference lies in our expression for $\omega(k)$. Debye only knew about sound where $\omega = vk$ is linear in the wavevector. We, on the other hand, have just calculated that for our microscopic mass and spring model ω is not linear in k (see Eq. 9.3). Other than this change in the dispersion relation, our calculation of heat capacity (which is exact for this model!) is identical to the approach of Debye.

In fact, Einstein's calculation of specific heat can also be phrased in exactly the same language—for Einstein's model the frequency ω is constant for all k (it is fixed at the Einstein frequency). We thus see Einstein's model, Debye's model, and our microscopic harmonic model in a very unified light. The only difference between the three is what we use for a dispersion relation.

One final comment is that it is frequently useful to replace integrals over k with integrals of a density of states over frequency (we did this when we studied the Debye model in Section 2.2.2). We obtain generally

$$\frac{Na}{2\pi} \int_{-\pi/a}^{\pi/a} dk = \int d\omega \, g(\omega)$$

where the density of states is given by[10]

$$g(\omega) = 2\frac{Na}{2\pi}|dk/d\omega|.$$

[10]The factor of 2 out front comes from the fact that each ω occurs for the two possible values of $\pm k$.

Recall again that the definition of density of states is that the number of modes with frequency between ω and $\omega + d\omega$ is given by $g(\omega)d\omega$.

Note that in the (one-dimensional) Debye model this density of states is constant from $\omega = 0$ to $\omega = \omega_{Debye} = v\pi/a$. In our model, as we have just calculated, the density of states is not a constant, but becomes zero at frequency above the maximum frequency $2\sqrt{\kappa/m}$ (in Exercise 9.2 we calculate this density of states explicitly). Finally in the Einstein model, this density of states is a delta function at the Einstein frequency.

9.4 Crystal Momentum

As mentioned in Section 9.1, the wavevector of a phonon is defined only modulo[11] the reciprocal lattice. In other words, k is the same as $k + G_m$ where $G_m = 2\pi m/a$ is a point in the reciprocal lattice. Now we are supposed to think of these phonons as particles—and we like to think of these particles as having energy $\hbar\omega$ and momentum $\hbar k$. But we cannot define a phonon's momentum this way because physically it is

[11]The word "modulo" or "mod" means "up to additive terms of". One can think of it also as "divide and take the remainder". For example, 15 modulo 7 = 1 since, up to additive terms of 7, the numbers 15 and 1 are the same. Equivalently we can say that dividing 15 by 7, you get a remainder of 1.

the same phonon whether we describe it as $\hbar k$ or $\hbar(k + G_m)$. We thus instead define a concept known as the *crystal momentum* which is the momentum modulo the reciprocal lattice—or equivalently we agree that we must always describe k within the first Brillouin zone.

In fact, this idea of crystal momentum is extremely powerful. Since we are thinking about phonons as being particles, it is actually possible for two (or more) phonons to bump into each other and scatter from each other—the same way particles do.[12] In such a collision, energy is conserved and *crystal momentum is conserved!* For example, three phonons each with crystal momentum $\hbar(2/3)\pi/a$ can scatter off of each other to produce three phonons each with crystal momentum $-\hbar(2/3)\pi/a$. This is allowed since the initial and final states have the same energy and

$$3 \times (2/3)\pi/a = 3 \times (-2/3)\pi/a \mod (2\pi/a)$$

During these collisions although momentum $\hbar k$ is not conserved, crystal momentum is.[13] In fact, the situation is similar when, for example, phonons scatter from electrons in a periodic lattice—crystal momentum becomes the conserved quantity rather than momentum. This is an extremely important principle which we will encounter again and again. In fact, it is a main cornerstone of solid state physics.

Aside: There is a very fundamental reason for the conservation of crystal momentum. Conserved quantities are results of symmetries—this is a deep and general statement known as Noether's theorem.[14] For example, conservation of momentum is a result of the translational invariance of space. If space is not the same from point to point, say there is a potential $V(x)$ which varies from place to place, then momentum is not conserved. The conservation of crystal momentum correspondingly results from space being invariant under translations of a, giving us momentum that is conserved modulo $2\pi/a$. Since the symmetry is not a continuous one, this is not a strict application of Noether's theorem, but it is very closely related.

[12] In the harmonic model we have considered phonons that do not scatter from each other. We know this because the phonons are eigenstates of the system, so their occupation does not change with time. However, if we add anharmonic (cubic and higher) terms to the interatomic potential, this corresponds to perturbing the phonon Hamiltonian and can be interpreted as allowing phonons to scatter from each other.

[13] This thing we have defined, $\hbar k$, has dimensions of momentum, but is not conserved. However, as we will discuss in Chapter 14, if a particle, like a photon, enters a crystal with a given momentum and undergoes a process that conserves crystal momentum but not momentum, when the photon exits the crystal we will find that total momentum of the system is indeed conserved, with the momentum of the entire crystal accounting for any momentum that is missing from the photon. See margin note 6 in Section 14.1.1.

[14] Emmy Noether has been described by Einstein, among others, as the most important woman in the history of mathematics. Being Jewish, she fled Germany in 1933 to take a job at Bryn Mawr College (she was also offered a job at Somerville College, Oxford, but she preferred the States). Sadly, she died suddenly only two years later at the relatively young age of 53.

Chapter Summary

A number of very crucial new ideas have been introduced in this section. Many of these will return again and again in later chapters.

- Normal modes are collective oscillations where all particles move at the same frequency.
- If a system is periodic in space with periodicity $\Delta x = a$, then in reciprocal space (k-space) the system is periodic with periodicity $\Delta k = 2\pi/a$.
- Values of k which differ by multiples of $2\pi/a$ (by an element of the reciprocal lattice) are physically equivalent. The set of points in k-space which are equivalent to $k = 0$ are known as the reciprocal lattice.

- Any value of k is equivalent to some k in the first Brillouin zone, $-\pi/a \leqslant k < \pi/a$ (in 1d).
- The sound velocity is the slope of the dispersion in the small k limit (group velocity = phase velocity in this limit).
- A classical normal mode of frequency ω gets translated into quantum-mechanical eigenstates with energies $E_n = \hbar\omega(n + \frac{1}{2})$. If a mode is in the nth eigenstate, we say that it is occupied by n phonons.
- Phonons can be thought of as particles, like photons, that obey Bose statistics.

References

Pretty much every solid state physics course will discuss the normal modes of a monatomic chain, and they all certainly discuss phonons as well. I think the following are fairly good.

- Kittel, beginning of chapter 4
- Goodstein, beginning of section 3.3
- Hook and Hall, section 2.3.1
- Burns, sections 12.1–12.2
- Ashcroft and Mermin, beginning of chapter 22
- Dove, section 8.3

Exercises

(9.1) **Classical Normal Modes to Quantum Eigenstates**

In Section 9.3 we stated without proof that a classical normal mode becomes a quantum eigenstate. Here we prove this fact for a simple diatomic molecule in a potential well (see Exercise 2.7 for a more difficult case, and see also Exercise 9.7 where this principle is proven in more generally).

Consider two particles, each of mass m in one dimension, connected by a spring (K), at the bottom of a potential well (with spring constant k). We write the potential energy as

$$U = \frac{k}{2}(x_1^2 + x_2^2) + \frac{K}{2}(x_1 - x_2)^2$$

▷ Write the classical equations of motion.

▷ Transform into relative $x_{rel} = (x_1 - x_2)$ and center of mass $x_{cm} = (x_1 + x_2)/2$ coordinates.

(a) Show that in these transformed coordinates, the system decouples, thus showing that the two normal modes have frequencies

$$\begin{aligned} \omega_{cm} &= \sqrt{k/m} \\ \omega_{rel} &= \sqrt{(k + 2K)/m} \end{aligned}$$

Note that since there are two initial degrees of freedom, there are two normal modes.

Now consider the quantum-mechanical version of the same problem. The Hamiltonian is

$$H = \frac{p_1^2}{2m} + \frac{p_2^2}{2m} + U(x_1, x_2)$$

▷ Again transform into relative and center of mass coordinates.

Define the corresponding momenta $p_{rel} = (p_1 - p_2)/2$ and $p_{cm} = (p_1 + p_2)$.

(b) Show that $[p_\alpha, x_\gamma] = -i\hbar\delta_{\alpha,\gamma}$ where α and γ take the values cm or rel.

(c) In terms of these new coordinates show that the Hamiltonian decouples into two independent harmonic oscillators with the same eigenfrequencies ω_{cm} and ω_{rel}. Conclude that the spectrum of this system is

$$E_{n_{rel},n_{cm}} = \hbar\omega_{rel}(n_{rel} + \frac{1}{2}) + \hbar\omega_{cm}(n_{cm} + \frac{1}{2})$$

where n_{cm} and n_{rel} are non-negative integers.

(d) At temperature T what is the expectation of the energy of this system?

(9.2) **Normal Modes of a One-Dimensional Monatomic Chain**

(a)‡ Explain what is meant by "normal mode" and by "phonon".

▷ Explain briefly why phonons obey Bose statistics.

(b)‡ Derive the dispersion relation for the longitudinal oscillations of a one-dimensional mass-and-spring crystal with N identical atoms of mass m, lattice spacing a, and spring constant κ (motion of the masses is restricted to be in one dimension).

(c)‡ Show that the mode with wavevector k has the same pattern of mass displacements as the mode with wavevector $k + 2\pi/a$. Hence show that the dispersion relation is periodic in reciprocal space (k-space).

▷ How many *different* normal modes are there.

(d)‡ Derive the phase and group velocities and sketch them as a function of k.

▷ What is the sound velocity?

▷ Show that the sound velocity is also given by $v_s = 1/\sqrt{\beta\rho}$ where ρ is the chain density and β is the compressibility.

(e) Find the expression for $g(\omega)$, the density of states of modes per angular frequency.

▷ Sketch $g(\omega)$.

(f) Write an expression for the heat capacity of this one-dimensional chain. You will inevitably have an integral that you cannot do analytically.

(g)* However, you can expand exponentials for high temperature to obtain a high-temperature approximation. It should be obvious that the high-temperature limit should give heat capacity $C/N = k_B$ (the law of Dulong–Petit in one dimension). By expanding to next non-trivial order, show that

$$C/N = k_B(1 - A/T^2 + \ldots)$$

where

$$A = \frac{\hbar^2\kappa}{6mk_B^2}.$$

(9.3) **More Vibrations**

Consider a one-dimensional spring and mass model of a crystal. Generalize this model to include springs not only between neighbors but also between second nearest neighbors. Let the spring constant between neighbors be called κ_1 and the spring constant between second neighbors be called κ_2. Let the mass of each atom be m.

(a) Calculate the dispersion curve $\omega(k)$ for this model.

(b) Determine the sound wave velocity. Show the group velocity vanishes at the Brillouin zone boundary.

(9.4) **Decaying Waves**

In the dispersion curve of the harmonic chain (Eq. 9.3), there is a maximum possible frequency of oscillation ω_{max}. If a vibration with frequency $\omega > \omega_{max}$ is forced upon the chain (say by a driving force) the "wave" will not propagate along the chain, but rather will decay as one moves away from the point where the oscillation is imposed (this is sometimes known as an "evanescent" wave). With $\omega > \omega_{max}$ solve Eq. 9.3 for a complex k to determine the decay length of this evanescent wave. What happens to this length as $\omega \to \omega_{max}$?

(9.5) **Reflection at an Interface***

Consider a harmonic chain of equally spaced identical masses of mass m where left of the $n = 0$ mass the spring constant is κ_L but right of the $n = 0$ mass, the spring constant is κ_R, as shown in this figure.

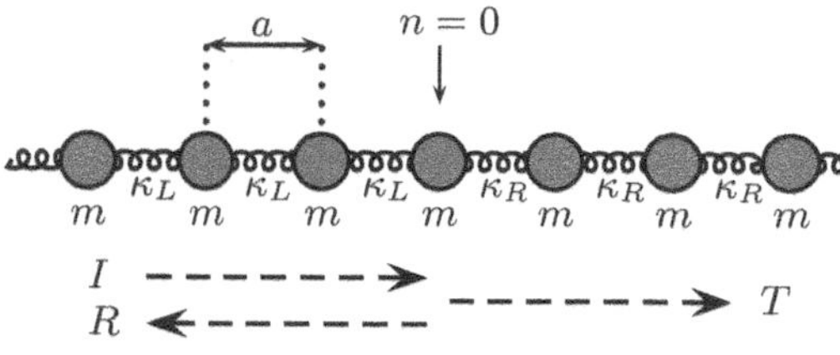

A wave with amplitude I is incident on this interface from the left, where it can be either transmitted with amplitude T or reflected with amplitude R. Using the following ansatz form

$$\delta x_n = \begin{cases} Te^{i\omega t - ik_R na} & n \geqslant 0 \\ Ie^{i\omega t - ik_L na} + Re^{i\omega t + ik_L na} & n < 0 \end{cases}$$

derive T/I and R/I given ω, κ_L, κ_R and m.

(9.6) **Impurity Phonon Mode***

Consider a harmonic chain where all spring constants have the same value κ and masses have value m, except for the mass at position $n = 0$ which instead has value $M < m$ as shown in this figure:

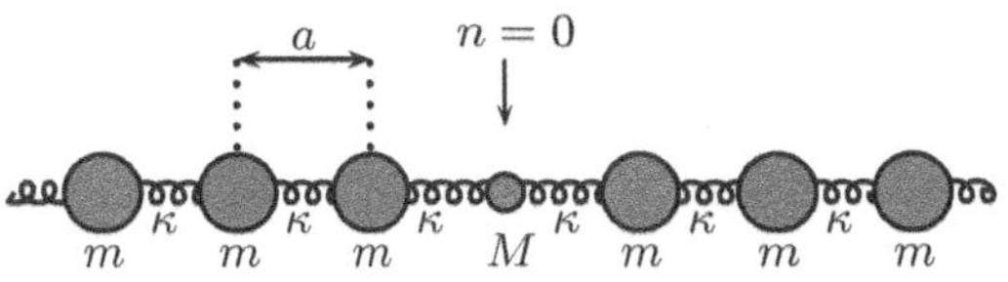

Along with traveling wave solutions, there can be a standing wave normal mode localized near the impurity. Use an ansatz of the form

$$\delta x_n = A e^{i\omega t - q|n|a}$$

with q complex to solve for the frequency of this impurity mode. Consider your result in the context of Exercise 9.4.

(9.7) **General Proof That Normal Modes Become Quantum Eigenstates***

This proof generalizes the argument given in Exercise 9.1. Consider a set of N particles $a = 1, \ldots N$ with masses m_a interacting via a potential

$$U = \frac{1}{2} \sum_{a,b} x_a \, V_{a,b} \, x_b$$

where x_a is the deviation of the position of particle a from its equilibrium position and V can be taken (without loss of generality) to be a symmetric matrix. (Here we consider a situation in 1d, however, we will see that to go to 3d we just need to keep track of three times as many coordinates.)

(i) Defining $y_a = \sqrt{m_a}\, x_a$, show that the classical equations of motion may be written as

$$\ddot{y}_a = -\sum_b S_{a,b} \, y_b$$

where

$$S_{a,b} = \frac{1}{\sqrt{m_a}} V_{a,b} \frac{1}{\sqrt{m_b}}$$

Thus show that the solutions are

$$y_a^{(m)} = e^{-i\omega_m t} s_a^{(m)}$$

where ω_m^2 is the mth eigenvalue of the matrix S with corresponding eigenvector $s_a^{(m)}$. These are the N normal modes of the system.

(ii) Recall the orthogonality relations for eigenvectors of hermitian matrices

$$\sum_a [s_a^{(m)}]^* [s_a^{(n)}] \;=\; \delta_{m,n} \qquad (9.8)$$

$$\sum_m [s_a^{(m)}]^* [s_b^{(m)}] \;=\; \delta_{a,b}. \qquad (9.9)$$

Since S is symmetric as well as hermitian, the eigenvectors can be taken to be real. Construct the transformed coordinates

$$Y^{(m)} \;=\; \sum_a s_a^{(m)} x_a \sqrt{m_a} \qquad (9.10)$$

$$P^{(m)} \;=\; \sum_a s_a^{(m)} p_a / \sqrt{m_a} \qquad (9.11)$$

show that these coordinates have canonical commutations

$$[P^{(m)}, Y^{(n)}] = -i\hbar \delta_{n,m} \qquad (9.12)$$

and show that in terms of these new coordinates the Hamiltonian is rewritten as

$$H = \sum_m \left[\frac{1}{2} [P^{(m)}]^2 + \frac{1}{2} \omega_m^2 [Y^{(m)}]^2 \right]. \qquad (9.13)$$

Conclude that the quantum eigenfrequencies of the system are also ω_m. (Can you derive this result from the prior two equations?)

(9.8) **Phonons in 2d***

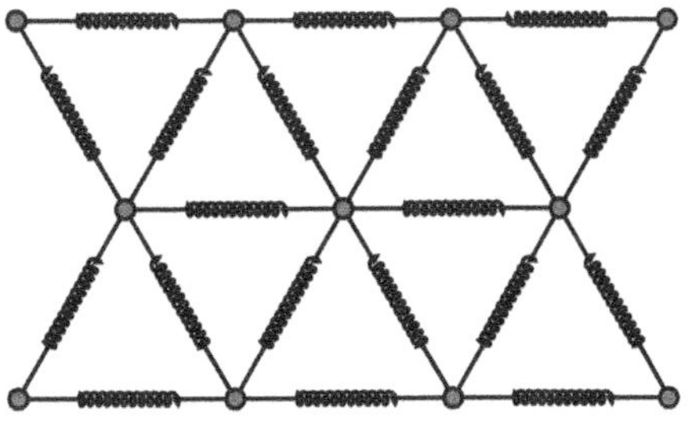

Consider a mass and spring model of a two-dimensional triangular lattice as shown in the figure (assume the lattice is extended infinitely in all directions). Assume that identical masses m are attached to each of their six neighbors by equal springs of equal length and spring constant κ. Calculate the dispersion curve $\omega(\mathbf{k})$. The two-dimensional structure is more difficult to handle than the one-dimensional examples given in this chapter. In Chapters 12 and 13 we study crystals in two and three dimensions, and it might be useful to read those chapters first and then return to try this exercise again.

Vibrations of a One-Dimensional Diatomic Chain

10

In the previous chapter we studied in detail a one-dimensional model of a solid where every atom is identical to every other atom. However, in real materials not every atom is the same (for example, in sodium chloride, NaCl, we have two types of atoms!). We thus intend to generalize our previous discussion of the one-dimensional solid to a one-dimensional solid with two types of atoms. We are not, however, just studying this for the sake of adding complexity. In fact we will see that several fundamentally new features will emerge in this more general situation.

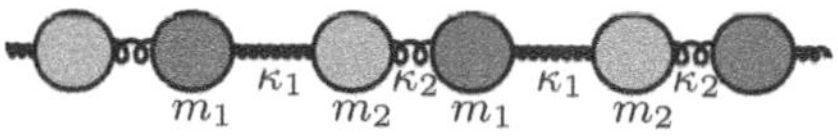

Fig. 10.1 A general diatomic chain with two different types of atoms (i.e., two different masses) and two different types of springs.

10.1 Diatomic Crystal Structure: Some Useful Definitions

Consider the model system shown in Fig. 10.1 which represents a periodic arrangement of two different types of atoms. Here we have given them two masses m_1 and m_2 which alternate along the one-dimensional chain. The springs connecting the atoms have spring constants κ_1 and κ_2 and also alternate.

In this circumstance with more than one type of atom, we first would like to identify the so-called *unit cell* which is the repeated motif in the arrangement of atoms. In Fig. 10.2, we have put a box around the unit cell. The length of the unit cell in one dimension is known as the *lattice constant* and it is labeled a.

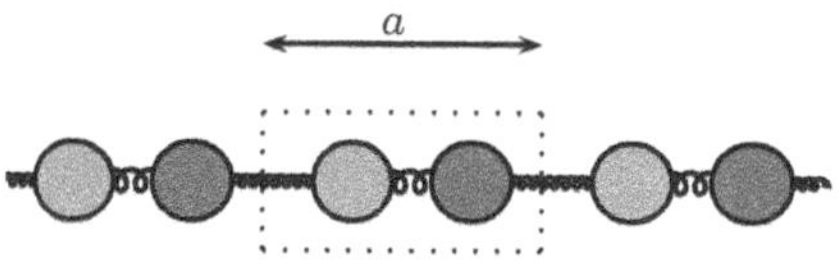

Fig. 10.2 A unit cell for the diatomic chain.

Note however, that the definition of the unit cell is extremely non-unique. We could just as well have chosen (for example) the unit cell to be that shown in Fig. 10.3.

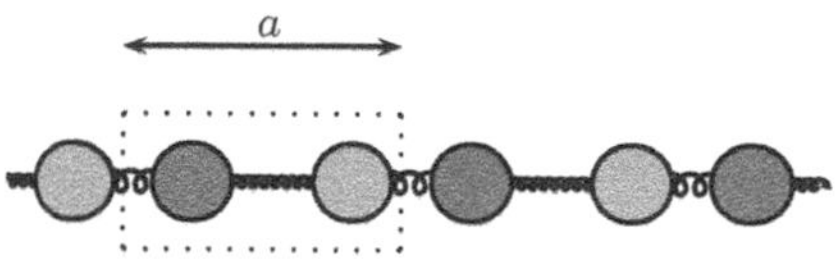

Fig. 10.3 Another possible unit cell for the diatomic chain.

The important thing in defining a periodic system is to choose *some* unit cell and then construct the full system by reproducing the same unit cell over and over (in other words, make a definition of the unit cell and stick with that definition!).

It is sometimes useful to pick some reference point inside each unit cell. This set of reference points makes a simple *lattice* (we will define the term "lattice" more closely in Chapter 12, but for now the idea is that a lattice has only one type of point in it—not two different types

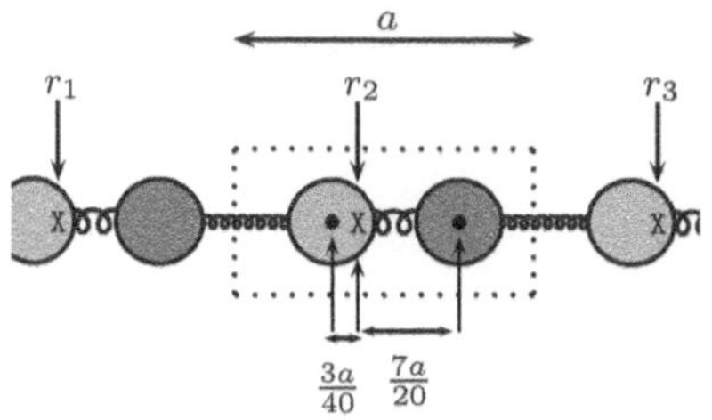

Fig. 10.4 The *basis* describes the objects in the crystal with respect to the positions of the reference lattice points. Here the reference point (at position r_n) is marked with an X.

of points). So in Fig. 10.4 we have marked our reference point in each unit cell (at positions r_n) with an X (again, the choice of this reference point is arbitrary).

Given the reference lattice point in the unit cell, the description of all of the atoms in the unit cell with respect to this reference point is known as a *basis*. In the case of Fig. 10.4 we might describe our basis as

light gray atom
at position $3a/40$ to the left of the reference lattice point,

dark gray atom
at position $7a/20$ to the right of the reference lattice point.

Thus if the reference lattice point in unit cell n is called r_n (and the spacing between the lattice points is a) we can set

$$r_n = an$$

with a the size of the unit cell. Then the (equilibrium) position of the light gray atom in the nth unit cell is

$$x_n^{eq} = an - 3a/40$$

whereas the (equilibrium) position of the dark gray atom in the nth unit cell is

$$y_n^{eq} = an + 7a/20.$$

10.2 Normal Modes of the Diatomic Solid

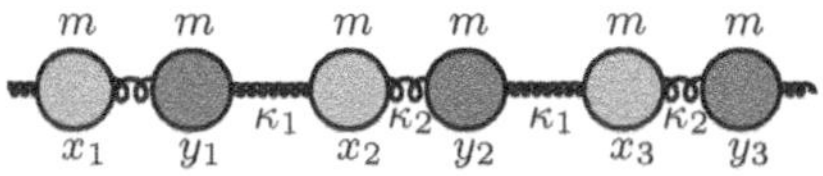

Fig. 10.5 The *alternating chain* has all masses the same but the values of the spring constants alternate. This might seem unnatural if each mass were a simple spherical atom, but if you imagine each mass is a small assymetric molecule, then it is natural to have different spring constants to the left versus right. (In fact, one can actually have an alternating chain of identical atoms, but this is a much more complicated situation. See Exercise 11.8.)

[1]This is the proper pluralization of ansatz. See margin note 1 from the previous chapter. "Ansätze" would be a great name for a heavy metal band.

For simplicity, let us focus on the case shown in Fig. 10.5 where all of the masses along our chain are the same $m_1 = m_2 = m$ but the two spring constants κ_1 and κ_2 are different (we still take the lattice constant to be a). In Exercise 10.1 we will consider the case where the masses are different, but the spring constants are the same. The physics of both cases are quite similar, but the case we address here is slightly easier algebraically. The current case is sometimes known as the *alternating* chain (presumably because the spring constants alternate).

Given the spring constants in the picture, we can write down Newton's equations of motion for the deviations of the positions of the masses from their equilibrium positions. We obtain

$$m\,\ddot{\delta x}_n = \kappa_2(\delta y_n - \delta x_n) + \kappa_1(\delta y_{n-1} - \delta x_n) \quad (10.1)$$
$$m\,\ddot{\delta y}_n = \kappa_1(\delta x_{n+1} - \delta y_n) + \kappa_2(\delta x_n - \delta y_n) \quad (10.2)$$

Analogous to the monatomic case we propose ansätze[1] for these quantities that have the form of a wave

$$\delta x_n = A_x e^{i\omega t - ikna} \quad (10.3)$$
$$\delta y_n = A_y e^{i\omega t - ikna} \quad (10.4)$$

where, as in the previous chapter, we implicitly mean to take the real part of the complex number. As such, we can always choose to take $\omega > 0$ as long as we consider k to be either positive and negative.

As we saw in the previous chapter, values of k that differ by $2\pi/a$ are physically equivalent. We can thus focus our attention to the first Brillouin zone $-\pi/a \leqslant k < \pi/a$. Any k outside the first Brillouin zone is redundant with some other k inside the zone. Note that the important length here is the unit cell length or lattice constant a.

As we found in the previous chapter, if our system has N unit cells (hence $L = Na$) then (putting periodic boundary conditions on the system) k will be quantized in units of $2\pi/(Na) = 2\pi/L$. Note that here the important quantity is N, the number of unit cells, not the number of atoms ($2N$).

Dividing the range of k in the first Brillouin zone by the spacing between neighboring k's, we obtain exactly N different possible values of k exactly as we did in Eq. 9.6. In other words, we have exactly one value of k per unit cell.

We might recall at this point the intuition that Debye used—that there should be exactly one possible excitation mode per degree of freedom of the system. Here we obviously have two degrees of freedom per unit cell, but we obtain only one possible value of k per unit cell. The resolution, as we will see in a moment, is that there will be two possible oscillation modes for each wavevector k.

We now proceed by plugging in our ansätze (Eqs. 10.3 and 10.4) into our equations of motion (Eqs. 10.1 and 10.2). We obtain

$$
\begin{aligned}
&-\omega^2 m A_x e^{i\omega t - ikna} = \\
&\qquad \kappa_2 A_y e^{i\omega t - ikna} + \kappa_1 A_y e^{i\omega t - ik(n-1)a} - (\kappa_1+\kappa_2) A_x e^{i\omega t - ikna}
\end{aligned}
$$

$$
\begin{aligned}
&-\omega^2 m A_y e^{i\omega t - ikna} = \\
&\qquad \kappa_1 A_x e^{i\omega t - ik(n+1)a} + \kappa_2 A_x e^{i\omega t - ikna} - (\kappa_1+\kappa_2) A_y e^{i\omega t - ikna}
\end{aligned}
$$

which simplifies to

$$
\begin{aligned}
-\omega^2 m A_x &= \kappa_2 A_y + \kappa_1 A_y e^{ika} - (\kappa_1+\kappa_2) A_x \\
-\omega^2 m A_y &= \kappa_1 A_x e^{-ika} + \kappa_2 A_x - (\kappa_1+\kappa_2) A_y \quad .
\end{aligned}
$$

This can be rewritten conveniently as an eigenvalue equation

$$
m\omega^2 \begin{pmatrix} A_x \\ A_y \end{pmatrix} = \begin{pmatrix} (\kappa_1+\kappa_2) & -\kappa_2 - \kappa_1 e^{ika} \\ -\kappa_2 - \kappa_1 e^{-ika} & (\kappa_1+\kappa_2) \end{pmatrix} \begin{pmatrix} A_x \\ A_y \end{pmatrix} . \qquad (10.5)
$$

The solutions of this are obtained by finding the zeros of the characteristic determinant[2]

$$
\begin{aligned}
0 &= \begin{vmatrix} (\kappa_1+\kappa_2) - m\omega^2 & -\kappa_2 - \kappa_1 e^{ika} \\ -\kappa_2 - \kappa_1 e^{-ika} & (\kappa_1+\kappa_2) - m\omega^2 \end{vmatrix} \\
&= \left|(\kappa_1+\kappa_2) - m\omega^2\right|^2 - \left|\kappa_2 + \kappa_1 e^{ika}\right|^2 ,
\end{aligned}
$$

[2]The characteristic determinant is sometimes called a "secular determinant". This old-style nomenclature does not have to do with religion (or lack thereof) but rather refers to secular astronomical phenomena—meaning on the time-scale of a century. Determinants were used to calculate the weak perturbative effects on the planetary orbits, hence explaining these secular phenomena.

the roots of which are clearly given by

$$m\omega^2 = (\kappa_1 + \kappa_2) \pm |\kappa_1 + \kappa_2 e^{ika}| \ .$$

The second term here can be simplified to

$$\begin{aligned} |\kappa_1 + \kappa_2 e^{ika}| &= \sqrt{(\kappa_1 + \kappa_2 e^{ika})(\kappa_1 + \kappa_2 e^{-ika})} \\ &= \sqrt{\kappa_1^2 + \kappa_2^2 + 2\kappa_1\kappa_2 \cos(ka)} \ , \end{aligned}$$

so we finally obtain

$$\begin{aligned} \omega_\pm &= \sqrt{\frac{\kappa_1 + \kappa_2}{m} \pm \frac{1}{m}\sqrt{\kappa_1^2 + \kappa_2^2 + 2\kappa_1\kappa_2 \cos(ka)}} \\ &= \sqrt{\frac{\kappa_1 + \kappa_2}{m} \pm \frac{1}{m}\sqrt{(\kappa_1 + \kappa_2)^2 - 4\kappa_1\kappa_2 \sin^2(ka/2)}} \qquad (10.6) \end{aligned}$$

Note in particular that for each k we find two normal modes—usually referred to as the two *branches* of the dispersion. Thus since there are N different k values, we obtain $2N$ modes total (if there are N unit cells in the entire system). This is in agreement with Debye's intuition that we should have exactly one normal mode per degree of freedom in our system. The dispersion of these two modes is shown in Fig. 10.6.

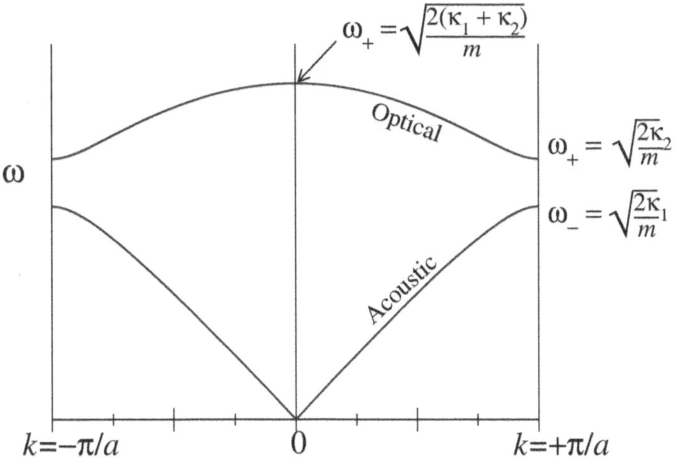

Fig. 10.6 Dispersion relation for vibrations of the one-dimensional diatomic chain. The dispersion is periodic in $k \to k + 2\pi/a$. Here the dispersion is shown for the case of $\kappa_2 = 1.5\kappa_1$. This scheme of plotting dispersions, putting all normal modes within the first Brillouin zone, is the *reduced zone scheme*. Compare this to Fig. 10.8.

A few things to note about this dispersion. First of all we note that there is a long-wavelength low-energy branch of excitations with linear dispersion (corresponding to ω_- in Eq. 10.6). This is the sound wave, or acoustic mode. Generally the definition of an *acoustic mode* is any mode that has linear dispersion as $k \to 0$.

By expanding Eq. 10.6 for small k it is easy to check that the sound velocity is

$$v_{sound} = \frac{d\omega_-}{dk} = \sqrt{\frac{a^2\kappa_1\kappa_2}{2m(\kappa_1 + \kappa_2)}} \qquad (10.7)$$

In fact, we could have calculated this sound velocity on general principles analogous to what we did in Eq. 8.2 and Eq. 8.3. The density of the chain is $2m/a$. The effective spring constant of two springs κ_1 and κ_2 in series is $\tilde{\kappa} = (\kappa_1\kappa_2)/(\kappa_1 + \kappa_2)$ so the compressibility of the chain is $\beta = 1/(\tilde{\kappa}a)$ (see Eq. 8.1). Then, plugging into Eq. 8.2 gives exactly the same sound velocity as we calculate here in Eq. 10.7.

The higher-energy branch of excitations is known as the *optical* mode. It is easy to check that in this case the optical mode goes to frequency $\sqrt{2(\kappa_1 + \kappa_2)/m}$ at $k = 0$, and also has zero group velocity at $k = 0$. The reason for the nomenclature "optical" has to do with how light scatters from solids (we will study scattering from solids in much more depth in Chapter 14). For now we give a very simplified description of why it is named this way. Consider a solid being exposed to light. It is possible for the light to be absorbed by the solid, but energy and momentum must both be conserved. However, light travels at a very high velocity c, so $\hbar\omega = \hbar ck$ is very large. Since phonons have a maximum frequency, this means that photons can only be absorbed for very small k. However, for small k, acoustic phonons have energy $\hbar vk \ll \hbar ck$ so that energy and momentum cannot be conserved. On the other hand, optical phonons have frequency $\omega_{optical}$ which is finite for small k so that at some value of small k, we have $\hbar\omega_{optical} = \hbar ck$ and one can match the frequency (energy) and momentum of the photon to that of the phonon.[3] Thus, whenever phonons interact with light, it is inevitably the optical phonons that are involved.

[3]From this naive argument, one might think that the process where one photon with frequency $\omega_{optical}$ is absorbed while emitting a phonon is an allowed process. This is not true since the photons carry spin while phonons do not, and spin must also be conserved. Much more typically the interaction between photons and phonons is one where a photon is absorbed and then re-emitted at a different frequency while emitting a phonon. I.e., the photon is inelastically scattered. We will discuss this later in Section 14.4.2.

Let us examine a bit more closely the acoustic and the optical mode as $k \to 0$. Examining our eigenvalue problem Eq. 10.5, we see that in this limit the matrix to be diagonalized takes the simple form

$$\omega^2 \begin{pmatrix} A_x \\ A_y \end{pmatrix} = \frac{\kappa_1 + \kappa_2}{m} \begin{pmatrix} 1 & -1 \\ -1 & 1 \end{pmatrix} \begin{pmatrix} A_x \\ A_y \end{pmatrix} . \quad (10.8)$$

The acoustic mode (which has frequency 0) corresponds to the eigenvector

$$\begin{pmatrix} A_x \\ A_y \end{pmatrix} = \begin{pmatrix} 1 \\ 1 \end{pmatrix} .$$

This tells us that the two masses in the unit cell (at positions x and y) move together for the case of the acoustic mode in the long wavelength limit. This is not surprising considering our understanding of sound waves as being very long wavelength compressions and rarefactions. This is depicted in Fig. 10.7. Note in the figure that the amplitude of the compression is slowly modulated, but always the two atoms in the unit cell move almost exactly the same way.

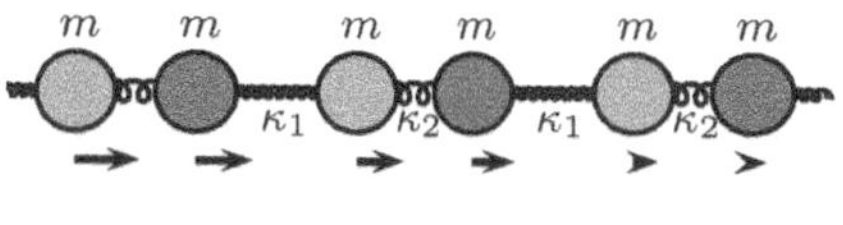

Fig. 10.7 A long wavelength acoustic mode for the alternating chain.

On the other hand, the optical mode at $k = 0$, having frequency $\omega^2 = \frac{2(\kappa_1+\kappa_2)}{m}$, has the eigenvector

$$\begin{pmatrix} A_x \\ A_y \end{pmatrix} = \begin{pmatrix} 1 \\ -1 \end{pmatrix}$$

which described the two masses in the unit cell moving in *opposite* directions, for the optical mode. This is depicted in Fig. 10.9. Note in the the

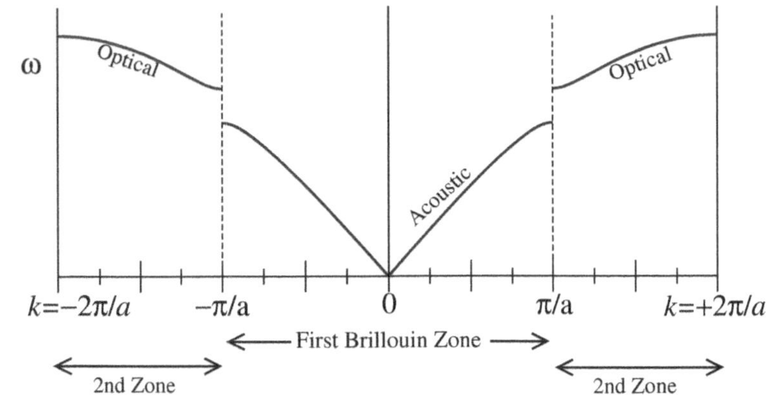

Fig. 10.8 Dispersion relation of vibrations of the one-dimensional diatomic chain in the extended zone scheme (again choosing $\kappa_2 = 1.5\kappa_1$). Compare this to Fig. 10.6. One can think of this as just unfolding the dispersion such that there is only one excitation plotted at each value of k. The first and second Brillouin zones are labeled here

figure that the amplitude of the compression is slowly modulated, but always the two atoms in the unit cell move almost exactly the opposite way.[4]

In order to get a better idea of how motion occurs for both the optical and acoustic modes, it is useful to see animations, which you can find on the web.[5] In this example we had two atoms per unit cell and we obtained two modes per distinct value of k. One of these modes is acoustic and one is optical. More generally, if there are M atoms per unit cell (in one dimension) we will have M modes per distinct value of k (i.e., M branches of the dispersion) of which one mode will be acoustic (goes to zero energy at $k = 0$) and all of the remaining modes are optical (do not go to zero energy at $k = 0$).

[4]As mentioned at the beginning of this section, many books discuss instead the diatomic chain where the two masses are different but the two spring constants are the same (see Exercise 10.1). Be warned that many references then *incorrectly* describe the motion of the masses—in that case the amplitude of motion of the two different masses is *not* the same except in the case of the very long wavelength acoustic mode!

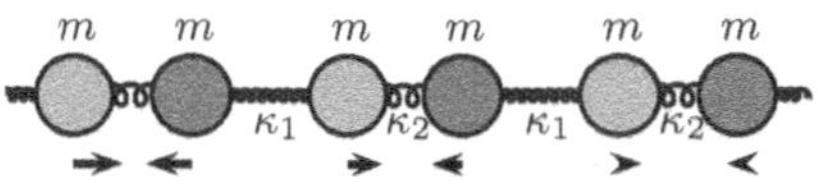

Fig. 10.9 A long-wavelength optical mode for the alternating chain.

Caution: We have been careful to discuss a true one-dimensional system, where the atoms are allowed to move only along the one-dimensional line. Thus each atom has only one degree of freedom. However, if we allow atoms to move in other directions (transverse to the one-dimensional line) we will have more degrees of freedom per atom. When we get to the 3d solid we should expect three degrees of freedom per atom—there should be three different acoustic modes at each k at long wavelength. In 3d, if there are n atoms per unit cell, there will be $3(n-1)$ optical modes but always three acoustic modes totalling $3n$ degrees of freedom per unit cell.

[5]Currently you can find links on my website, but there are plenty of other places to look as well!

One thing that we should study closely is the behavior at the Brillouin zone boundary. It is also easy to check that the frequencies $\omega_\pm$ at the zone boundary ($k = \pm\pi/a$) are $\sqrt{2\kappa_1/m}$ and $\sqrt{2\kappa_2/m}$, the larger of the two being ω_+. We can also check that the group velocity $d\omega/dk$ of both modes goes to zero at the zone boundary (similarly the optical mode has zero group velocity at $k = 0$).

In Fig. 10.6 we have shown both modes at each value of k, such that we only need to show k within the first Brillouin zone. This is known as the *reduced zone scheme*. Another way to plot exactly the same dispersions is shown in Fig. 10.8, and is known as the *extended zone scheme*. Essentially you can think of this as "unfolding" the dispersions such that there is only one mode at each value of k.

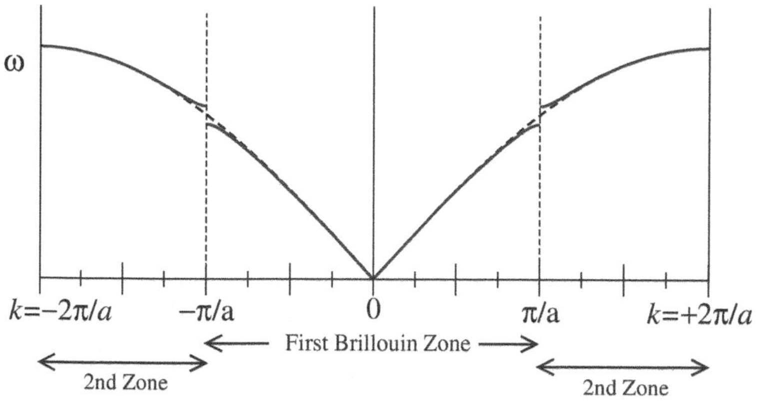

Fig. 10.10 How a diatomic dispersion becomes a monatomic dispersion when the two different atoms become the same. **Solid:** Dispersion relation of vibrations of the one-dimensional diatomic chain in the extended zone scheme with κ_2 not too different from κ_1 ($\kappa_2 = 1.25\kappa_1$ here). **Dashed:** Dispersion relation when $\kappa_2 = \kappa_1$. In this case, the two atoms become exactly the same, and we have a monatomic chain with lattice spacing $a/2$. This single band dispersion precisely matches that calculated in Chapter 9, only with the lattice constant redefined to $a/2$.

In Fig. 10.8 we have defined (for the first time) the *second Brillouin zone*. Recall the first zone in 1d is defined as $|k| \leqslant \pi/a$. Analogously the second Brillouin zone is now $\pi/a \leqslant |k| \leqslant 2\pi/a$. In Chapter 13 we will define the Brillouin zones more generally.

Here is an example where it is very useful to think using the extended zone scheme. We have been considering cases with $\kappa_2 > \kappa_1$; now let us consider what would happen if we take the limit of $\kappa_2 \to \kappa_1$. When the two spring constants become the same, then in fact the two atoms in the unit cell become identical, and we have a simple monatomic chain (which we discussed at length in the previous chapter). As such we should define a new smaller unit cell with lattice constant $a/2$, and the dispersion curve is now just a simple $|\sin|$ as it was in Chapter 9 (see Eq. 9.3).

Thus it is frequently useful, if the two atoms in a unit cell are not too different from each other, to think about the dispersion as being a small perturbation to a situation where all atoms are identical. When the atoms are made slightly different, a small gap opens up at the zone boundary, but the rest of the dispersion continues to look mostly as if it is the dispersion of the monatomic chain. This is illustrated in Fig. 10.10.

Chapter summary

A number of key concepts are introduced in this chapter:

- A unit cell is the repeated motif that comprises a crystal.
- The basis is the description of the unit cell with respect to a reference lattice.
- The lattice constant is the size of the unit cell (in 1d).

- If there are M atoms per unit cell we will find M normal modes at each wavevector k (for one-dimensional motion).
- One of these modes is an acoustic mode, meaning that it has linear dispersion at small k, whereas the remaining $M-1$ are optical, meaning they have finite frequency at $k = 0$.
- For the acoustic mode, all atoms in the unit cell move in-phase with each other (at $k = 0$), whereas for optical modes they move out of phase with each other (at $k = 0$).
- If all of the dispersion curves are plotted within the first Brillouin zone $|k| \leqslant \pi/a$ we call this the reduced zone scheme. If we "unfold" the curves such that there is only one excitation plotted per k, but we use more than one Brillouin zone, we call this the extended zone scheme.
- For a diatomic chain, if the two atoms in the unit cell become identical, the new unit cell is half the size of the old unit cell. It is convenient to describe this limit in the extended zone scheme.

References

- Ashcroft and Mermin, chapter 22 (but not the 3d part)
- Ibach and Luth, section 4.3
- Kittel, chapter 4
- Hook and Hall, sections 2.3.2, 2.4, 2.5
- Burns, section 12.3
- Dove, section 8.5

Exercises

(10.1) **Normal modes of a One-Dimensional Diatomic Chain**

(a) What is the difference between an acoustic mode and an optical mode.

▷ Describe how particles move in each case.

(b) Derive the dispersion relation for the longitudinal oscillations of a one-dimensional diatomic mass-and-spring crystal where the unit cell is of length a and each unit cell contains one atom of mass m_1 and one atom of mass m_2 connected together by springs with spring constant κ, as shown in the figure (all springs are the same, and motion of particles is in one dimension only).

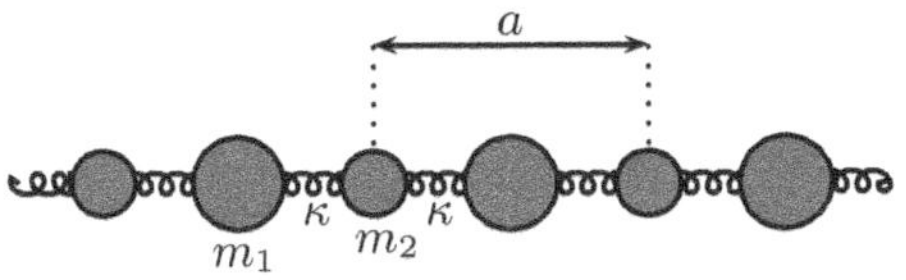

(c) Determine the frequencies of the acoustic and optical modes at $k = 0$ as well as at the Brillouin zone boundary.

▷ Describe the motion of the masses in each case (see margin note 4 of this chapter!).

▷ Determine the sound velocity and show that the group velocity is zero at the zone boundary.

▷ Show that the sound velocity is also given by $v_s = \sqrt{\beta^{-1}/\rho}$ where ρ is the chain density and β is the compressibility.

(d) Sketch the dispersion in both reduced and extended zone scheme.

▷ If there are N unit cells, how many different normal modes are there?

▷ How many *branches* of excitations are there? I.e., in reduced zone scheme, how many modes are there there at each k?

(e) What happens when $m_1 = m_2$?

(10.2) **Decaying Waves**

Consider the alternating diatomic chain dispersion as discussed in the text Eq. 10.6 and shown in Fig. 10.6. For frequencies above $\omega_+(k=0)$ there are no propagating wave modes, and similarly for frequencies between $\omega_-(k=\pi/a)$ and $\omega_+(k=\pi/a)$ there are no propagating wave modes. As in Exercise 9.4, if this chain is driven at a frequency ω for which there are no propagating wave modes, then there will be a decaying, or evanescent, wave instead. By solving Eq. 10.6 for a complex k, find the length scale of this decaying wave.

(10.3) **General Diatomic Chain***

Consider a general diatomic chain as shown in Fig. 10.1 with two different masses m_1 and m_2 as well as two different spring constants κ_1 and κ_2 and lattice constant a.

(a) Calculate the dispersion relation for this system.

(b) Calculate the acoustic mode velocity and compare it to $v_s = \sqrt{\beta^{-1}/\rho}$ where ρ is the chain density and β is the compressibility.

(10.4) **Second Neighbor Diatomic Chain***

Consider the diatomic chain from Exercise 10.1. In addition to the spring constant κ between neighboring masses, suppose that there is also a next nearest-neighbor coupling with spring constant κ' connecting equivalent masses in adjacent unit cells. Determine the dispersion relation for this system. What happens if $\kappa' \gg \kappa$?

(10.5) **Triatomic Chain***

Consider a mass-and-spring model with three different masses and three different springs per unit cell as shown in this diagram.

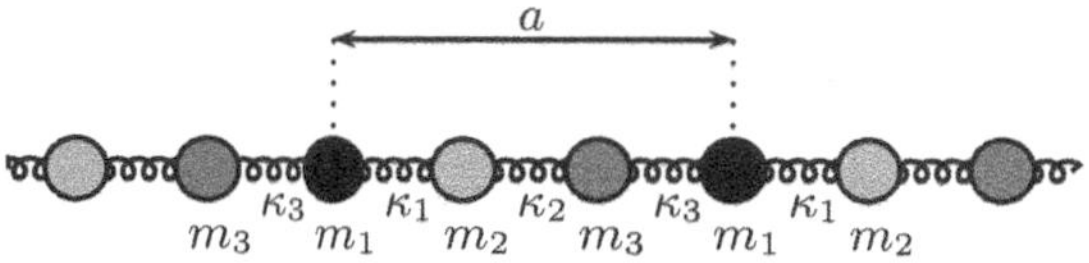

As usual, assume that the masses move only in one dimension.

(a) At $k=0$ how many optical modes are there? Calculate the frequencies of these modes. Hint: You will get a cubic equation. However, you already know one of the roots since it is the frequency of the acoustic mode at $k=0$

(b)* If all the masses are the same and $\kappa_1 = \kappa_2$ determine the frequencies of all three modes at the zone boundary $k=\pi/a$. You will have a cubic equation, but you should be able to guess one root which corresponds to a particularly simple normal mode.

(c)* If all three spring constants are the same, and $m_1 = m_2$ determine the frequencies of all three modes at the zone boundary $k=\pi/a$. Again you should be able to guess one of the roots.

(10.6) **Einstein and Debye Again*&**

As a simplification of the dispersion relation of diatomic chain, let us approximate the optical mode as having a fixed frequency ω_E and approximate the acoustic mode as having frequency $\omega = v|k|$. Calculate the heat capacity of this system. This is a more precise description of what was meant in Exercise 2.9.

Tight Binding Chain (Interlude and Preview)

11

In the previous two chapters we considered the properties of vibrational waves (phonons) in a one-dimensional system. At this point, we are going to make a bit of an excursion to consider electrons in solids again. The point of this excursion, besides being a preview of much of the physics that will reoccur later on, is to make the point that all waves in periodic environments (in crystals) are similar. In the previous two chapters we considered vibrational waves, whereas in this chapter we will consider electron waves (remember that in quantum mechanics particles are just as well considered to be waves!).

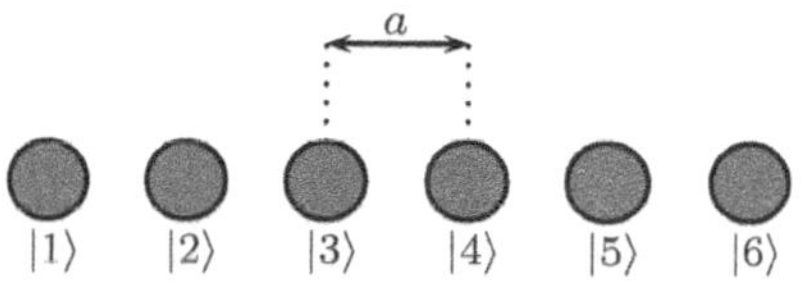

Fig. 11.1 The tight binding chain. There is one orbital on each atom, and electrons are allowed to hop from one atom to the neighboring atom.

11.1 Tight Binding Model in One Dimension

We described the molecular orbital, tight binding, or LCAO picture for molecules previously in Section 6.2.2. Here we will consider a chain of such molecular orbitals to represent orbitals in a macroscopic (one-dimensional) solid as shown in Fig. 11.1.

In this picture, there is a single orbital on atom n which we call $|n\rangle$. For convenience we will assume that the system has periodic boundary conditions (i.e., there are N sites, and site N is the same as site 0). Further, we assume that all of the orbitals are orthogonal to each other[1]

$$\langle n|m\rangle = \delta_{n,m} \quad . \tag{11.1}$$

Let us now take a general trial wavefunction of the form

$$|\Psi\rangle = \sum_n \phi_n |n\rangle \quad .$$

As we discussed for the tight-binding model (see Eq. 6.4 and Exercise 6.2) the effective Schroedinger equation can be written as[2]

$$\sum_m H_{nm}\phi_m = E\phi_n \tag{11.2}$$

where H_{nm} is the matrix element of the Hamiltonian

$$H_{nm} = \langle n|H|m\rangle \quad .$$

[1] As in Section 6.2.2 this is not a great approximation, particularly when the atoms get close to each other. Doing it more correctly, however, only adds algebraic complexity and is not all that enlightening. See Exercise 6.5 and 11.3, where we work through the calculation more correctly.

[2] Another way to get this effective equation is to start with the real Schroedinger equation $H|\psi\rangle = E|\psi\rangle$, insert a complete set $1 = \sum_m |m\rangle\langle m|$ between H and $|\psi\rangle$ and then apply $\langle n|$ from the left on both sides to obtain Eq. 11.2 where $\phi_n = \langle n|\psi\rangle$. If the set $|m\rangle$ were really complete set (which it is not) this would be a good derivation. However, to the extent that these orbitals *approximate* a complete set, this is an approximate derivation. More precisely one should interpret this as a variational approximation as discussed in Exercise 6.2.

As mentioned previously when we studied the molecular orbital model (Section 6.2.2), this Schroedinger equation is actually a variational approximation. For example, instead of finding the exact ground state, it finds the best possible ground state made up of the orbitals that we have put in the model.

One can make the variational approach increasingly better by expanding the Hilbert space and putting more orbitals into the model. For example, instead of having only one orbital $|n\rangle$ at a given site, one could consider many $|n, \alpha\rangle$ where α runs from 1 to some number p. As p is increased the approach becomes increasingly more accurate and eventually is essentially exact. This method of using tight-binding-like orbitals to increasingly well approximate the exact Schroedinger equation is known as LCAO (linear combination of atomic orbitals). However, one complication (which we treat in Exercise 11.3) is that when we add many more orbitals we typically have to give up our nice orthogonality assumption, i.e., $\langle n, \alpha | m, \beta \rangle = \delta_{nm}\delta_{\alpha\beta}$ no longer holds. This makes the effective Schroedinger equation a bit more complicated, but not fundamentally different (see comments in Section 6.2.2).

At any rate, in the current chapter we will work with only one orbital per site, and we assume the orthogonality Eq. 11.1.

We write the Hamiltonian as

$$H = K + \sum_j V_j$$

where $K = \mathbf{p}^2/(2m)$ is the kinetic energy and V_j is the Coulomb interaction of the electron at position $\mathbf{r}$ with the nucleus at site j,

$$V_j = V(\mathbf{r} - \mathbf{R_j})$$

where $\mathbf{R_j}$ is the position of the jth nucleus.

With these definitions we have

$$H|m\rangle = (K + V_m)|m\rangle + \sum_{j \neq m} V_j |m\rangle \quad .$$

Now, we should recognize that $K + V_m$ is the Hamiltonian which we would have if there were only a single nucleus (the mth nucleus) and no other nuclei in the system. Thus, if we take the tight-binding orbitals $|m\rangle$ to be the atomic orbitals, then we have

$$(K + V_m)|m\rangle = \epsilon_{atomic}|m\rangle$$

where ϵ_{atomic} is the energy of an electron on nucleus m in the absence of any other nuclei. Thus we can write

$$H_{n,m} = \langle n|H|m\rangle = \epsilon_{atomic}\, \delta_{n,m} + \sum_{j \neq m} \langle n|V_j|m\rangle \quad .$$

We now have to figure out what the final term of this equation is. The meaning of this term is that, via the interaction with some nucleus which is not the mth, an electron on the mth atom can be transferred (can "hop") to the nth atom. Generally this can only happen if n and m are very close to each other. Thus, we write

$$\sum_{j \neq m} \langle n|V_j|m\rangle = \begin{cases} V_0 & n = m \\ -t & n = m \pm 1 \\ 0 & \text{otherwise} \end{cases} \tag{11.3}$$

which defines both V_0 and t. (The V_0 term here does not hop an electron from one site to another, but rather just shifts the energy on a given site.) Note that by translational invariance of the system, we expect that the result should depend only on $n - m$, which this form does. These two types of terms V_0 and t are entirely analogous to the two types of terms V_{cross} and t that we met in Section 6.2.2 when we studied covalent bonding of two atoms.[3] The situation here is similar except that now there are many nuclei instead of just two.

With the above matrix elements we obtain

$$H_{n,m} = \epsilon_0 \delta_{n,m} - t\left(\delta_{n+1,m} + \delta_{n-1,m}\right) \tag{11.4}$$

where we have now defined[4]

$$\epsilon_0 = \epsilon_{atomic} + V_0 \quad .$$

This Hamiltonian is a very heavily studied model, known as the tight binding chain. Here t is known as the hopping term, as it allows the Hamiltonian (which generates time evolution) to move the electron from one site to another, and it has dimensions of energy. It stands to reason that the magnitude of t depends on how close together the orbitals are—becoming large when the orbitals are close together and decaying exponentially when they are far apart.

11.2 Solution of the Tight Binding Chain

The solution of the tight binding model in one dimension (the tight binding chain) is very analogous to what we did to study vibrations (and hence the point of presenting the tight binding model at this point!). We propose an ansatz solution[5]

$$\phi_n = \frac{e^{-ikna}}{\sqrt{N}} \tag{11.5}$$

where the denominator is included for normalization where there are N sites in the system. We now plug this ansatz into the Schroedinger equation Eq. 11.2. Note that in this case (as compared to the vibrational chains) there is no frequency in the exponent of our ansatz. This is simply because we are trying to solve the time-independent Schroedinger equation. Had we used the time-dependent Schroedinger equation, we would need a factor of $e^{i\omega t}$ as well!

[3] Just to be confusing, atomic physicists sometimes use J where I have used t here.

[4] Once again ϵ_0 is *not* a dielectric constant or the permittivity of free space, but rather just the energy of having an electron sit on a site.

[5] Recall that in Section 4.5 we ran into a puzzle that the mean free path of electrons in metals seems unreasonably long. The fact that electrons hopping between orbitals form eigenstates which are plane waves (i.e., are delocalized across the entire system) hints towards the solution to this puzzle. We will return to reconsider this issue in more detail in Section 15.2.

As with vibrations, it is obvious that $k \to k + 2\pi/a$ gives the same solution. Further, if we consider the system to have periodic boundary conditions with N sites (length $L = Na$), the allowed values of k are quantized in units of $2\pi/L$. As with Eq. 9.6 there are precisely N possible different solutions of the form of Eq. 11.5.

Plugging the ansatz into the left side of the Schroedinger equation 11.2 and then using Eq. 11.4 gives us

$$\sum_m H_{n,m}\phi_m = \epsilon_0 \frac{e^{-ikna}}{\sqrt{N}} - t\left(\frac{e^{-ik(n+1)a}}{\sqrt{N}} + \frac{e^{-ik(n-1)a}}{\sqrt{N}}\right)$$

which we set equal to the right side of the Schroedinger equation

$$E\phi_n = E\frac{e^{-ikna}}{\sqrt{N}}$$

to obtain the spectrum

$$E = \epsilon_0 - 2t\cos(ka) \qquad (11.6)$$

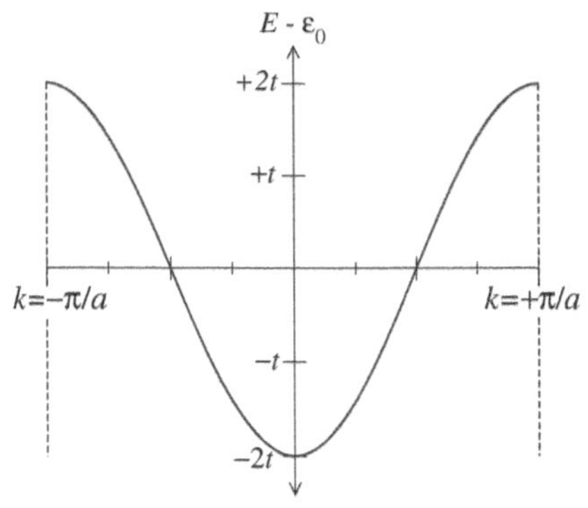

Fig. 11.2 Dispersion of the tight binding chain. Energy is plotted versus wavevector in the first Brillouin zone.

which looks rather similar to the *phonon* spectrum of the one-dimensional monatomic chain which was (see Eq. 9.2)

$$\omega^2 = 2\frac{\kappa}{m} - 2\frac{\kappa}{m}\cos(ka) \quad .$$

Note however, that in the electronic case one obtains the energy whereas in the phonon case one obtains the *square* of the frequency.[6]

The dispersion curve of the tight binding chain (Eq. 11.6) is shown in Fig. 11.2. Analogous to the phonon case, it is periodic in $k \to k + 2\pi/a$. Further, analogous to the phonon case, the dispersion always has zero group velocity (is flat) for $k = n\pi/a$ for n any integer (i.e., at the Brillouin zone boundary).

Note that unlike free electrons, the electron dispersion here has a maximum energy as well as a minimum energy. Electrons only have eigenstates within a certain energy *band*. The word "band" is used both to describe the energy range for which eigenstates exist, as well as to describe one connected branch of the dispersion curve. (In this picture there is only a single mode at each k, hence one branch, hence a single band.)

The energy difference from the bottom of the band to the top is known as the *bandwidth*. Within this bandwidth (between the top and bottom of the band) for any energy there exists at least one k state having that energy. For energies outside the bandwidth there are no k-states with that energy.

The bandwidth (which in this model is $4t$) is determined by the magnitude of the hopping, which depends on the distance between nuclei.[7] As a function of the interatomic spacing the bandwidth changes roughly as shown in Fig 11.3. On the right of this diagram there are N states, each one being an atomic orbital $|n\rangle$. On the left of the diagram these N states form a band, yet as discussed above in this section, there remain

[6]This difference is due to the fact that the time dependent Schroedinger equation has one time derivative, but Newton's equation of motion ($F = ma$) has two.

[7]Since the hopping t depends on an overlap between orbitals on adjacent atoms (see Eq. 11.3 and comments thereafter), in the limit that the atoms are well separated, the bandwidth will decrease exponentially as the atoms are pulled further apart.

precisely N states (this should not surprise us, being that we have not changed the dimension of the Hilbert state, we have just expressed it in terms of the complete set of eigenstates of the Hamiltonian). Note that the average energy of a state in this band remains always ϵ_0.

By allowing hopping between orbitals, some of the eigenstates in the band have decreased in energy from the energy ϵ_0 of the atomic eigenstate and some of the eigenstates have increased in energy. This is entirely analogous to what we found in Section 6.2.2 when we found bonding and antibonding orbitals form when we allow hopping between two atoms. In both cases, the hopping splits the energy levels (originally ϵ_0) into some higher energy states and some lower energy states.

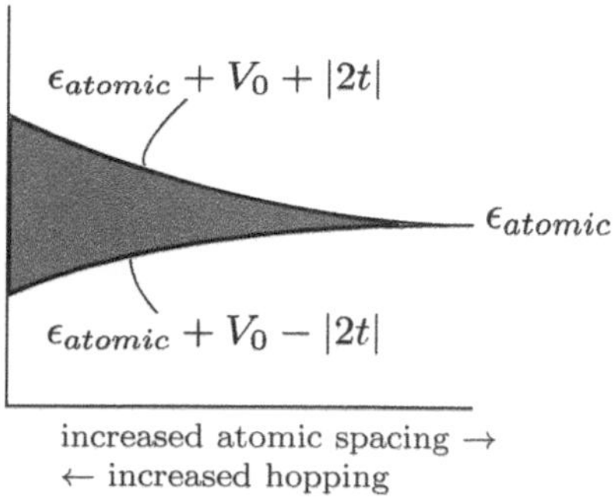

Fig. 11.3 Caricature of the dependence of bandwidth on interatomic spacing. On the far right there is no hopping and the energy of every state in the band is ϵ_0. As hopping increases (towards the left) the energies of states in the band spread out. At each value of hopping there are eigenstates with energies within the shaded region, but not outside the shaded region.

Aside: Note that if the band is not completely filled, the *total* energy of all of the electrons *decreases* as the atoms are moved together and the band width increases (since the average energy remains zero, but some of the higher energy states are not filled). This decrease in energy is precisely the binding force of a "metallic bond" which we discussed in Section 6.4.[8] We also mentioned previously that one property of metals is that they are typically soft and malleable. This is a result of the fact that the electrons that hold the atoms together are mobile—in essence, because they are mobile, they can readjust their positions somewhat as the crystal is deformed.

Near the bottom of the band, the dispersion is parabolic. For the dispersion Eq. 11.6, expanding for small k, we obtain

$$E(k) = \text{Constant} + ta^2k^2 \quad .$$

(Note that for $t < 0$, the energy minimum is at the Brillouin zone boundary $k = \pi/a$. In this case we would expand for k close to π/a instead of for k close to 0.) The resulting parabolic behavior is similar to that of free electrons which have a dispersion

$$E_{free}(k) = \frac{\hbar^2k^2}{2m} \quad .$$

We can therefore view the bottom of the band as being almost like free electrons, except that we have to define a new *effective mass* which we call m^* such that

$$\frac{\hbar^2k^2}{2m^*} = ta^2k^2$$

which gives us

$$m^* = \frac{\hbar^2}{2ta^2} \quad .$$

In other words, the *effective mass* m^* is defined such that the dispersion of the bottom of the band is exactly like the dispersion of free particles of mass m^*. (We will discuss effective mass in much more depth in Chapter 17. This is just a quick first look at it.) Note that this mass has nothing to do with the actual mass of the electron, but rather depends on the hopping matrix element t. Further, we should keep in mind that the k that enters into the dispersion relationship is actually the crystal momentum, not the actual momentum of the electron (recall that crystal

[8] Of course, we have not considered the repulsive force between neighboring nuclei, so the nuclei do not get too close together. As in the case of the covalent bond considered in Section 6.2.2, some of the Coulomb repulsion between nuclei will be canceled by V_{cross} (here V_0), the attraction of the electron on a given site to other nuclei.

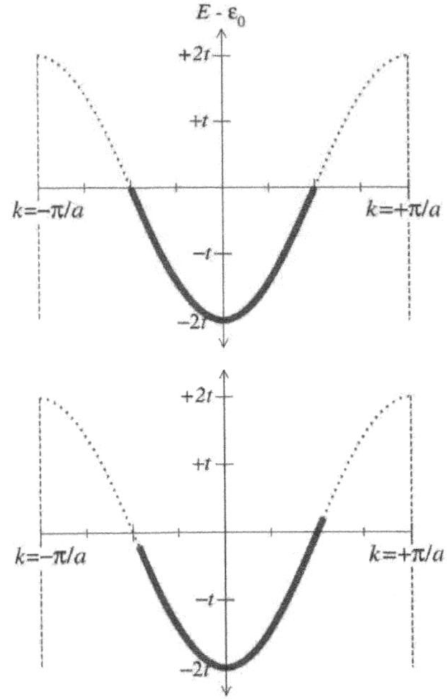

Fig. 11.4 Top: If each atom has valence 1, then the band is half-filled. The states that are shaded are filled with both up and down spin electrons. The Fermi surface is the boundary between the filled and unfilled states. **Bottom:** When a small electric field is applied, at only a small cost of energy, the Fermi sea can shift slightly (moving a few electrons from the right side to the left side) thus allowing current to flow.

momentum is defined only modulo $2\pi/a$). However, so long as we stay at very small k, then there is no need to worry about the periodicity of k which occurs. Nonetheless, we should keep in mind that if electrons scatter off of other electrons, or off of phonons, it is crystal momentum that is conserved (see the discussion in Section 9.4).

11.3 Introduction to Electrons Filling Bands

We now imagine that our tight binding model is actually made up of atoms and each atom "donates" one electron into the band (i.e., the atom has *valence* one). Since there are N possible k-states in the band, and electrons are fermions, you might guess that this would precisely fill the band. However, there are two possible spin states for an electron at each k, so in fact, this then only half-fills the band. This is depicted in the top of Fig. 11.4. The filled states (shaded) in this picture are filled with both up and down spins.

It is crucial in this picture that there is a Fermi surface—the points where the shaded meets the unshaded region. If a small electric field is applied to the system, it only costs a very small amount of energy to shift the Fermi surface as shown in the bottom of Fig. 11.4, populating a few k-states moving right and depopulating some k-states moving left. In other words, the state of the system responds by changing a small bit and a current is induced. As such, this system is a *metal* in that it conducts electricity. Indeed, crystals of atoms that are monovalent are very frequently metals!

On the other hand, if each atom in our model were di-valent (donates two electrons to the band) then the band would be entirely full of electrons. In fact, it does not matter if we think about this as being a full band where every k-state $|k\rangle$ is filled with two electrons (one up and one down), or a filled band where every site $|n\rangle$ is filled—these two statements describe the same multi-electron wavefunction. In fact, there is a single unique wavefunction that describes this completely filled band.

In the case of the filled band, were one to apply a small electric field to this system, the system cannot respond at all. There is simply no freedom to repopulate the occupation of k-states because every state is already filled. Thus we conclude an important principle.

Principle: A filled band carries no current.[9]

Thus our example of a divalent tightbinding model is an insulator (this type of insulator is known as a *band insulator*). Indeed, many systems of divalent atoms are insulators (although in a moment we will discuss how divalent atoms can also form metals).

[9] In one dimension this principle is absolutely correct. In higher dimensions, it is sometimes possible to have Hall effect current (but not longitudinal current) from bands that are entirely filled. While this situation, a so-called "Chern band", is of great current research interest, it is far beyond the scope of this book. As such, I'd recommend ignoring this unusual possibility and just viewing the principle as being *almost* always true for most practical situations.

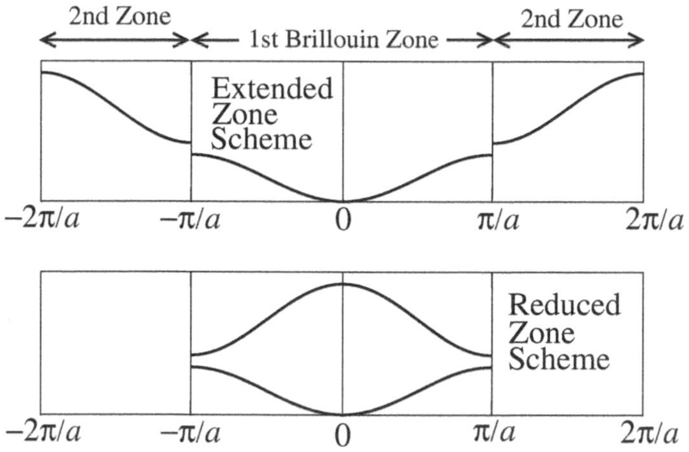

Fig. 11.5 Diatomic tight binding dispersion in one dimension. **Bottom:** Reduced zone scheme. **Top:** Extended zone scheme. Note that in obtaining the extended zone scheme from the reduced zone scheme, one simply translates pieces of the dispersion curve by appropriate reciprocal lattice vectors.

11.4 Multiple Bands

In the tight binding chain considered in this chapter, we considered only the case where there is a single atom in the unit cell and a single orbital per atom. However, more generally we might consider a case where we have multiple orbitals per unit cell.

One possibility is to consider one atom per unit cell, but several orbitals per atom.[10] Analogous to what we found for the tight binding model having only one orbital per atom, when the atoms are very far apart, one has only the atomic orbitals on each atom. However, as the atoms are moved closer together, the orbitals merge together and the energies spread to form bands.[11] Analogous to Fig. 11.3 we have shown how this occurs for the two band case in Fig. 11.6.

A very similar situation occurs when we have two atoms per unit cell but only one orbital per atom (see Exercises 11.2 and 11.4.) The general result will be quite analogous to what we found for vibrations of a diatomic chain in Chapter 10.

In Fig. 11.5 we show the spectrum of a tight-binding model with two different atoms per unit cell—each having a single orbital. We have shown results here in both the reduced and extended zone schemes.

As for the case of vibrations, we see that there are now two possible energy eigenstates at each value of k. In the language of electrons, we say that there are two bands (we do not use the words "acoustic" and "optical" for electrons, but the idea is similar). Note that there is a gap between the two bands where there are simply no energy eigenstates.

Let us think for a second about what might result in this situation where there are two atoms per unit cell and one orbital per atom. If each atom (of either type) were divalent, then the two electrons donated per atom would completely fill the single orbital on each atom. In this case, both bands would be completely filled with both spin-up and spin-down electrons.

[10] Each atom actually has an infinite number of orbitals to be considered at higher and higher energy. But only a small number of them are filled, and within our level of approximation we can only consider very few of them.

[11] This picture of atomic orbitals in the weak hopping limit merging together to form bands does not depend on the fact that the crystal of atoms is ordered. Glasses and amorphous solids can have this sort of band structure as well!

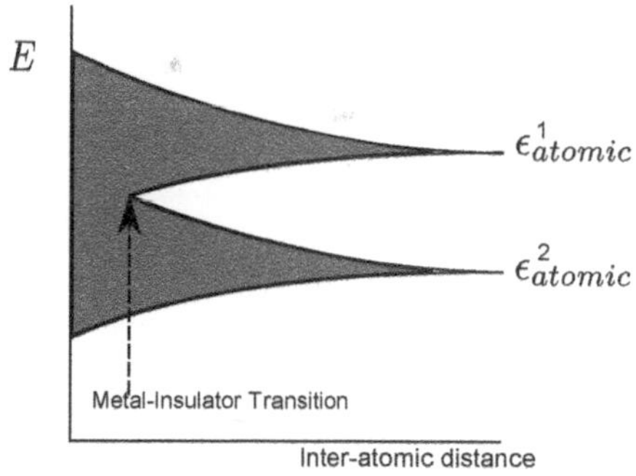

Fig. 11.6 Caricature of bands for a two-band model as a function of inter-atomic spacing. In the atomic limit, the orbitals have energies ϵ^1_{atomic} and ϵ^2_{atomic}. If the system has valence two per unit cell, then in the atomic limit the lower orbital is filled and the upper orbital is empty. When the atoms are pushed together, the lower band remains filled, and the upper remains empty, until the bands start to overlap, whereupon we have two bands both partially filled, which becomes a metal.

On the other hand, if each atom (of either type) is monovalent, then this means exactly half of the states of the system should be filled. However, here, when one fills half of the states of the system, then all of the states of the lower band are completely filled (with both spins) but all of the states in the upper band are completely empty. In the extended zone scheme it appears that a gap has opened up precisely where the Fermi surface is (at the Brillouin zone boundary!).

In the situation where a lower band is completely filled but an upper band is completely empty, if we apply a weak electric field to the system can current flow? In this case, one cannot rearrange electrons within the lower band, but one can remove an electron from the lower band and put it in the upper band in order to change the overall (crystal) momentum of the system. However, moving an electron from the lower band requires a finite amount of energy—one must overcome the gap between the bands. As a result, for small enough electric fields (and at low temperature), this cannot happen. We conclude that a filled band is an insulator as long as there is a finite gap to any higher empty bands.

As with the single-band case, one can imagine the magnitude of hopping changing as one changes the distance between atoms. When the atoms are far apart, then one is in the atomic limit, but these atomic states spread into bands as the atoms get closer together, as shown in Fig. 11.6.

For the case where each of the two atoms is monovalent, in the atomic limit, half of the states are filled—that is, the lower-energy atomic orbital is filled with both spin-up and spin-down electrons, whereas the higher-energy orbital is completely empty (i.e., an electron is transferred from the higher-energy atom to the lower-energy atom and this completely fills the lower-energy band). As the atoms are brought closer together, the atomic orbitals spread into bands (the hopping t increases). However, at some point the bands get so wide that their energies overlap[12]—in which case there is no gap to transfer electrons between bands, and the system becomes a metal as marked in Fig. 11.6. (If it is not clear how bands may overlap, consider, for example, the right side of Fig. 16.2. In fact band overlap of this type is very common in real materials!)

[12] See Exercise 11.4.

Chapter Summary

- Solving the tight-binding Schroedinger equation for electron waves is very similar to solving Newton's equations for vibrational (phonon) waves. The structure of the reciprocal lattice and the Brillouin zone remains the same.
- We obtain energy bands where energy eigenstates exist, and gaps between bands.
- Zero hopping is the atomic limit. As hopping increases, atomic orbitals spread into bands.

- Energies are parabolic in k near bottom of band—like free electrons, but with a modified effective mass.
- A filled band with a gap to the next band is an insulator (a *band* insulator), a partially filled band has a Fermi surface and is a metal.
- Whether a band is filled depends on the valence of the atoms.
- As we found for phonons, gaps open at Brillouin zone boundaries. Group velocities are also zero at zone boundaries.

References

No book has an approach to tight binding that is exactly like what we have here. The books that come closest do essentially the same thing, but in three dimensions (which complicates life a bit). These books are:

- Ibach and Luth, section 7.3
- Kittel, chapter 9, section on tight-binding
- Burns, sections 10.9–10.10
- Singleton, chapter 4

Possibly the nicest (albeit short) description is given by
- Dove, section 5.5.5.

Also a nice short description of the physics (without detail) is given by
- Rosenberg, section 8.19.

Finally, an alternative approach to tight binding is given by
- Hook and Hall, section 4.3.

This is a good discussion, but they insist on using time-dependent Schroedinger equation, which is annoying.

Exercises

(11.1) **Monatomic Tight Binding Chain**

Consider a one-dimensional tight binding model of electrons hopping between atoms. Let the distance between atoms be called a, and here let us label the atomic orbital on atom n as $|n\rangle$ for $n = 1 \ldots N$ (you may assume periodic boundary conditions, and you may assume orthonormality of orbitals, i.e., $\langle n|m\rangle = \delta_{nm}$). Suppose there is an on-site energy ϵ and a hopping matrix element $-t$. In other words, suppose $\langle n|H|m\rangle = \epsilon$ for $n = m$ and $\langle n|H|m\rangle = -t$ for $n = m \pm 1$.

▷ Derive and sketch the dispersion curve for electrons. (Hint: Use the effective Schroedinger equations of Exercise 6.2a. The resulting equation should look very similar to that of Exercise 9.2.)

▷ How many different eigenstates are there in this system?

▷ What is the effective mass of the electron near the bottom of this band?

▷ What is the density of states?

▷ If each atom is monovalent (it donates a single electron) what is the density of states at the Fermi surface?

▷ Give an approximation of the heat capacity of the system (see Exercise 4.3).

▷ What is the heat capacity if each atom is divalent?

(11.2) **Diatomic Tight Binding Chain**

We now generalize the calculation of the previous exercise to a one-dimensional diatomic solid which might look as follows:

$$-A-B-A-B-A-B-$$

Suppose that the onsite energy of type A is different from the onsite energy of type B. I.e, $\langle n|H|n\rangle$ is ϵ_A for n being on a site of type A and is ϵ_B for n being on a site of type B. (All hopping matrix elements $-t$ are still identical to each other.)

▷ Calculate the new dispersion relation. (This is extremely similar to Exercise 10.1. If you are stuck, try studying that exercise again.)

▷ Sketch this dispersion relation in both the reduced and extended zone schemes.

▷ What happens if $\epsilon_A = \epsilon_B$?

▷ What happens in the "atomic" limit when t becomes very small.

▷ What is the effective mass of an electron near the bottom of the lower band?

▷ If each atom (of either type) is monovalent, is the system a metal or an insulator?

▷ *Given the results of this exercise, explain why LiF (which has very ionic bonds) is an extremely good insulator.

(11.3) **Tight Binding Chain Done Right**

Let us reconsider the one-dimensional tight binding model as in Exercise 11.1. Again we assume an on-site energy ϵ and a hopping matrix element $-t$. In other words, suppose $\langle n|H|m\rangle = \epsilon$ for $n = m$ and $\langle n|H|m\rangle = -t$ for $n = m \pm 1$. However, now, let us no longer assume that orbitals are orthonormal. Instead, let us assume $\langle n|m\rangle = A$ for $n = m$ and $\langle n|m\rangle = B$ for $n = m+1$ with $\langle n|m\rangle = 0$ for $|n-m| > 1$.

▷ Why is this last assumption (the $|n-m| > 1$ case) reasonable?

Treating the possible non-orthogonality of orbitals here is very similar to what we did in Exercise 6.5. Go back and look at that exercise.

▷ Use the effective Schroedinger equation from Exercise 6.5 to derive the dispersion relation for this one-dimensional tight binding chain.

(11.4) **Two Orbitals per Atom**

(a) Consider an atom with two orbitals, A and B having eigenenergies ϵ^A_{atomic} and ϵ^B_{atomic}. Now suppose we make a one-dimensional chain of such atoms and let us assume that these orbitals remain orthogonal. We imagine hopping amplitudes t_{AA} which allows an electron on orbital A of a given atom to hop to orbital A on the neighboring atom. Similarly we imagine a hopping amplitude t_{BB} that allows an electron on orbital B of a given atom to hop to orbital B on the neighboring atom. (We assume that V_0, the energy shift of the atomic orbital due to neighboring atoms, is zero).

▷ Calculate and sketch the dispersion of the two resulting bands.

▷ If the atom is divalent, derive a condition on the quantities $\epsilon^A_{atomic} - \epsilon^B_{atomic}$, as well as t_{AA} and t_{BB} which determines whether the system is a metal or an insulator.

(b)* Now suppose that there is in addition a hopping term t_{AB} which allows an electron on one atom in orbital A to hop to orbital B on the neighboring atom (and vice versa). What is the dispersion relation now?

(11.5) **Electronic Impurity State***

Consider the one-dimensional tight binding Hamiltonian given in Eq. 11.4. Now consider the situation where one of the atoms in the chain (atom $n = 0$) is an impurity such that it has an atomic orbital energy which differs by Δ from all the other atomic orbital energies. In this case the Hamiltonian becomes

$$H_{n,m} = \epsilon_0 \delta_{n,m} - t(\delta_{n+1,m} + \delta_{n-1,m}) + \Delta \delta_{n,m}\delta_{n,0}.$$

(a) Using an ansatz

$$\phi_n = Ae^{-qa|n|}$$

with q real, and a the lattice constant, show that there is a localized eigenstate for any negative Δ, and find the eigenstate energy. This exercise is very similar to Exercise 9.6.

(b) Consider instead a continuum one-dimensional Hamiltonian with a delta-function potential

$$H = -\frac{\hbar^2}{2m^*}\partial_x^2 + (a\Delta)\delta(x).$$

Similarly show that there is a localized eigenstate for any negative Δ and find its energy. Compare your result to that of part (a).

(11.6) **Reflection from an Impurity***

Consider the tight binding Hamiltonian from the previous exercise representing a single impurity in a chain. Here the intent is to see how this impurity scatters a plane wave incoming from the left with unit amplitude (this is somewhat similar to Exercise 9.5). Use an ansatz wavefunction

$$\phi_n = \begin{cases} Te^{-ikna} & n \geqslant 0 \\ e^{-ikna} + Re^{+ikna} & n < 0 \end{cases}$$

to determine the transmission T and reflection R as a function of k.

(11.7) **Transport in One Dimension***

(a) Consider the one-dimensional tight binding chain discussed in this chapter at (or near) zero temperature ($k_B T \ll \mu$). Suppose the right end of this chain is attached to a reservoir at chemical potential μ_R and the left end of the chain is attached to a reservoir at chemical potential μ_L and let us assume $\mu_L > \mu_R$. The electrons moving towards the left will be filled up to chemical potential μ_R, whereas the electrons moving towards the right will be filled up to chemical potential μ_L, as shown in the bottom of Fig. 11.4, and also diagrammed schematically in the following figure

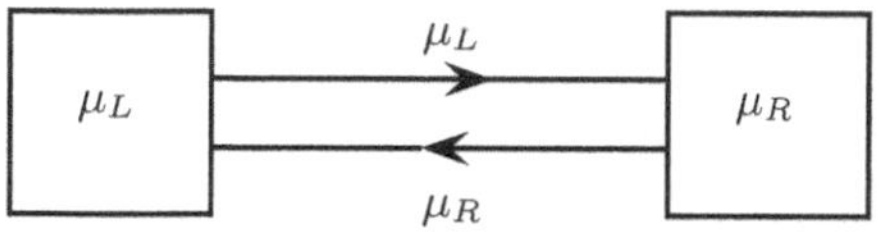

(i) Argue that the total electrical current of all the electrons moving to the right is

$$j_R = -e \int_0^{\infty} \frac{dk}{\pi} v(k) n_F(\beta(E(k) - \mu_L))$$

with $v(k) = (1/\hbar) dE(k)/dk$ the group velocity and n_F the Fermi occupation factor; and an analogous equation holds for left moving current.

(ii) Calculate the conductance G of this wire, defined as

$$J_{total} = GV$$

where $J_{total} = j_L - j_R$ and $eV = \mu_L - \mu_R$, and show $G = 2e^2/h$ with h Planck's constant. This "quantum" of conductance is routinely measured in disorder free one-dimensional electronic systems.

(iii) In the context of Exercise 11.6, imagine that an impurity is placed in this chain between the two reservoirs to create some backscattering. Argue that the conductance is reduced to $G = 2e^2|T|^2/h$. This is known as the Landauer formula and is a pillar of nano-scale electronics.

(b) Now suppose that the chemical potentials at both reservoirs are the same, but the temperatures are T_L and T_R respectively.

(i) Argue that the heat current j^q of all the electrons moving to the right is

$$j_R^q = \int_0^{\infty} \frac{dk}{\pi} v(k)\ (E(k) - \mu)\ n_F(\beta_L(E(k) - \mu))$$

and a similar equation holds for left-moving heat current.

(ii) Define the *thermal* conductance K to be

$$J^q = K(T_L - T_R)$$

where $J^q = j_L^q - j_R^q$ and $T_L - T_R$ is assumed to be small. Derive that the thermal conductance can be rewritten as

$$K = \frac{-2}{hT} \int_{-\infty}^{\infty} dE (E - \mu)^2 \frac{\partial}{\partial E} n_F(\beta(E - \mu)).$$

Evaluating this expression, confirm the Wiedemann–Franz ratio for clean one-dimensional systems

$$\frac{K}{TG} = \frac{\pi^2 k_B^2}{3e^2}$$

(Note that this is a relationship between conductance and thermal conductance rather than between conductivity and thermal conductivity.) In evaluating the above integral you will want to use

$$\int_{-\infty}^{\infty} dx\ x^2 \frac{\partial}{\partial x} \frac{1}{e^x + 1} = -\frac{\pi^2}{3}.$$

If you are very adventurous, you can prove this nasty identity using the techniques analogous to those mentioned in footnote 20 of Chapter 2, as well as the fact that the Riemann zeta function takes the value $\zeta(2) = \pi^2/6$ which you can prove analogous to the appendix of that chapter.

(11.8) **Peierls Distortion*&**

Consider a chain made up of all the same type of atom, but in such a way that the spacing between atoms alternates as long-short-long-short as follows

$$-A = A - A = A - A = A-$$

In a tight binding model, the shorter bonds (marked with =) will have hopping matrix element $t_{short} = t(1 + \epsilon)$ whereas the longer bonds (marked with $-$) have hopping matrix element $t_{long} = t(1 - \epsilon)$.

(a) Calculate the tight-binding energy $E(k)$ spectrum of this chain. (The onsite energy E_0 is the same on every atom).

(b) Suppose the lower band is filled and the upper band is empty (what is the valence of each atom in this case?). The total ground-state energy is the integral of $E(k)$ over the filled states. Unfortunately this is a messy integral. Instead, let us approximate this integral by splitting it into two parts. Focus on the Brillouin zone boundary at π/a. Define $\kappa = \pi/a - k$. Split the relevant integral at $\kappa \approx \epsilon/a$, and show that the energy decrease for having small but finite ϵ (compared to $\epsilon = 0$) has a term of the form

$$\delta E_{total} \approx |2tL| \int_{\kappa=\epsilon/a}^{\text{cutoff}} d\kappa \left[\sqrt{\epsilon^2 + (\kappa a)^2/4} - \kappa a/2\right]$$

where the location of the cutoff of the integral will not be important.

(c) Show that the decrease in energy for having finite ϵ has a leading behavior of the form $\epsilon^2 \log \epsilon$. Check that the other parts of the integration that we have thrown out are smaller than the pieces we have kept.

(d) Now consider a chain of equally spaced identical A atoms connected together with identical springs with spring constant κ. Show that making a distortion whereby alternating springs are shorter/longer by δx costs energy proportional to $(\delta x)^2$. Conclude that for a chain with the valence as in part (b), a distortion of this sort will occur spontaneously. This is known as a Peierls distortion.

(11.9) **Tight Binding in 2d***

Consider a rectangular lattice in two dimensions as shown in the figure. Now imagine a tight binding model where there is one orbital at each lattice site, and where the hopping matrix element is $\langle n|H|m\rangle = t_1$ if sites n and m are neighbors in the horizontal direction and is $= t_2$ if n and m are neighbors in the vertical direction. Calculate the dispersion relation for this tight binding model. What does the dispersion relation look like near the bottom of the band? (The two-dimensional structure is more difficult to handle than the one-dimensional examples given in this chapter. In Chapters 12 and 13 we study crystals in two and three dimensions, and it might be useful to read those chapters first and then return to try this exercise again.)

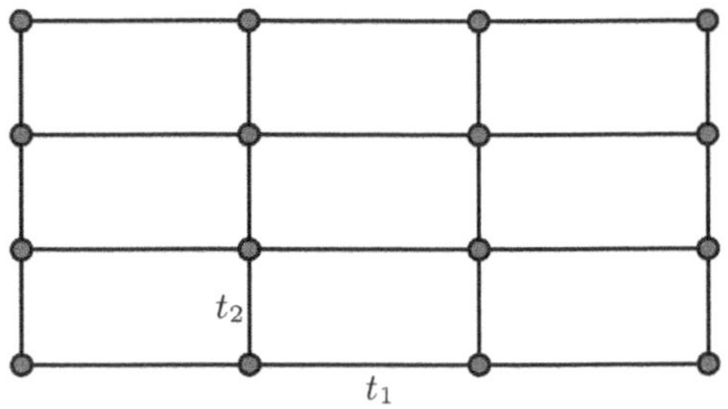

(11.10) **Finite Chains and Complex Hopping*&**

(a) Consider a tight binding chain of N sites (numbered $1 \ldots N$) in a straight (not periodic) chain, with real hopping amplitude t. You can model the ends of the chain by imagining two additional sites just beyond the ends of the chain (numbered 0 and $N+1$) and imposing the boundary conditions that the wavefunction must always vanish on these two additional sites. By superposing left-going and right-going waves to satisfy the boundary condiditions, find the eigenstates and eigenenergies for this system.

(b) Consider a case where the hopping is complex, $t = |t|e^{i\phi}$ for hopping from left to right. Make sure when you write down the Schroedinger equation you have a hermitian Hamiltonian! Show that for the finite chain described above, no physical quantity is dependent on the phase ϕ.

(c) Now consider putting the N sites in a ring (i.e., it is now periodic). With complex hopping, what are the eigenstates and eigenenergies now? In the large N limit, what (if any) physical quantities can depend on the hopping phase? For small N what is the effect of the hopping phase?

(d)** When might such a nontrivial phase occur in a hopping amplitude?

Part IV

Geometry of Solids

12 Crystal Structure

Having introduced a number of important ideas in one dimension, we must now deal with the fact that our world is actually spatially three-dimensional. While this adds a bit of complication, really the important concepts are no harder in three dimensions than they were in one dimension. Some of the most important ideas we have already met in one dimension, but we will reintroduce them more generally here.

There are two things that might be difficult here. First, we do need to wrestle with a bit of geometry. Hopefully most will not find this too hard. Secondly we will also need to establish a language in order to describe structures in two and three dimensions intelligently. As such, much of this chapter is just a list of definitions to be learned, but unfortunately this is necessary in order to be able to continue further at this point.

12.1 Lattices and Unit Cells

Definition 12.1 *A* ***lattice***[1] *is an infinite set of points defined by integer sums of a set of linearly independent primitive lattice*[2] *vectors.*

For example, in two dimensions, as shown in Fig. 12.1 the lattice points are described as

$$\mathbf{R}_{[n_1\, n_2]} = n_1\mathbf{a_1} + n_2\mathbf{a_2} \qquad n_1, n_2 \in \mathbb{Z} \qquad (2d)$$

with $\mathbf{a_1}$ and $\mathbf{a_2}$ being the primitive lattice vectors and n_1 and n_2 being integers. In three dimensions points of a lattice are analogously indexed by three integers:

$$\mathbf{R}_{[n_1\, n_2\, n_3]} = n_1\mathbf{a_1} + n_2\mathbf{a_2} + n_3\mathbf{a_3} \qquad n_1, n_2, n_3 \in \mathbb{Z} \qquad (3d). \qquad (12.1)$$

Note that in one dimension this definition of a lattice fits with our previous description of a lattice as being the points $R = na$ with n an integer.

It is important to point out that in two and three dimensions, the choice of primitive lattice vectors is not unique,[3] as shown in Fig. 12.2. (In one dimension, the single primitive lattice vector is unique up to the sign, or direction, of a.)

[1]Warning: Some books (Ashcroft and Mermin in particular) refer to this as a *Bravais lattice*. This enables them to use the term *lattice* to describe other things that we would not call a lattice (e.g., the honeycomb). However, the definition we use here is more common among crystallographers, and more correct mathematically as well.

[2]Very frequently "primitive lattice vectors" are called "primitive basis vectors" (not the same use of the word "basis" as in Section 10.1) or "primitive translation vectors".

[3]Given a set of primitive lattice vectors $\mathbf{a_i}$ a new set of primitive lattice vectors may be constructed as $\mathbf{b_i} = \sum_j m_{ij}\mathbf{a_j}$ so long as m_{ij} is an invertible matrix with integer entries *and* the inverse matrix $[m^{-1}]_{ij}$ also has integer entries. Equivalently m_{ij} are integers and $\det[m] = \pm 1$.

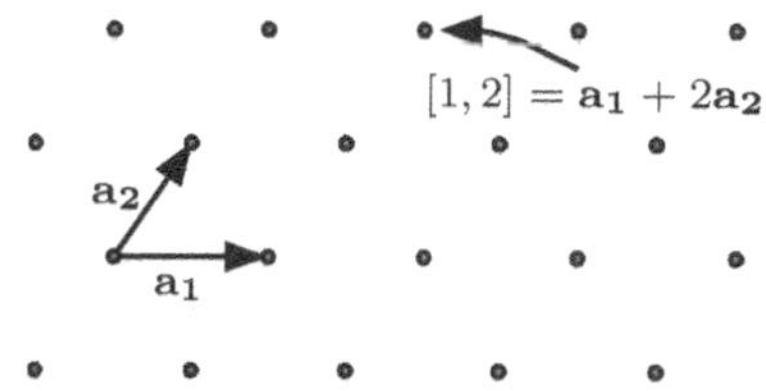

Fig. 12.1 A lattice is defined as integer sums of primitive lattice vectors.

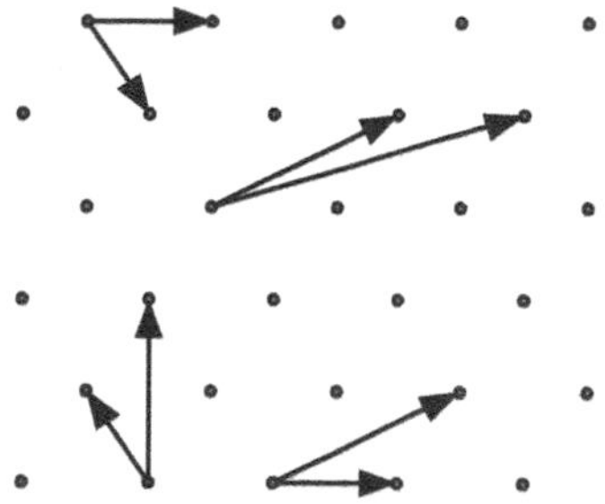

Fig. 12.2 The choice of primitive lattice vectors for a lattice is not unique. (Four possible sets of primitive lattice vectors are shown, but there are an infinite number of possibilities!)

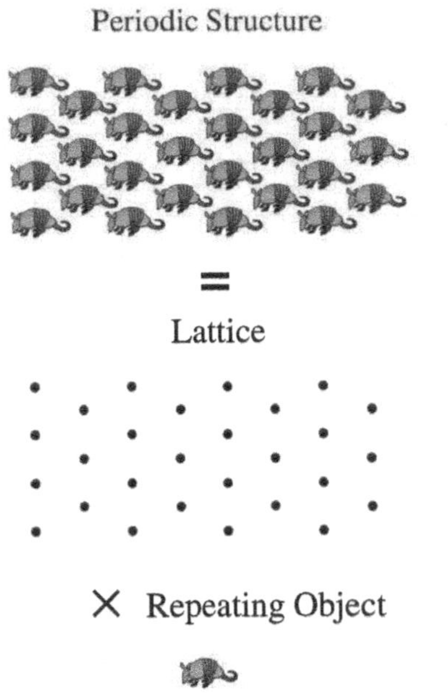

Fig. 12.3 Any periodic structure can be represented as a lattice of repeating motifs.

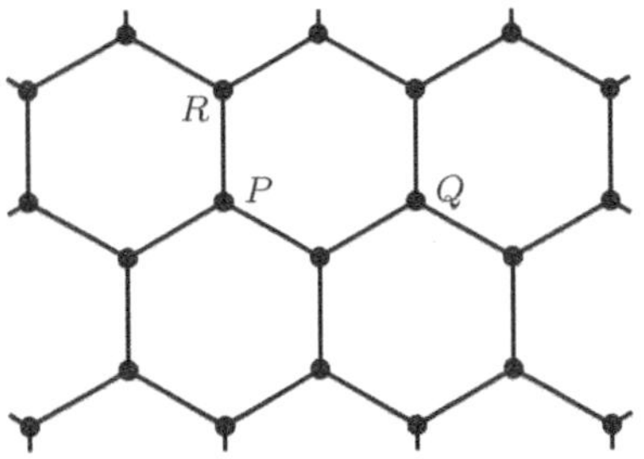

Fig. 12.4 The honeycomb is not a lattice. Points P and R are inequivalent (points P and Q *are* equivalent).

[4]One should be very careful *not* to call the honeycomb a hexagonal lattice. First of all, by our definition it is not a lattice at all since all points do not have the same environment. Secondly, some people (perhaps confusingly) use the term "hexagonal" to mean what the rest of us call a triangular lattice: a lattice of triangles where each point has six nearest neighbor points (see Fig. 12.6).

It turns out that there are several definitions that are entirely equivalent to the one we have just given:

Equivalent Definition 12.1.1 *A* ***lattice*** *is an infinite discrete set of vectors where addition or subtraction of any two vectors in the set gives a third vector in the set.*

It is easy to see that our first definition 12.1 implies the second one 12.1.1. Here is a less crisply defined, but sometimes more useful definition.

Equivalent Definition 12.1.2 *A* ***lattice*** *is an infinite discrete set of points where the environment of any given point is equivalent to the environment of any other given point.*

It turns out that any periodic structure can be expressed as a lattice of repeating motifs. A cartoon of this statement is shown in Fig. 12.3. One should be cautious however, that not all periodic arrangements of points are lattices. The honeycomb[4] shown in Fig. 12.4 is *not* a lattice. This is obvious from the third definition 12.1.2: The environment of point P and point R are actually different—point P has a neighbor directly above it (the point R), whereas point R has no neighbor directly above.

In order to describe a honeycomb (or other more complicated arrangements of points) we have the idea of a unit cell, which we have met before in Section 10.1. Generally we have

Definition 12.2 *A* ***unit cell*** *is a region of space such that when many identical units are stacked together it tiles (completely fills) all of space and reconstructs the full structure.*

An equivalent (but less rigorous) definition is

Equivalent Definition 12.2.1 *A* ***unit cell*** *is the repeated motif which is the elementary building block of the periodic structure.*

To be more specific we frequently want to work with the smallest possible unit cell:

Definition 12.3 *A* ***primitive unit cell*** *for a periodic crystal is a unit cell containing exactly one lattice point.*

As mentioned in Section 10.1 the definition of the unit cell is never unique. This is shown, for example, in Fig. 12.5.

Sometimes it is useful to define a unit cell which is not primitive in order to make it simpler to work with. This is known as a *conventional unit cell.* Almost always these conventional unit cells are chosen so as to have orthogonal axes.

Some examples of possible unit cells are shown for the triangular lattice in Fig. 12.6. In this figure the conventional unit cell (upper left) is chosen to have orthogonal axes—which is often easier to work with than axes which are non-orthogonal.

A note about counting the number of lattice points in the unit cell. It is frequently the case that we will work with unit cells where the lattice points live at the corners (or edges) of the cells. When a lattice point is on the boundary of the unit cell, it should only be counted fractionally depending on what fraction of the point is actually in the cell. So for example in the conventional unit cell shown in Fig. 12.6, there are two lattice points within this cell. There is one point in the center, then four points at the corners—each of which is one quarter inside the cell, so we obtain $2 = 1 + 4(\frac{1}{4})$ points in the cell. (Since there are two lattice points in this cell, it is by definition not primitive.) Similarly for the primitive cell shown in Fig. 12.6 (upper right), the two lattice points at the far left and the far right have a 60° degree slice (which is 1/6 of a circle) inside the cell. The other two lattice points each have 1/3 of the lattice point inside the unit cell. Thus this unit cell contains $2(\frac{1}{3}) + 2(\frac{1}{6}) = 1$ point, and is thus primitive. Note however, that we can just imagine shifting the unit cell a tiny amount in almost any direction such that a single lattice point is completely inside the unit cell and the others are completely outside the unit cell. This sometimes makes counting much easier.

Also shown in Fig. 12.6 is a so-called *Wigner–Seitz unit cell*

Definition 12.4 *Given a lattice point, the set of all points in space which are closer to that given lattice point than to any other lattice point constitute the* ***Wigner–Seitz cell*** *of the given lattice point.*[5]

There is a rather simple scheme for constructing such a Wigner–Seitz cell: choose a lattice point and draw lines to all of its possible near neighbors (not just its nearest neighbors). Then draw perpendicular bisectors of all of these lines. The perpendicular bisectors bound the Wigner–Seitz cell. It is always true that the Wigner–Seitz construction for a lattice gives a primitive unit cell. In Fig. 12.7 we show another example of the Wigner–Seitz construction for a two-dimensional lattice.

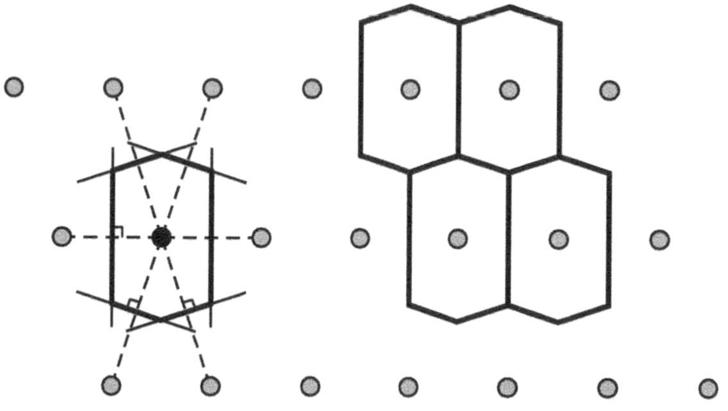

Fig. 12.7 The Wigner–Seitz construction for a lattice in two dimensions. On the left perpendicular bisectors are added between the darker point and each of its neighbors. The area bounded defines the Wigner–Seitz cell. On the right it is shown that the Wigner–Seitz cell is a primitive unit cell. (The cells on the right are exactly the same shape as the bounded area on the left!)

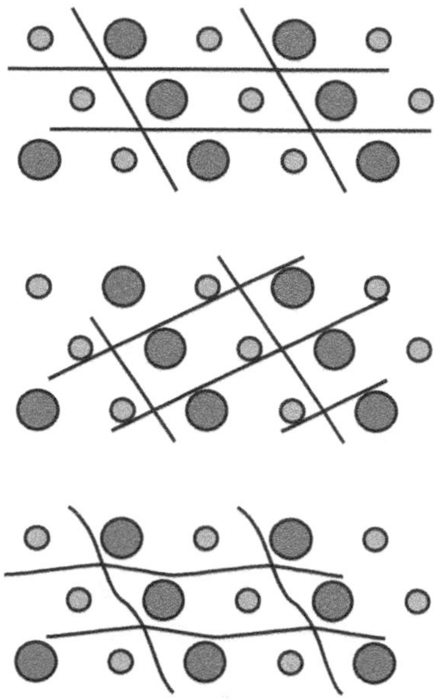

Fig. 12.5 The choice of a unit cell is not unique. All of these unit cells can be used as "tiles" to perfectly reconstruct the full crystal.

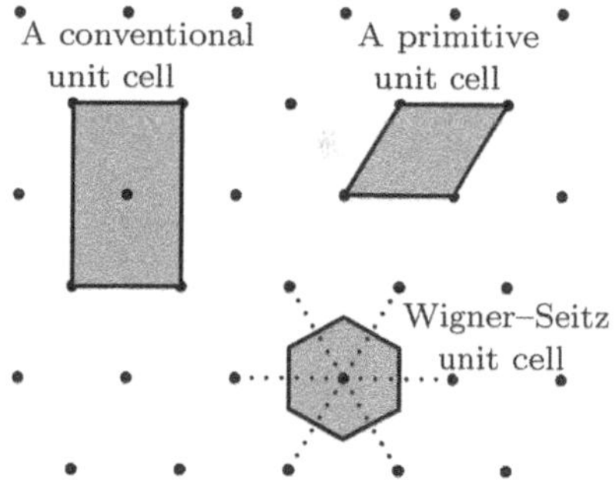

Fig. 12.6 Some unit cells for the triangular lattice.

[5] A construction analogous to Wigner–Seitz can be performed on an irregular collection of points as well as on a periodic lattice. For such an irregular set of point the region closer to one particular point than to any other of the points is known as a Voronoi cell.

A similar construction can be performed in three dimensions in which case one must construct perpendicular-bisecting planes to bound the Wigner–Seitz cell.[6] See for example, Figs. 12.13 and 12.16.

[6]Eugene Wigner was yet another Nobel laureate who was one of the truly great minds of the last century of physics. Perhaps as important to physics was the fact that his sister, Margit, married Dirac. It was often said that Dirac could be a physicist only because Margit handled everything else. Fredrick Seitz was far less famous, but gained notoriety in his later years by being a consultant for the tobacco industry, a strong proponent of the Regan-era Star Wars missile defense system, and a prominent sceptic of global warming. He passed away in 2007.

Definition 12.5 *The description of objects in the unit cell with respect to the reference lattice point in the unit cell is known as a* ***basis****.*

This is the same definition of "basis" that we used in Section 10.1. In other words, we think of reconstructing the entire crystal by associating with each lattice point a basis of atoms.

In Fig. 12.8 (top) we show a periodic structure in two dimensions made of two types of atoms. On the bottom we show a primitive unit cell (expanded) with the position of the atoms given with respect to the reference point of the unit cell which is taken to be the lower left-hand corner. We can describe the basis of this crystal as follows:

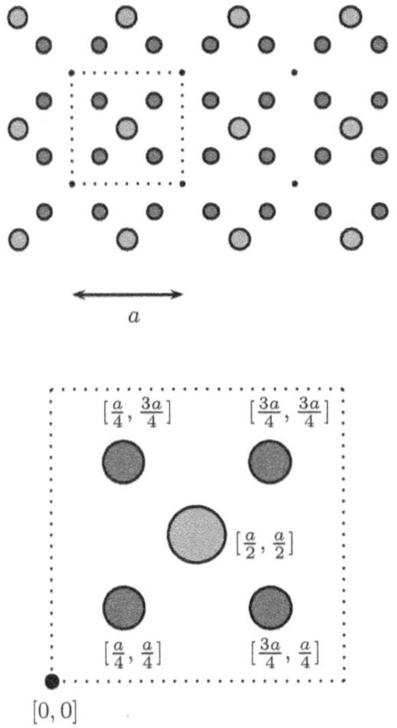

Fig. 12.8 Top: A periodic structure in two dimensions. A unit cell is marked with the dotted lines. **Bottom:** A blow-up of the unit cell with the coordinates of the objects in the unit cell with respect to the reference point in the lower left-hand corner. The basis is the description of the atoms along with these positions.

Basis for crystal in Fig. 12.8 =		
Large Light Gray Atom	Position=	$[a/2, a/2]$
Small Dark Gray Atoms	Position=	$[a/4, a/4]$
		$[a/4, 3a/4]$
		$[3a/4, a/4]$
		$[3a/4, 3a/4]$

The reference points (the small black dots in the figure) forming the square lattice have positions

$$\mathbf{R}_{[n_1\, n_2]} = [a\,n_1, a\,n_2] = a\,n_1\hat{\boldsymbol{x}} + a\,n_2\hat{\boldsymbol{y}} \tag{12.2}$$

with n_1, n_2 integers so that the large light gray atoms have positions

$$\mathbf{R}^{light-gray}_{[n_1\, n_2]} = [a\,n_1, a\,n_2] + [a/2, a/2]$$

whereas the small dark gray atoms have positions

$$\begin{aligned}
\mathbf{R}^{dark-gray1}_{[n_1\, n_2]} &= [a\,n_1, a\,n_2] + [a/4, a/4] \\
\mathbf{R}^{dark-gray2}_{[n_1\, n_2]} &= [a\,n_1, a\,n_2] + [a/4, 3a/4] \\
\mathbf{R}^{dark-gray3}_{[n_1\, n_2]} &= [a\,n_1, a\,n_2] + [3a/4, a/4] \\
\mathbf{R}^{dark-gray4}_{[n_1\, n_2]} &= [a\,n_1, a\,n_2] + [3a/4, 3a/4].
\end{aligned}$$

In this way you can say that the positions of the atoms in the crystal are "the lattice plus the basis".

We can now return to the case of the honeycomb shown in Fig. 12.4. The same honeycomb is shown in Fig. 12.9 with the lattice and the basis explicitly shown. Here, the reference points (small black dots) form a (triangular) lattice, where we can write the primitive lattice vectors as

$$\begin{aligned} \mathbf{a}_1 &= a\,\hat{\boldsymbol{x}} \\ \mathbf{a}_2 &= (a/2)\,\hat{\boldsymbol{x}} + (a\sqrt{3}/2)\,\hat{\boldsymbol{y}}. \end{aligned} \tag{12.3}$$

In terms of the reference points of the lattice, the basis for the primitive unit cell, i.e., the coordinates of the two larger circles with respect to the reference point, are given by $\frac{1}{3}(\mathbf{a}_1 + \mathbf{a}_2)$ and $\frac{2}{3}(\mathbf{a}_1 + \mathbf{a}_2)$.

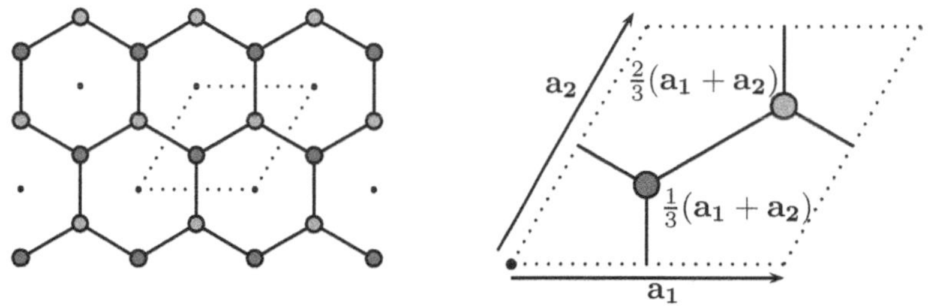

Fig. 12.9 Left: The honeycomb from Fig. 12.4 is shown with the two inequivalent points of the unit cell given different shades. The unit cell is outlined dotted and the corners of the unit cell are marked with small black dots (which form a triangular lattice). **Right:** The unit cell is expanded and coordinates are given with respect to the reference point at the lower left corner.

12.2 Lattices in Three Dimensions

The simplest lattice in three dimensions is the simple cubic lattice shown in Fig. 12.10 (sometimes known as cubic "P" or cubic-primitive lattice). The primitive unit cell in this case can most conveniently be taken to be a single cube—which includes 1/8 of each of its eight corners (see Fig. 12.11).

Only slightly more complicated than the simple cubic lattice are the *tetragonal* and *orthorhombic* lattices where the axes remain perpendicular, but the primitive lattice vectors may be of different lengths (shown in Fig. 12.11). The orthorhombic unit cell has three different lengths of its perpendicular primitive lattice vectors, whereas the tetragonal unit cell has two lengths the same and one different.

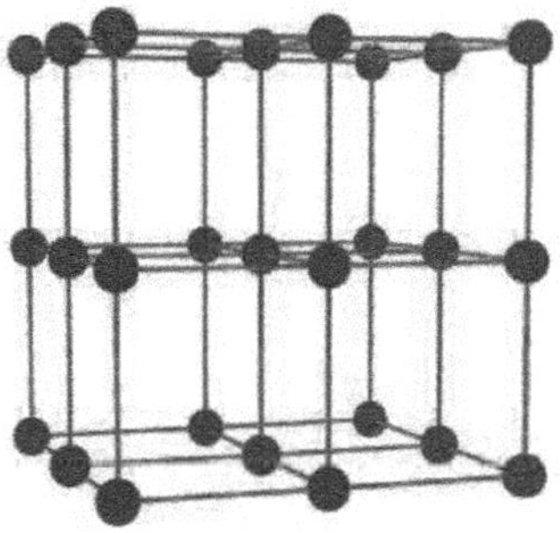

Fig. 12.10 A cubic lattice, otherwise known as cubic "P" or cubic primitive.

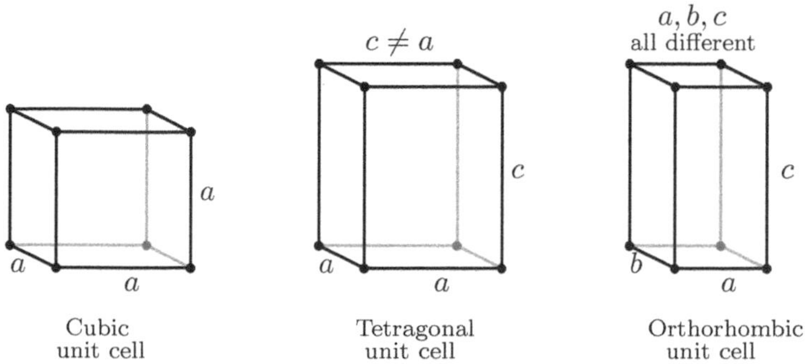

Fig. 12.11 Unit cells for cubic, tetragonal, and orthorhombic lattices.

Conventionally, to represent a given vector amongst the infinite number of possible lattice vectors in a lattice, one writes

$$[uvw] = u\mathbf{a_1} + v\mathbf{a_2} + w\mathbf{a_3} \tag{12.4}$$

where u,v, and w are integers. For cases where the lattice vectors are orthogonal, the basis vectors $\mathbf{a}_1$, $\mathbf{a}_2$, and $\mathbf{a}_3$ are assumed to be in the $\hat{\boldsymbol{x}}$, $\hat{\boldsymbol{y}}$, and $\hat{\boldsymbol{z}}$ directions. We have seen this notation before,[7] for example, in the subscripts of the equations after definition 12.1.

[7] This notation is also sometimes abused, as in Eq. 12.2 or Fig. 12.8, where the brackets enclose not integers, but distances. The notation can also be abused to specify points which are not members of the lattice, by choosing, u, v, or w to be non-integers. We will sometimes engage in such abuse. Note also that crystallographers often use the square brackets to indicate a *direction* in real space, rather than a position.

Lattices in three dimensions also exist where axes are not orthogonal. We will not cover all of these more complicated lattices in detail in this book. (In Section 12.2.4 we will briefly look through these other cases, but only at a very cursory level.) The principles we learn in the more simple cases (with orthogonal axes) generalize fairly easily, and just add further geometric and algebraic complexity without illuminating the physics much further.

Two particular lattices (with orthogonal axes) which we will cover in some detail are body-centered cubic (bcc) lattices and face-centered cubic (fcc) lattices.

12.2.1 The Body-Centered Cubic (bcc) Lattice

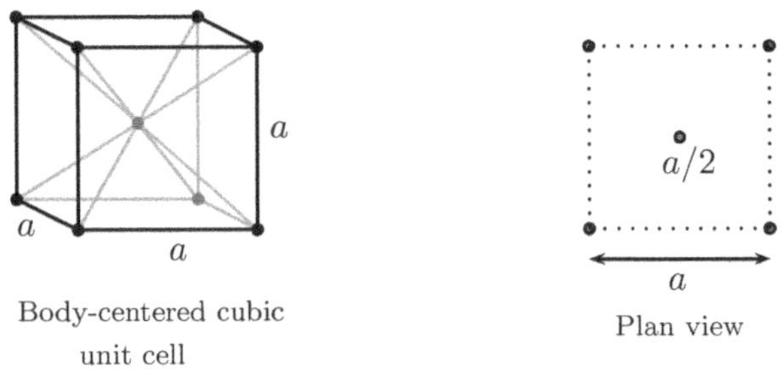

Fig. 12.12 Conventional unit cell for the body-centered cubic (I) lattice. **Left:** 3D view. **Right:** A plan view of the conventional unit cell. Unlabeled points are both at heights 0 and a.

The body-centered cubic (bcc) lattice is a simple cubic lattice where there is an additional lattice point in the very center of the cube (this is sometimes known[8] as cubic-I.) The unit cell is shown in the left of Fig. 12.12. Another way to show this unit cell, which does not rely on showing a three-dimensional picture, is to use a so-called *plan view* of the unit cell, shown in the right of Fig. 12.12. A plan view (a term used in engineering and architecture) is a two-dimensional projection from the top of an object where heights are labeled to show the third dimension.

[8] Cubic-I comes from "Innenzentriert" (inner-centered). This notation was introduced by Bravais in his 1848 treatise (Interestingly, Europe was burning in 1848, but obviously that didn't stop science from progressing.)

In the picture of the bcc unit cell, there are eight lattice points on the corners of the cell (each of which is 1/8 inside of the conventional unit cell) and one point in the center of the cell. Thus the conventional unit cell contains exactly two $(= 8 \times 1/8 + 1)$ lattice points.

Packing together these unit cells to fill space, we see that the lattice points of a full bcc lattice can be described as being points having coordinates $[x, y, z]$ where either all three coordinates are integers $[uvw]$ times the lattice constant a, or all three are half-odd-integers times the lattice constant a.

It is often convenient to think of the bcc lattice as a simple cubic lattice with a basis of two atoms per conventional cell. The simple cubic lattice contains points $[x, y, z]$ where all three coordinates are integers in units of the lattice constant. Within the conventional simple-cubic unit cell we put one point at position $[0, 0, 0]$ and another point at the position $[\frac{1}{2}, \frac{1}{2}, \frac{1}{2}]$ in units of the lattice constant. Thus the points of the bcc lattice are written in units of the lattice constant as

$$\begin{aligned} \mathbf{R}_{corner} &= [n_1, n_2, n_3] \\ \mathbf{R}_{center} &= [n_1, n_2, n_3] + [\tfrac{1}{2}, \tfrac{1}{2}, \tfrac{1}{2}] \end{aligned}$$

as if the two different types of points were two different types of atoms, although all points in this lattice should be considered equivalent (they only look inequivalent because we have chosen a conventional unit cell with two lattice points in it). From this representation we see that we can also think of the bcc lattice as being two interpenetrating simple cubic lattices displaced from each other by $[\frac{1}{2}, \frac{1}{2}, \frac{1}{2}]$. (See also Fig. 12.14.)

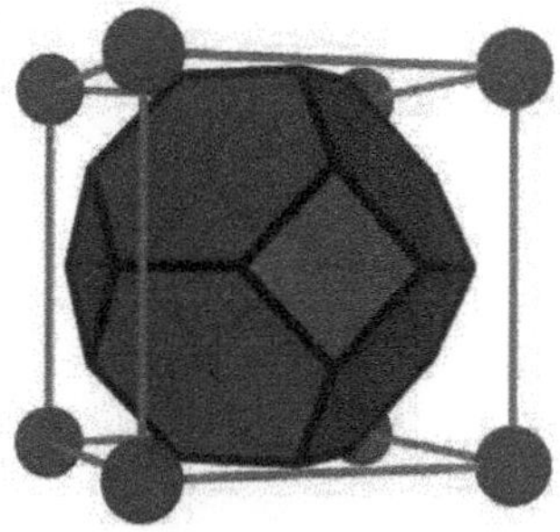

Fig. 12.13 The Wigner–Seitz cell of the bcc lattice (this shape is a "truncated octahedron"). The hexagonal face is the perpendicular bisecting plane between the lattice point (shown as a sphere) in the center and the lattice point (also a sphere) on the corner. The square face is the perpendicular bisecting plane between the lattice point in the center of the unit cell and a lattice point in the center of the neighboring unit cell.

We may ask why it is that this set of points forms a lattice. In terms of our first definition of a lattice (definition 12.1) we can write the primitive lattice vectors of the bcc lattice as

$$\begin{aligned} \mathbf{a}_1 &= [1, 0, 0] \\ \mathbf{a}_2 &= [0, 1, 0] \\ \mathbf{a}_3 &= [\tfrac{1}{2}, \tfrac{1}{2}, \tfrac{1}{2}] \end{aligned}$$

in units of the lattice constant. It is easy to check that any combination

$$\mathbf{R} = n_1\mathbf{a}_1 + n_2\mathbf{a}_2 + n_3\mathbf{a}_3 \tag{12.5}$$

with n_1, n_2, and n_3 integers gives a point within our definition of the bcc lattice (that the three coordinates are either all integers or all half-odd integers times the lattice constant). Further, one can check that any point satisfying the conditions for the bcc lattice can be written in the form of Eq. 12.5.

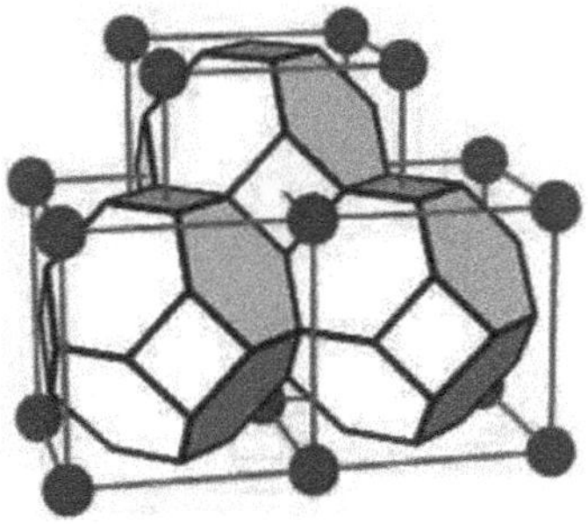

Fig. 12.14 The Wigner–Seitz cells of the bcc lattice pack together to tile all of space. Note that the structure of the bcc lattice is that of two interpenetrating simple cubic lattices.

We can also check that our description of a bcc lattice satisfies our second description of a lattice (definition 12.1.1) that addition of any two points of the lattice (given by Eq. 12.5) gives another point of the lattice.

More qualitatively we can consider definition 12.1.2 of the lattice—that the local environment of every point in the lattice should be the same. Examining the point in the center of the unit cell, we see that it has precisely eight nearest neighbors in each of the possible diagonal directions. Similarly, any of the points in the corners of the unit cells will have eight nearest neighbors corresponding to the points in the center of the eight adjacent unit cells.

The *coordination number* of a lattice (frequently called Z or z) is the number of nearest neighbors any point of the lattice has. For the bcc lattice the coordination number is $Z = 8$.

As in two dimensions, a Wigner–Seitz cell can be constructed around each lattice point which encloses all points in space that are closer to that lattice point than to any other point in the lattice. This Wigner–Seitz unit cell for the bcc lattice is shown in Fig. 12.13. Note that this cell is bounded by the perpendicular bisecting planes between lattice points. These Wigner–Seitz cells, being primitive, can be stacked together to fill all of space as shown in Fig. 12.14.

12.2.2 The Face-Centered Cubic (fcc) Lattice

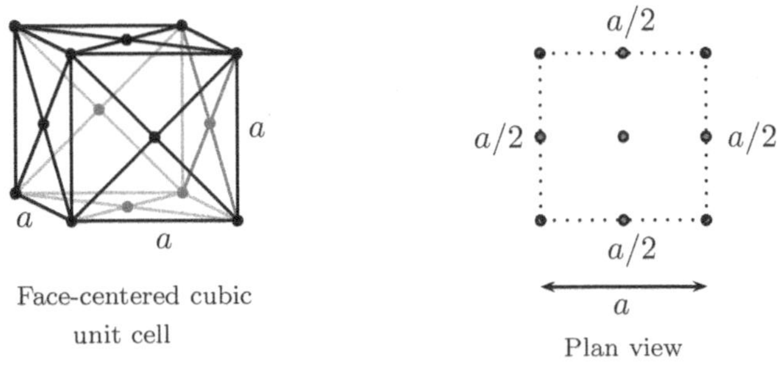

Fig. 12.15 Conventional unit cell for the face-centered cubic (F) lattice. **Left:** 3D view. **Right:** A plan view of the conventional unit cell. Unlabeled points are both at heights 0 and a.

The face-centered (fcc) lattice is a simple cubic lattice where there is an additional lattice point in the center of every face of every cube (this is sometimes known as cubic-F, for "face-centered"). The unit cell is shown in the left of Fig. 12.15. A plan view of the unit cell is shown on the right of Fig. 12.15 with heights labeled to indicate the third dimension.

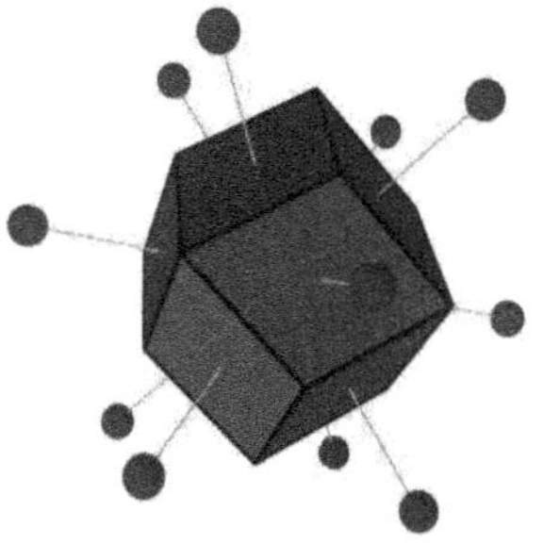

Fig. 12.16 The Wigner–Seitz cell of the fcc lattice (this shape is a "rhombic dodecahedron"). Each face is the perpendicular bisector between the central point and one of its 12 nearest neighbors.

In the picture of the fcc unit cell, there are eight lattice points on the corners of the cell (each of which is 1/8 inside of the conventional unit cell) and one point in the center of each of the six faces (each of which is 1/2 inside the cell). Thus the conventional unit cell contains exactly four ($= 8 \times 1/8 + 6 \times 1/2$) lattice points. Packing together these unit cells to fill space, we see that the lattice points of a full fcc lattice can be described as being points having coordinates (x, y, z) where either all three coordinates are integers times the lattice constant a, or two of the three coordinates are half-odd integers times the lattice constant a and the remaining one coordinate is an integer times the lattice constant a. Analogous to the bcc case, it is sometimes convenient to think of the fcc lattice as a simple cubic lattice with a basis of four atoms per conventional unit cell. The simple cubic lattice contains points $[x, y, z]$ where all three coordinates are integers in units of the lattice constant a. Within the conventional simple-cubic unit cell we put one point at position $[0, 0, 0]$ and another point at the position $[\frac{1}{2}, \frac{1}{2}, 0]$ another point at $[\frac{1}{2}, 0, \frac{1}{2}]$ and another point at $[0, \frac{1}{2}, \frac{1}{2}]$. Thus the lattice points of the fcc lattice are written in units of the lattice constant as

$$\begin{aligned} \mathbf{R}_{corner} &= [n_1, n_2, n_3] \\ \mathbf{R}_{face-xy} &= [n_1, n_2, n_3] + [\tfrac{1}{2}, \tfrac{1}{2}, 0] \\ \mathbf{R}_{face-xz} &= [n_1, n_2, n_3] + [\tfrac{1}{2}, 0, \tfrac{1}{2}] \\ \mathbf{R}_{face-yz} &= [n_1, n_2, n_3] + [0, \tfrac{1}{2}, \tfrac{1}{2}]. \end{aligned} \qquad (12.6)$$

Again, this expresses the points of the lattice as if they were four different types of points but they only look inequivalent because we have chosen a conventional unit cell with four lattice points in it. Since the conventional unit cell has four lattice points in it, we can think of the fcc lattice as being four interpenetrating simple cubic lattices.

Again we can check that this set of points forms a lattice. In terms of our first definition of a lattice (definition 12.1) we write the primitive lattice vectors of the fcc lattice as

$$\begin{aligned} \mathbf{a}_1 &= [\tfrac{1}{2}, \tfrac{1}{2}, 0] \\ \mathbf{a}_2 &= [\tfrac{1}{2}, 0, \tfrac{1}{2}] \\ \mathbf{a}_3 &= [0, \tfrac{1}{2}, \tfrac{1}{2}] \end{aligned}$$

in units of the lattice constant. Again it is easy to check that any combination

$$\mathbf{R} = n_1\mathbf{a}_1 + n_2\mathbf{a}_2 + n_3\mathbf{a}_3$$

with n_1, n_2, and n_3 integers gives a point within our definition of the fcc lattice (that the three coordinates are either all integers, or two of three are half-odd integers and the remaining is an integer in units of the lattice constant a).

We can also similarly check that our description of a fcc lattice satisfies our other two definitions of (definition 12.1.1 and 12.1.2) of a lattice. The Wigner–Seitz unit cell for the fcc lattice is shown in Fig. 12.16. In Fig. 12.17 it is shown how these Wigner–Seitz cells pack together to fill all of space.

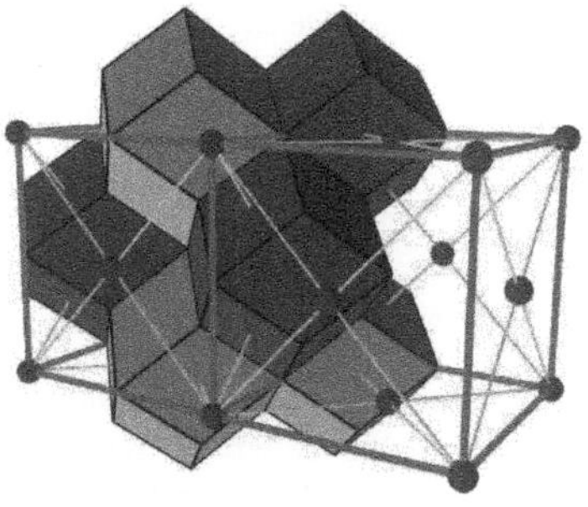

Fig. 12.17 The Wigner–Seitz cells of the fcc lattice pack together to tile all of space. Also shown in the picture are two conventional (cubic) unit cells.

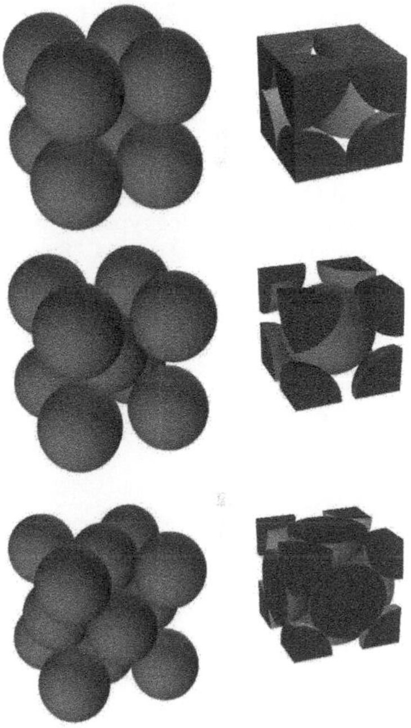

Fig. 12.18 Top: Simple cubic, **Middle:** bcc, **Bottom:** fcc. The left shows packing of spheres into these lattices. The right shows a cutaway of the conventional unit cell exposing how the fcc and bcc lattices leave much less empty space than the simple cubic.

12.2.3 Sphere Packing

Although the simple cubic lattice (see Fig. 12.10) is conceptually the simplest of all lattices, in fact, real crystals of atoms are rarely simple cubic.[9] To understand why this is so, think of atoms as small spheres that weakly attract each other and therefore try to pack close together. When you assemble spheres into a simple cubic lattice you find that it is a very inefficient way to pack the spheres together—you are left with a lot of empty space in the center of the unit cells, and this turns out to be energetically unfavorable in most cases. Packings of spheres into simple cubic, bcc, and fcc lattices are shown in Fig. 12.18. It is easy to see that the bcc and fcc lattices leave much less open space between

[9]Of all of the chemical elements, polonium is the only one which can form a simple cubic lattice with a single atom basis. (It can also form another crystal structure depending on how it is prepared.)

the spheres than packing the spheres in a simple cubic lattice[10] (see also Exercise 12.4). Correspondingly, bcc and fcc lattices are realized much more frequently in nature than simple cubic (at least in the case of a single atom basis). For example, the elements Al, Ca, Au, Pb, Ni, Cu, Ag (and many others) are fcc whereas the elements Li, Na, K, Fe, Mo, Cs (and many others) are bcc.

[10]In fact it is impossible to pack spheres more densely than you would get by placing the spheres at the vertices of an fcc lattice. This result (known empirically to people who have tried to pack oranges in a crate) was first officially conjectured by Johannes Kepler in 1611, but was not mathematically proven until 1998! Note however that there is another lattice, the *hexagonal close packed* lattice which achieves precisely the same packing density for spheres as the fcc lattice.

12.2.4 Other Lattices in Three Dimensions

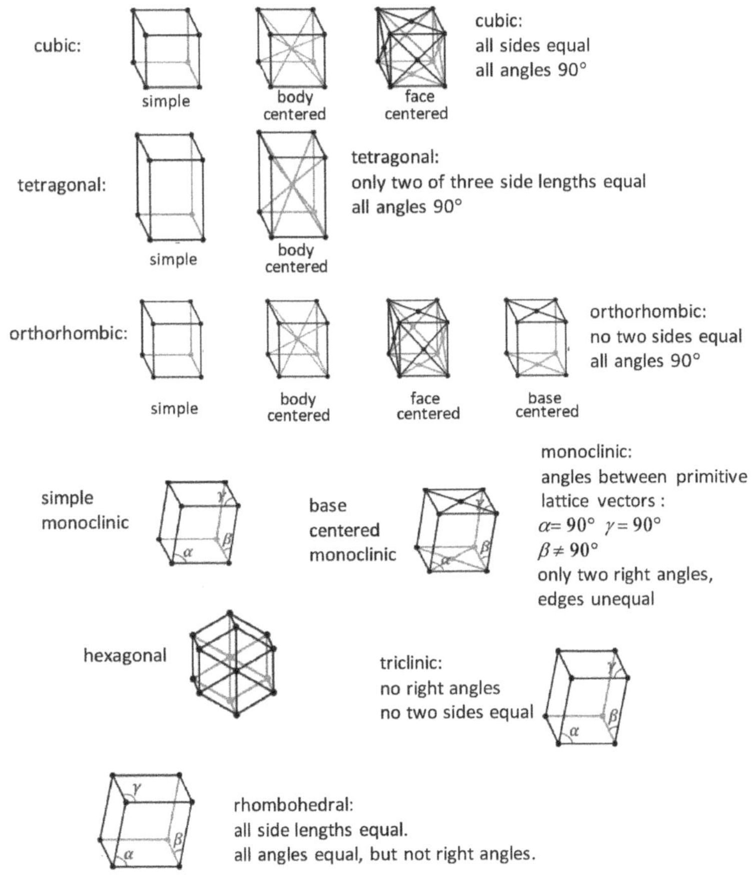

Fig. 12.19 Conventional unit cells for the fourteen Bravais lattice types. Note that if you tried to construct a "face-centered tetragonal" lattice, you would find that by turning the axes at 45 degrees it would actually be equivalent to a body-centered tetragonal lattice. Hence face-centered tetragonal is not listed as a Bravais lattice type (nor is base-centered tetragonal for a similar reason, etc.).

In addition to the simple cubic, orthorhombic, tetragonal, fcc, and bcc lattices, there are nine other types of lattices in three dimensions. These are known as the fourteen *Bravais lattice types*.[11] Although the study of all of these lattice types is beyond the scope of this book, it is probably a good idea to know that they exist.

[11]Named after Auguste Bravais who classified all the three-dimensional lattices in 1848. Actually they should be named after Moritz Frankenheim who studied the same thing over ten years earlier—although he made a minor error in his studies, and therefore missed getting his name associated with them.

Figure 12.19 shows the full variety of Bravais lattice types in three dimensions. While it is an extremely deep fact that there are only fourteen lattice types in three dimensions, the precise statement of this theorem, as well of the proof of it, are beyond the scope of this book. The key result is that any crystal, no matter how complicated, has a lattice which is one of these fourteen types.[12]

12.2.5 Some Real Crystals

Once we have discussed lattices we can combine a lattice with a basis to describe any periodic structure—and in particular, we can describe any crystalline structure. Several examples of real (and reasonably simple) crystal structures are shown in Figs. 12.20 and 12.21.

[12]There is a real subtlety here in classifying a crystal as having a particular lattice type. There are only these fourteen lattice types, but in principle a crystal could have one lattice, but have the *symmetry* of another lattice. An example of this would be if the a lattice were cubic, but the unit cell did not look the same from all six sides. Crystallographers would not classify this as being a cubic material even if the lattice happened to be cubic. The reason for this is that if the unit cell did not look the same from all six sides, there would be no particular reason that the three primitive lattice vectors should have the same length—it would be an insane coincidence were this to happen, and almost certainly in any real material the primitive lattice vector lengths would actually have slightly different values if measured more closely.

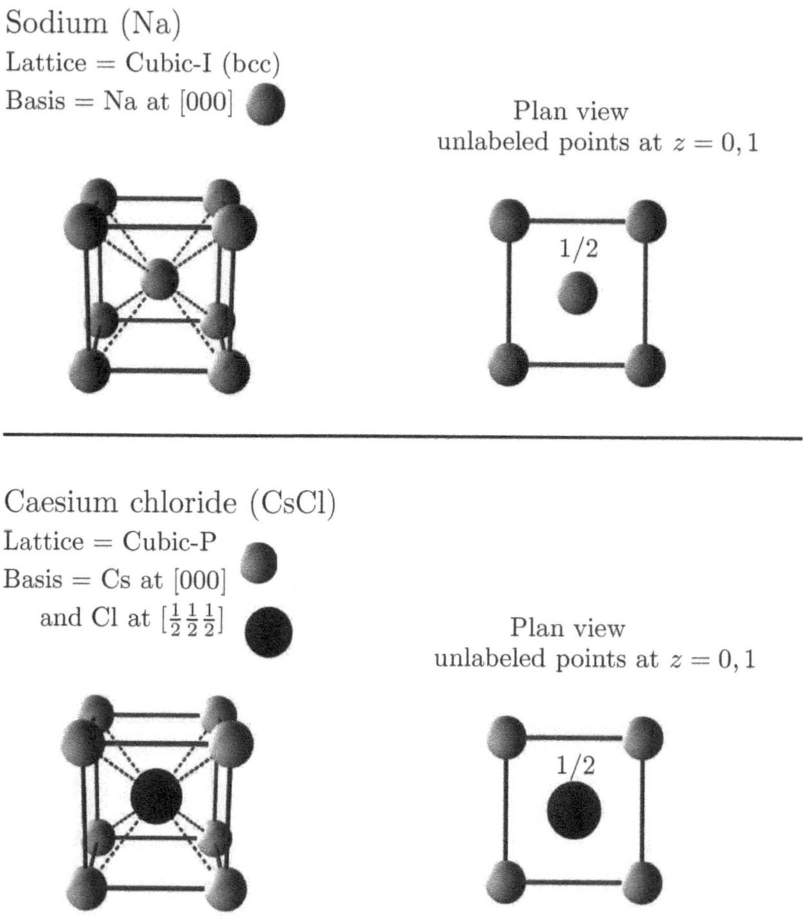

Fig. 12.20 Top: Sodium forms a bcc lattice. **Bottom:** Caesium chloride forms a cubic lattice with a two atom basis. Note carefully: CsCl is *not* bcc! In a bcc lattice all of the points (including the body center) must be identical. For CsCl, the point in the center is Cl whereas the points in the corner are Cs.

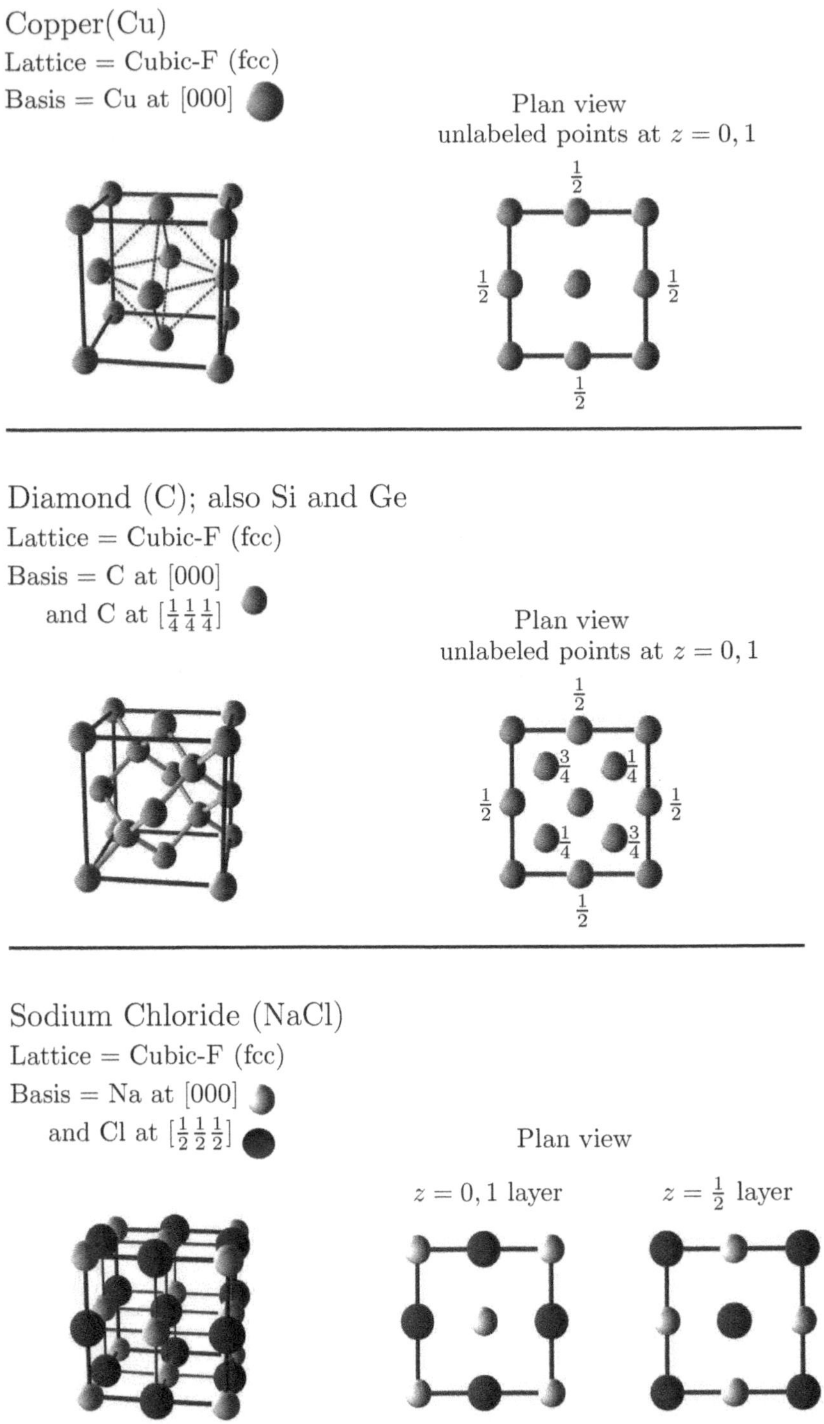

Fig. 12.21 Some crystals based on the

fcc lattice. **Top:** Copper forms an fcc lattice. **Middle:** Diamond (carbon) is

an fcc lattice with a two-atom basis. **Bottom:** NaCl (salt) is also an fcc lattice with a two atom basis. Note that in every case, a conventional unit cell is shown but the basis is given for the

primitive unit cell.

Copper(Cu)Lattice = Cubic

Lattice = Cubic-F (fcc)

Basis = Cu at [000]

Plan view unlabeled points at z = 0, 1

Diamond (C); also Si and Ge

Lattice = Cubic-F (fcc)

Basis = C at [000]

and C at $\left[\frac{1}{4}\frac{1}{4}\frac{1}{4}\right]$

Plan view unlabeled points at z = 0, 1

Sodium Chloride (NaCl)

Lattice = Cubic-F (fcc)

Basis = Na at [000]

and Cl at $\left[\frac{1}{2},\frac{1}{2},\frac{1}{2}\right]$

Plan view

z = 0, 1 layer

 $z = \frac{1}{2}$ layer

Chapter summary

This chapter introduced a plethora of new definitions, aimed at describing crystal structure in three dimensions. Here is a list of some of the concepts that one should know:

- Definition of a *lattice* in three different ways. See definitions 12.1, 12.1.1, 12.1.2.
- Definition of a *unit cell* for a periodic structure, and definition of a *primitive unit cell* and a *conventional unit cell.*
- Definition and construction of the *Wigner–Seitz (primitive) unit cell.*
- One can write any periodic structure in terms of a lattice and a basis (see examples in Fig. 12.20 and 12.21).
- In 3d, know the simple cubic lattice, the fcc lattice and the bcc lattices in particular. Orthorhombic and tetragonal lattices are also very useful to know.
- The fcc and bcc lattices can be thought of as simple cubic lattices with a basis.
- Know how to read a *plan view* of a structure.

References

All solid state books cover crystals. Some books give *way* too much detail. I recommend the following as giving not too much and not too little:

- Kittel, chapter 1
- Ashcroft and Mermin, chapter 4 (Caution of the nomenclature issue, see margin note 1 of this chapter.)
- Hook and Hall, sections 1.1–1.3 (probably not enough detail here!)

For greater detail about crystal structure see the following:

- Glazer, chapters 1–3
- Dove, sections 3.1–3.2 (brief but good)

Exercises

(12.1) **Crystal Structure of NaCl**

Consider the NaCl crystal structure shown in Fig. 12.21. If the lattice constant is $a = 0.563$ nm, what is the distance from a sodium atom to the nearest chlorine? What is the distance from a sodium atom to the nearest other sodium atom?

(12.2) **Neighbors in the Face-Centered Lattice.**

(a) Show that each lattice point in an fcc lattice has twelve nearest neighbors, each the same distance from the initial point. What is this distance if the conventional unit cell has lattice constant a?

(b)* Now stretch the side lengths of the fcc lattice such that you obtain a face-centered *orthorhombic* lattice where the conventional unit cell has sides of length a, b, and c which are all different. What are the distances to these twelve neighboring points now? How many nearest neighbors are there?

(12.3) **Crystal Structure**

The diagram of Fig. 12.22 shows a plan view of a structure of cubic ZnS (zincblende) looking down the z axis. The numbers attached to some atoms represent the heights of the atoms above the $z = 0$ plane expressed as a fraction of the cube edge a. Unlabeled atoms are at $z = 0$ and $z = a$.

(a) What is the Bravais lattice type?

(b) Describe the basis.

(c) Given that $a = 0.541$ nm, calculate the nearest-neighbor Zn–Zn, Zn–S, and S–S distances.

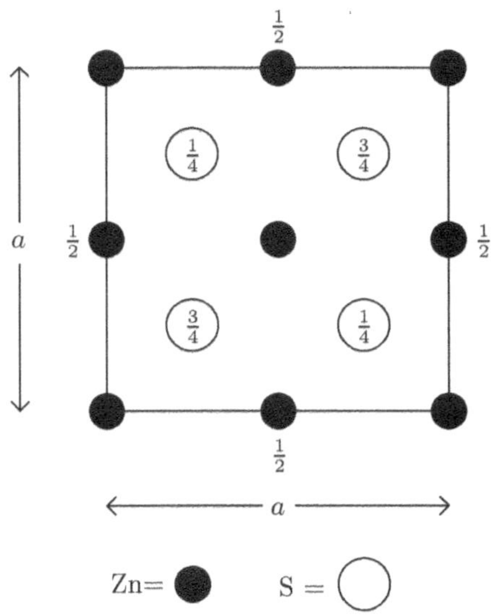

Fig. 12.22 Plan view of conventional unit cell of zincblende.

(12.4) **Packing Fractions**

Consider a lattice with a sphere at each lattice point. Choose the radius of the spheres to be such that neighboring spheres just touch (see for example, Fig. 12.18). The packing fraction is the fraction of the volume of all of space which is enclosed by the union of all the spheres (i.e., the ratio of the volume of the spheres to the total volume).

(a) Calculate the packing fraction for a simple cubic lattice.

(b) Calculate the packing fraction for a bcc lattice.

(c) Calculate the packing fraction for an fcc lattice.

(12.5) **Fluorine Beta Phase**

Fluorine can crystalize into a so-called beta-phase at temperatures between 45 and 55 Kelvin. Fig. 12.23 shows the cubic conventional unit cell for beta phase fluorine in three-dimensional form along with a plan view.

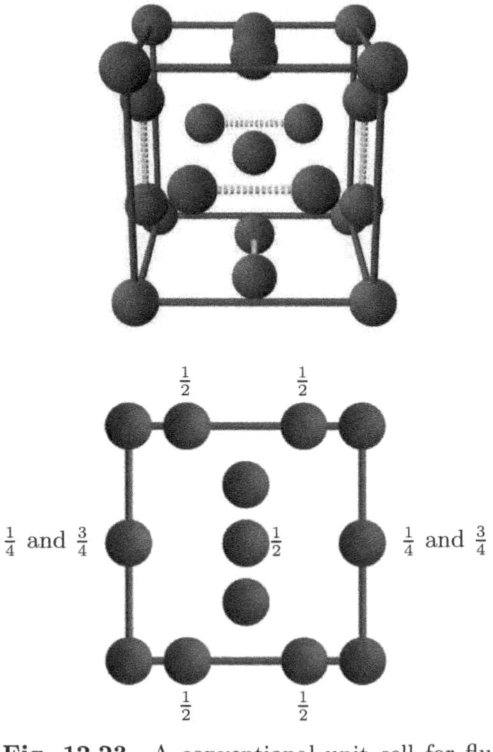

Fig. 12.23. A conventional unit cell for fluorine beta phase. All atoms in the picture are fluorine. Lines are drawn for clarity **Top:** Three-dimensional view. **Bottom:** Plan view. Unlabeled atoms are at height 0 and 1 in units of the lattice constant.

▷ How many atoms are in this conventional unit cell?

▷ What is the lattice and the basis for this crystal?

(12.6) **Different Primitive Lattice Vectors&**

Prove that the two conditions stated in footnote 3 of this chapter are equivalent.

Reciprocal Lattice, Brillouin Zone, Waves in Crystals

13

In the last chapter we explored lattices and crystal structure. However, as we saw in Chapters 9–11, the important physics of waves in solids (whether they are vibrational waves, or electron waves) is best described in reciprocal space. This chapter thus introduces reciprocal space in three dimensions. As with the previous chapter, there is some tricky geometry in this chapter, and a few definitions to learn as well. As a result this material is a bit tough to slog through, but stick with it because soon we will make substantial use of what we learn here. At the end of this chapter we will finally have enough definitions to describe the dispersions of phonons and electrons in three-dimensional systems.

13.1 The Reciprocal Lattice in Three Dimensions

13.1.1 Review of One Dimension

Let us first recall some results from our study of one dimension. We consider a simple lattice in one dimension $R_n = na$ with n an integer. Recall that two points in k-space (reciprocal space) were defined to be equivalent to each other if $k_1 = k_2 + G_m$ where $G_m = 2\pi m/a$ with m an integer. The points G_m form the reciprocal lattice.

Recall that the reason that we identified different k values with each other was because we were considering waves of the form

$$e^{ikx_n} = e^{ikna}$$

with n an integer. Because of this form of the wave, we find that shifting $k \to k + G_m$ leaves this functional form unchanged since

$$e^{i(k+G_m)x_n} = e^{i(k+G_m)na} = e^{ikna}e^{i(2\pi m/a)na} = e^{ikx_n}$$

where we have used

$$e^{i2\pi mn} = 1$$

in the last step. Thus, so far as the wave is concerned, k is the same as $k + G_m$.

13.1.2 Reciprocal Lattice Definition

Generalizing this one-dimensional result, we define

Definition 13.1 *Given a (direct) lattice of points* $\mathbf{R}$*, a point* $\mathbf{G}$ *is a point in the* ***reciprocal lattice*** *if and only if*

$$e^{i\mathbf{G}\cdot\mathbf{R}} = 1 \tag{13.1}$$

for all points $\mathbf{R}$ *of the direct lattice.*

To construct the reciprocal lattice, let us first write the points of the direct lattice in the form[1] (here we specialize to the three-dimensional case)

$$\mathbf{R} = n_1\mathbf{a_1} + n_2\mathbf{a_2} + n_3\mathbf{a_3} \tag{13.2}$$

with n_1, n_2, and n_3 integers, and with $\mathbf{a_1}$, $\mathbf{a_2}$, and $\mathbf{a_3}$ being primitive lattice vectors of the direct lattice.

We now make two key claims:

(1) We claim that the reciprocal lattice (defined by Eq. 13.1) is a lattice in reciprocal space (thus explaining its name).

(2) We claim that the primitive lattice vectors of the reciprocal lattice (which we will call $\mathbf{b_1}$, $\mathbf{b_2}$, and $\mathbf{b_3}$) are defined to have the following property:

$$\mathbf{a_i}\cdot\mathbf{b_j} = 2\pi\delta_{ij} \tag{13.3}$$

where δ_{ij} is the Kronecker delta.[2]

We can certainly construct vectors $\mathbf{b_i}$ to have the desired property of Eq. 13.3, as follows:

$$\begin{aligned}
\mathbf{b_1} &= \frac{2\pi\,\mathbf{a_2}\times\mathbf{a_3}}{\mathbf{a_1}\cdot(\mathbf{a_2}\times\mathbf{a_3})}\\
\mathbf{b_2} &= \frac{2\pi\,\mathbf{a_3}\times\mathbf{a_1}}{\mathbf{a_1}\cdot(\mathbf{a_2}\times\mathbf{a_3})}\\
\mathbf{b_3} &= \frac{2\pi\,\mathbf{a_1}\times\mathbf{a_2}}{\mathbf{a_1}\cdot(\mathbf{a_2}\times\mathbf{a_3})}.
\end{aligned}$$

It is easy to check that Eq. 13.3 is satisfied. For example,

$$\begin{aligned}
\mathbf{a_1}\cdot\mathbf{b_1} &= \frac{2\pi\,\mathbf{a_1}\cdot(\mathbf{a_2}\times\mathbf{a_3})}{\mathbf{a_1}\cdot(\mathbf{a_2}\times\mathbf{a_3})} = 2\pi\\
\mathbf{a_2}\cdot\mathbf{b_1} &= \frac{2\pi\,\mathbf{a_2}\cdot(\mathbf{a_2}\times\mathbf{a_3})}{\mathbf{a_1}\cdot(\mathbf{a_2}\times\mathbf{a_3})} = 0.
\end{aligned}$$

Now, given vectors $\mathbf{b_1}$, $\mathbf{b_2}$, and $\mathbf{b_3}$ satisfying Eq. 13.3 we have claimed that these are in fact primitive lattice vectors for the reciprocal lattice. To prove this, let us write an *arbitrary* point in reciprocal space as

$$\mathbf{G} = m_1\mathbf{b_1} + m_2\mathbf{b_2} + m_3\mathbf{b_3} \tag{13.4}$$

[1] There are certainly other ways to specify the points of a direct lattice. For example, it is sometimes convenient to choose $\mathbf{a_i}$'s to describe the edges vectors of a *conventional* unit cell, but then the n_i's are not simply described as all integers. This is done in section 13.1.5, and is relevant for the **Important Comment** there.

[2] Leopold Kronecker was a mathematician who is famous (among other things) for the sentence "God made the integers, everything else is the work of man". In case you don't already know this, the Kronecker delta is defined as $\delta_{ij} = 1$ for $i = j$ and is zero otherwise. (Kronecker did a lot of other interesting things as well.)

and for the moment, let us not require m_1, m_2, and m_3 to be integers. (We are about to discover that for $\mathbf{G}$ to be a point of the reciprocal lattice, they must be integers, but this is what we want to prove!)

To find points of the reciprocal lattice we must show that Eq. 13.1 is satisfied for all points $\mathbf{R} = n_1\mathbf{a_1} + n_2\mathbf{a_2} + n_3\mathbf{a_3}$ of the direct lattice with n_1, n_2, and n_3 integers. We thus write

$$e^{i\mathbf{G}\cdot\mathbf{R}} = e^{i(m_1\mathbf{b_1}+m_2\mathbf{b_2}+m_3\mathbf{b_3})\cdot(n_1\mathbf{a_1}+n_2\mathbf{a_2}+n_3\mathbf{a_3})} = e^{2\pi i(n_1m_1+n_2m_2+n_3m_3)}.$$

In order for $\mathbf{G}$ to be a point of the reciprocal lattice, this must equal unity for all points $\mathbf{R}$ of the direct lattice, i.e., for all integer values of n_1, n_2 and n_3. Clearly this can only be true if m_1, m_2 and m_3 are also integers. Thus, we find that the points of the reciprocal lattice are precisely those of the form of Eq. 13.4 with m_1, m_2 and m_3 integers. This further proves our claim that the reciprocal lattice is in fact a lattice!

13.1.3 The Reciprocal Lattice as a Fourier Transform

Quite generally one can think of the reciprocal lattice as being a Fourier transform of the direct lattice. It is easiest to start by thinking in one dimension. Here the direct lattice is given again by $R_n = an$. If we want to describe a "density" of lattice points in one dimension, we might put a delta function at each lattice points and write the density as[3]

$$\rho(r) = \sum_n \delta(r - an).$$

Fourier transforming this function gives[4]

$$\begin{aligned}\mathcal{F}[\rho(r)] = \int dr e^{ikr}\rho(r) &= \sum_n \int dr e^{ikr}\delta(r-an) = \sum_n e^{ikan} \\ &= \frac{2\pi}{|a|}\sum_m \delta(k - 2\pi m/a).\end{aligned}$$

The last step here is a bit non-trivial.[5] Here e^{ikan} is clearly unity if $k = 2\pi m/a$, i.e., if k is a point on the reciprocal lattice. In this case, each term of the sum contributes unity to the sum and one obtains an infinite result.[6] If k is not such a reciprocal lattice point, then the terms of the sum oscillate and the sum comes out to be zero.

This principle generalizes to the higher (two- and three-)dimensional cases. Generally

$$\mathcal{F}[\rho(\mathbf{r})] = \sum_{\mathbf{R}} e^{i\mathbf{k}\cdot\mathbf{R}} = \frac{(2\pi)^D}{v}\sum_{\mathbf{G}} \delta^D(\mathbf{k} - \mathbf{G}) \tag{13.5}$$

where in the middle term, the sum is over lattice points $\mathbf{R}$ of the direct lattice, and in the last term it is a sum over points $\mathbf{G}$ of the reciprocal lattice and v is the volume of the unit cell. Here D is the number of dimensions (1, 2 or 3) and the δ^D is a D-dimensional delta function.[7]

[3] Since the sums are over all lattice points they should go from $-\infty$ to $+\infty$. Alternatively, one uses periodic boundary conditions and sums over all points.

[4] With Fourier transforms there are several different conventions about where one puts the factors of 2π. Possibly in your mathematics class you learned to put $1/\sqrt{2\pi}$ with each integral. However, in solid state physics, conventionally $1/(2\pi)$ comes with each k integral, and no factor of 2π comes with each r integral. See Section 2.2.1 to see why this is used.

[5] This is sometimes known as the Poisson resummation formula, after Siméon Denis Poisson, the same guy after whom Poisson's equation $\nabla^2\phi = -\rho/\epsilon_0$ is named, as well as other mathematical things such as the Poisson random distribution. His last name means "fish" in French.

[6] Getting the prefactor right is a bit harder. But actually, the prefactor isn't going to be too important for us.

[7] For example, in two dimensions $\delta^2(\mathbf{r} - \mathbf{r_0}) = \delta(x - x_0)\delta(y - y_0)$ where $\mathbf{r} = (x, y)$

The equality in Eq. 13.5 is similar to the one-dimensional case. If $\mathbf{k}$ is a point of the reciprocal lattice, then $e^{i\mathbf{k}\cdot\mathbf{R}}$ is always unity and the sum is infinite. However, if $\mathbf{k}$ is not a point on the reciprocal lattice then the summands oscillate, and the sum comes out to be zero. Thus one obtains delta-function peaks precisely at the positions of reciprocal lattice vectors.

[8]See Exercise 13.1.

Aside: It is an easy exercise to show[8] that the reciprocal lattice of an fcc direct lattice is a bcc lattice in reciprocal space. Conversely, the reciprocal lattice of a bcc direct lattice is an fcc lattice in reciprocal space.

Fourier Transform of Any Periodic Function

In the prior section we considered the Fourier transform of a function $\rho(\mathbf{r})$ which is just a set of delta functions at lattice points. However, it is not too different to consider the Fourier transform of *any* function with the periodicity of the lattice (and this will be quite important in Chapter 14). We say a function $\rho(\mathbf{r})$ has the periodicity of a lattice if $\rho(\mathbf{r}) = \rho(\mathbf{r}+\mathbf{R})$ for any lattice vector $\mathbf{R}$. We then want to calculate

$$\mathcal{F}[\rho(\mathbf{r})] = \int \mathbf{dr}\; e^{i\mathbf{k}\cdot\mathbf{r}}\rho(\mathbf{r}).$$

The integral over all of space can be broken up into a sum of integrals over each unit cell. Here we write any point in space $\mathbf{r}$ as the sum of a lattice point $\mathbf{R}$ and a vector $\mathbf{x}$ within the unit cell

$$\mathcal{F}[\rho(\mathbf{r})] = \sum_{\mathbf{R}} \int_{unit-cell} \mathbf{dx}\; e^{i\mathbf{k}\cdot(\mathbf{x}+\mathbf{R})}\rho(\mathbf{x}+\mathbf{R}) = \sum_{\mathbf{R}} e^{i\mathbf{k}\cdot\mathbf{R}} \int_{unit-cell} \mathbf{dx}\; e^{i\mathbf{k}\cdot\mathbf{x}}\rho(\mathbf{x}).$$

where here we have used the invariance of ρ under lattice translations $\mathbf{x} \to \mathbf{x}+\mathbf{R}$. The sum of exponentials, as in Eq. 13.5, just gives a sum of delta functions yielding

$$\mathcal{F}[\rho(\mathbf{r})] = \frac{(2\pi)^D}{v}\sum_{\mathbf{G}} \delta^D(\mathbf{k}-\mathbf{G})S(\mathbf{k})$$

where

$$S(\mathbf{k}) = \int_{unit-cell} \mathbf{dx}\; e^{i\mathbf{k}\cdot\mathbf{x}}\rho(\mathbf{x}) \tag{13.6}$$

is known as the *structure factor* and will become very important in the next chapter.

13.1.4 Reciprocal Lattice Points as Families of Lattice Planes

Another way to understand the reciprocal lattice is via families of lattice planes of the direct lattice.

Definition 13.2 *A* ***lattice plane*** *(or* ***crystal plane****) is a plane containing at least three non-collinear (and therefore an infinite number of) points of a lattice.*

Definition 13.3 *A **family of lattice planes** is an infinite set of equally separated parallel lattice planes which taken together contain all points of the lattice.*

In Fig. 13.1, several examples of families of lattice planes are shown. Note that the planes are parallel and equally spaced, and every point of the lattice is included in exactly one lattice plane.

I now make the following claim:

Claim 13.1 *The families of lattice planes are in one-to-one correspondence*[9] *with the possible directions of reciprocal lattice vectors, to which they are normal. Further, the spacing between these lattice planes is* $d = 2\pi/|\mathbf{G}_{\mathbf{min}}|$ *where* $\mathbf{G}_{\mathbf{min}}$ *is the minimum length reciprocal lattice vector in this normal direction.*

This correspondence is made as follows. First we consider the set of planes defined by points $\mathbf{r}$ such that for some integer m,

$$\mathbf{G} \cdot \mathbf{r} = 2\pi m. \tag{13.7}$$

This defines an infinite set of parallel planes normal to $\mathbf{G}$. Since $e^{i\mathbf{G}\cdot\mathbf{r}} = 1$ we know that every lattice point is a member of one of these planes (since this is the definition of $\mathbf{G}$ in Eq. 13.1). However, for the planes defined by Eq. 13.7, not every plane needs to contain a lattice point (so generically this is a family of parallel equally spaced planes, but not a family of lattice planes). For this larger family of planes, the spacing between planes is given by

$$d = \frac{2\pi}{|\mathbf{G}|}. \tag{13.8}$$

To prove this we simply note (from Eq. 13.7) that two adjacent planes must have

$$\mathbf{G} \cdot (\mathbf{r_1} - \mathbf{r_2}) = 2\pi.$$

Thus in the direction parallel to $\mathbf{G}$, the spacing between planes is $2\pi/|\mathbf{G}|$ as claimed.

Clearly different values of $\mathbf{G}$ that happen to point in the same direction, but have different magnitudes, will define parallel sets of planes. As we increase the magnitude of $\mathbf{G}$, we add more and more planes. For example, examining Eq. 13.7 we see that when we double the magnitude of $\mathbf{G}$ we correspondingly double the density of planes, which we can see from the spacing formula Eq. 13.8. However, whichever $\mathbf{G}$ we choose, all of the lattice points will be included in one of the defined planes. If we choose the maximally possible spaced planes, hence the smallest possible value of $\mathbf{G}$ allowed in any given direction which we call $\mathbf{G}_{\mathbf{min}}$, then in fact every defined plane will include lattice points and therefore be

(010) family of lattice planes

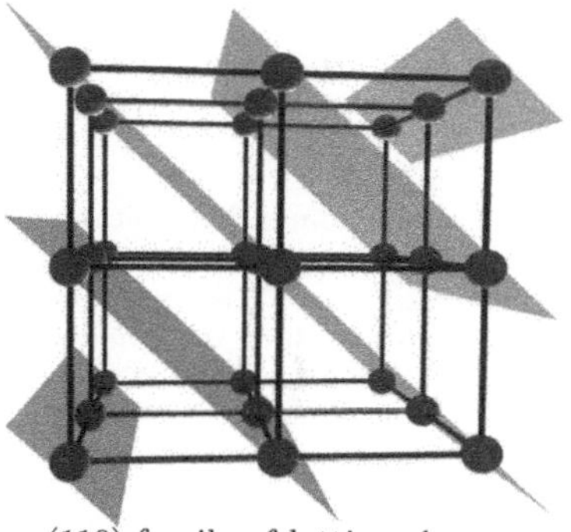

(110) family of lattice planes

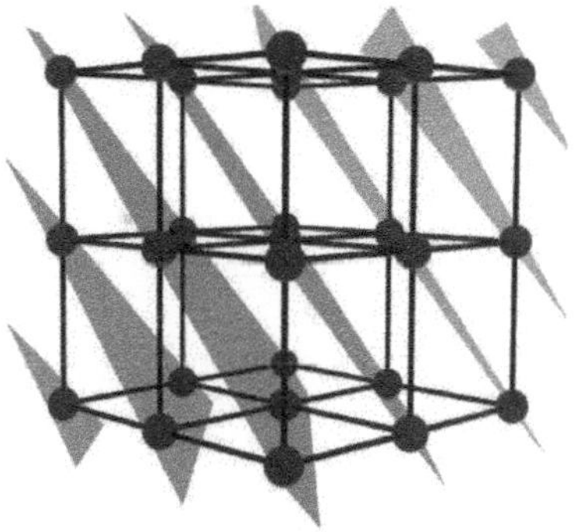

(111) family of lattice planes

Fig. 13.1 Examples of families of lattice planes on the cubic lattice. Each of these planes is a lattice plane because it intersects at least three non-collinear lattice points. Each picture is a family of lattice planes since every lattice point is included in one of the parallel lattice planes. The families are labeled in Miller index notation. **Top** (010); **Middle** (110); **Bottom** (111). In the top and middle the x-axis points to the right and the y-axis points up. In the bottom figure the axes are rotated for clarity.

[9]For this one-to-one correspondence to be precisely true we must define $\mathbf{G}$ and $-\mathbf{G}$ to be the same direction. If this sounds like a cheap excuse, we can say that "oriented" families of lattice planes are in one-to-one correspondence with the directions of reciprocal lattice vectors, thus keeping track of the two possible normals of the family of lattice planes.

lattice planes, and the spacing between these planes is correspondingly $2\pi/|\mathbf{G_{min}}|$. This proves[10] Claim 13.1.

13.1.5 Lattice Planes and Miller Indices

There is a useful notation for describing lattice planes (or reciprocal lattice vectors) known as *Miller indices*.[11] One first chooses edge vectors $\mathbf{a_i}$ for a unit cell in direct space (which may be primitive or non-primitive). One then constructs reciprocal space vectors $\mathbf{b_i}$ to satisfy $\mathbf{a_i} \cdot \mathbf{b_j} = 2\pi\delta_{ij}$ (see Eq. 13.3). In terms of these vectors $\mathbf{b_i}$, one writes (h, k, l) or (hkl) with integers h, k and l, to mean the reciprocal space vector[12]

$$\mathbf{G}_{(h,k,l)} = h\mathbf{b_1} + k\mathbf{b_2} + l\mathbf{b_3}. \qquad (13.9)$$

Note that Miller indices can be negative, such as $(1, -1, 1)$. Conventionally, the minus sign is denoted with an over-bar rather than a minus sign, so we write $(1\bar{1}1)$ instead.[13]

Note that if one chooses $\mathbf{a_i}$ to be the real (direct) space primitive lattice vectors, then $\mathbf{b_i}$ will be the primitive lattice vectors for the reciprocal lattice. In this case, any set of integer Miller indices (hkl) represents a reciprocal lattice vector. To represent a family of lattice planes, one should take the shortest reciprocal lattice vector in the given direction (see Claim 13.1), meaning h, k, and l should have no common divisors. If (hkl) are not the shortest reciprocal lattice vector in a given direction, then they represent a family of planes that is not a family of lattice planes (i.e., there are some planes that do not intersect lattice points).

On the other hand, if one chooses $\mathbf{a_i}$ to describe the edges of some non-primitive (conventional) unit cell, the corresponding $\mathbf{b_i}$ will not be primitive reciprocal lattice vectors. As a result not all integer sets of Miller indices will be reciprocal lattice vectors.

Important Comment: For any cubic lattice (simple cubic, fcc, or bcc) it is conventional to choose $\mathbf{a_i}$ to be $a\hat{\mathbf{x}}, a\hat{\mathbf{y}}$, and $a\hat{\mathbf{z}}$ with a the cube edge length. I.e., one chooses the orthogonal edge vectors of the conventional (cube) unit cell. Correspondingly, $\mathbf{b_i}$ are the vectors $2\pi\hat{\mathbf{x}}/a$, $2\pi\hat{\mathbf{y}}/a$, and $2\pi\hat{\mathbf{z}}/a$. For the primitive (simple) cubic case these are primitive reciprocal lattice vectors, but for the fcc and bcc case, they are not.[14] So in the fcc and bcc cases not all integer sets of Miller indices (hkl) are reciprocal lattice vectors.

To illustrate this point, consider the (010) family of planes for the cubic lattice, shown in the top of Fig. 13.1. This family of planes intersects every corner of the cubic unit cell. However, if we were discussing a bcc lattice, there would also be another lattice point in the center of every conventional unit cell which the (010) lattice planes would not intersect (see top of Fig. 13.2). However, the (020) planes *would* intersect these

[11] These are named after the nineteenth century mineralogist William Hallowes Miller.

[12] We have already used the corresponding notation $[uvw]$ to represent lattice points of the direct lattice. See for example, Eq. 12.1 and Eq. 12.4.

[13] How $(1\bar{1}1)$ is pronounced is a bit random. Some people say "one-(bar-one)-one" and others say "one-(one-bar)-one". I have no idea how the community got so confused as to have these two different conventions. I think in Europe the former is more prevalent whereas in America the latter is more prevalent. At any rate, it is always clear when it is written.

[14] Although this convention of working with non-primitive vectors $\mathbf{b_i}$ makes some things very complicated our only other option would be to work with the non-orthogonal coordinate axes of the primitive lattice vectors—which would complicate life even more!

[10] More rigorously, if there is a family of lattice planes in direction $\hat{\mathbf{G}}$ with spacing between planes d, then $\mathbf{G} = 2\pi\hat{\mathbf{G}}/d$ is necessarily a reciprocal lattice vector. To see this note that $e^{i\mathbf{G}\cdot\mathbf{R}} = 1$ will be unity for all lattice points. Further, in a family of lattice planes, all lattice points are included within the planes, so $e^{i\mathbf{G}\cdot\mathbf{R}} = 1$ for all $\mathbf{R}$ a lattice point, which implies $\mathbf{G}$ is a reciprocal lattice vector. Furthermore, $\mathbf{G}$ is the shortest reciprocal lattice vector in the direction of $\hat{\mathbf{G}}$ since increasing $\mathbf{G}$ will result in a smaller spacing of lattice planes and some planes will not intersect lattice points $\mathbf{R}$.

central points as well, so in this case (020) represents a true family of lattice planes (and hence a reciprocal lattice vector) for the bcc lattice whereas (010) does not! (See Fig. 13.2.) In Section 14.2 we will discuss the "selection rules" for knowing when a set of Miller indices represents a true family of lattice planes in the fcc and bcc cases.

From Eq. 13.8 one can write the spacing between adjacent planes of a family of planes specified by Miller indices (h, k, l)

$$d_{(hkl)} = \frac{2\pi}{|\mathbf{G}|} = \frac{2\pi}{\sqrt{h^2|\mathbf{b_1}|^2 + k^2|\mathbf{b_2}|^2 + l^2|\mathbf{b_3}|^2}} \tag{13.10}$$

where we have assumed that the coordinate axes of the lattice vectors $\mathbf{b_i}$ are orthogonal. Recall that in the case of orthogonal axes $|b_i| = 2\pi/|a_i|$ where a_i are the lattice constants in the three orthogonal directions. Thus we can equivalently write

$$\frac{1}{|d_{(hkl)}|^2} = \frac{h^2}{a_1^2} + \frac{k^2}{a_2^2} + \frac{l^2}{a_3^2}. \tag{13.11}$$

Note that for a cubic lattice this simplifies to

$$d^{cubic}_{(hkl)} = \frac{a}{\sqrt{h^2 + k^2 + l^2}}. \tag{13.12}$$

A useful shortcut for figuring out the geometry of lattice planes is to look at the intersection of a plane with the three coordinate axes. The intersections x_1, x_2, x_3 with the three coordinate axes (in units of the three lattice constants) are related to the Miller indices via

$$\frac{1}{x_1} : \frac{1}{x_2} : \frac{1}{x_3} = h : k : l.$$

This construction is illustrated in Fig. 13.3.

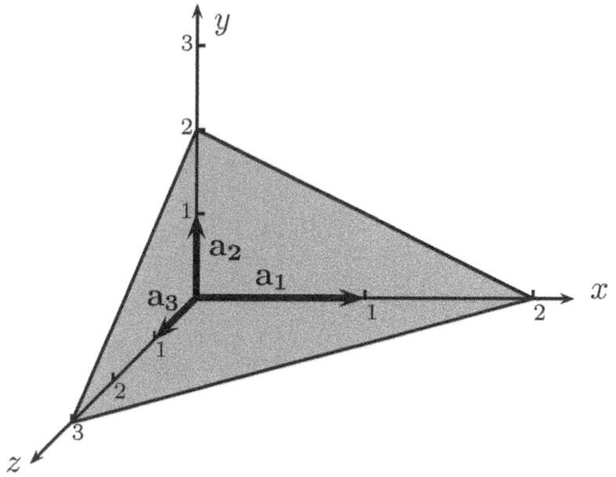

Fig. 13.3 Determining Miller indices from the intersection of a plane with the coordinate axes. This plane intersects the coordinate axes at $x = 2$, $y = 2$ and $z = 3$ in units of the lattice constants. The reciprocals of these intercepts are $\frac{1}{2}, \frac{1}{2}, \frac{1}{3}$. The smallest integers having these ratios are 3, 3, 2. Thus the Miller indices of this family of lattice planes are (332). The spacing between lattice planes in this family would be $1/|d_{(233)}|^2 = 3^2/a_1^2 + 3^2/a_2^2 + 2^2/a_3^2$ (assuming orthogonal axes).

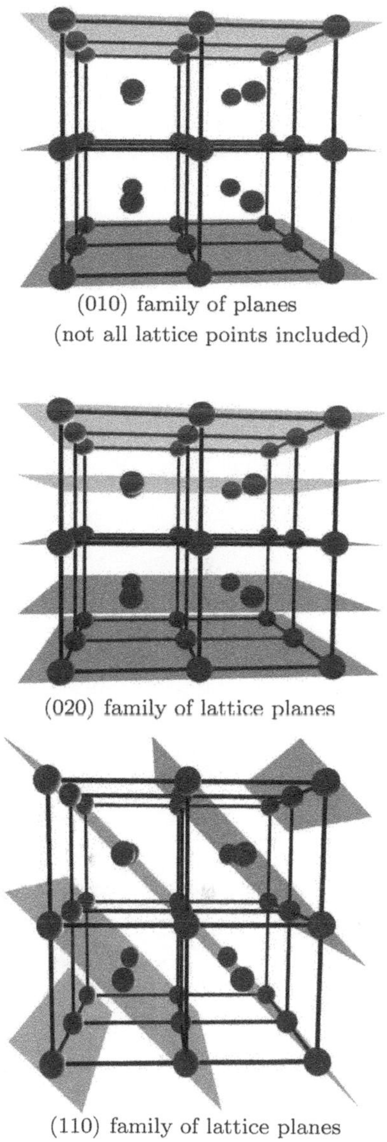

Fig. 13.2 Top: For the bcc lattice, the (010) planes are not a true family of lattice planes since the (010) planes do not intersect the lattice points in the middle of the cubes. **Middle:** The (020) planes are a family of lattice planes since they intersect all of the lattice points. **Bottom** The (110) planes are also a family of lattice planes.

Finally, we note that different lattice planes may be the same under a symmetry of the crystal. For example, in a cubic lattice, (111) looks the same as $(1\bar{1}1)$ after rotation (and possibly reflection) of the axes of the crystal (but would never look like (122) under any rotation or reflection since the spacing between planes is different!).[15] If we want to describe all lattice planes that are equivalent in this way, we write $\{111\}$ instead.

It is interesting that lattice planes in crystals were well understood long before people even knew for sure there was such a thing as atoms. By studying how crystals cleave along certain planes, scientists like Miller and Bravais could reconstruct a great deal about how these materials must be assembled.[16]

[15] It can sometimes be subtle to figure out if a crystal looks the same from two different directions: one needs to check that the *basis* of the crystal looks the same from the two directions!

[16] There is a law known as "Bravais' law", which states that crystals cleave most readily along faces having the highest density of lattice points, or equivalently the largest distance between lattice planes. To a large extent this means that crystals cleave on lattice planes with small Miller indices.

13.2 Brillouin Zones

The whole point of going into such gross detail about the structure of reciprocal space is in order to describe waves in solids. In particular, it will be important to understand the structure of the Brillouin zone.

13.2.1 Review of One-Dimensional Dispersions and Brillouin Zones

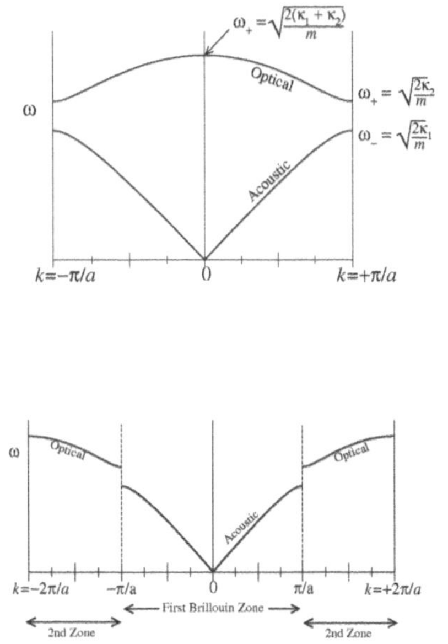

Fig. 13.4 Phonon spectrum of a diatomic chain in one dimension. **Top:** Reduced zone scheme. **Bottom:** Extended zone scheme. (See Figs. 10.6 and 10.8.) We can display the dispersion in either form due to the fact that wavevector is only defined modulo $2\pi/a$, that is, it is periodic in the Brillouin zone.

As we learned in Chapters 9–11, the Brillouin zone is extremely important in describing the excitation spectrum of waves in periodic media. As a reminder, in Fig. 13.4 we show the excitation spectrum of vibrations of a diatomic chain (Chapter 10) in both the reduced, and extended zone schemes. Since waves are physically equivalent under shifts of the wavevector k by a reciprocal lattice vector $2\pi/a$, we can always express every excitation within the first Brillouin zone, as shown in the reduced zone scheme (top of Fig. 13.4). In this example, since there are two atoms per unit cell, there are precisely two excitation modes per wavevector. On the other hand, we can always unfold the spectrum and put the lowest (acoustic) excitation mode in the first Brillouin zone and the higher-energy excitation mode (optical) in the second Brillouin zone, as shown in the extended zone scheme (bottom of Fig. 13.4). Note that there is a jump in the excitation spectrum at the Brillouin zone boundary.

13.2.2 General Brillouin Zone Construction

Definition 13.4 *A **Brillouin zone** is any primitive unit cell of the reciprocal lattice.*

Entirely equivalent to the one-dimensional situation, physical waves in crystals are unchanged if their wavevector is shifted by a reciprocal lattice vector $\mathbf{k} \to \mathbf{k} + \mathbf{G}$. Alternately, we realize that the physically relevant quantity is the crystal momentum. Thus, the Brillouin zone has been defined to include each physically different crystal momentum exactly once (each $\mathbf{k}$ point within the Brillouin zone is physically different, and all physically different points occur once within the zone).

While the most general definition of Brillouin zone allows us to choose any shape primitive unit cell for the reciprocal lattice, there are some definitions of unit cells which are more convenient than others.

We define the *first Brillouin zone* in reciprocal space quite analogously to the construction of the Wigner–Seitz cell for the direct lattice.

Definition 13.5 *Start with the reciprocal lattice point* $\mathbf{G} = \mathbf{0}$. *All* $\mathbf{k}$ *points which are closer to* $\mathbf{0}$ *than to any other reciprocal lattice point define the* ***first Brillouin zone****. Similarly all* $\mathbf{k}$ *points where the point* $\mathbf{0}$ *is the second closest reciprocal lattice point to that point constitute the* ***second Brillouin zone****, and so forth. Zone boundaries are defined in terms of this definition of Brillouin zones.*

As with the Wigner–Seitz cell, there is a simple algorithm to construct the Brillouin zones. Draw the perpendicular bisector between the point $\mathbf{0}$ and each of the reciprocal lattice vectors. These bisectors form the Brillouin zone boundaries. Any point that you can get to from $\mathbf{0}$ without crossing a perpendicular bisector is in the first Brillouin zone. If you cross only one perpendicular bisector, you are in the second Brillouin zone, and so forth.

In Fig. 13.5, we show the Brillouin zones of the square lattice. A few general principles to note:

(1) The first Brillouin zone is necessarily connected, but the higher Brillouin zones typically are made of disconnected pieces.

(2) A point on a Brillouin zone boundary lies on the perpendicular bisector between the point $\mathbf{0}$ and some reciprocal lattice point $\mathbf{G}$. Adding the vector $-\mathbf{G}$ to this point necessarily results in a point (the same distance from $\mathbf{0}$) which is on another Brillouin zone boundary (on the bisector of the segment from $\mathbf{0}$ to $-\mathbf{G}$). This means that Brillouin zone boundaries occur in parallel pairs symmetric around the point $\mathbf{0}$ which are separated by a reciprocal lattice vector (see Fig. 13.5).

(3) Each Brillouin zone has exactly the same total area (or volume in three dimensions). This must be the case since there is a one-to-one mapping of points in each Brillouin zone to the first Brillouin zone. Finally, as in one dimension, we claim that there are exactly as many k-states within the first Brillouin zone as there are primitive unit cells in the entire system.[17]

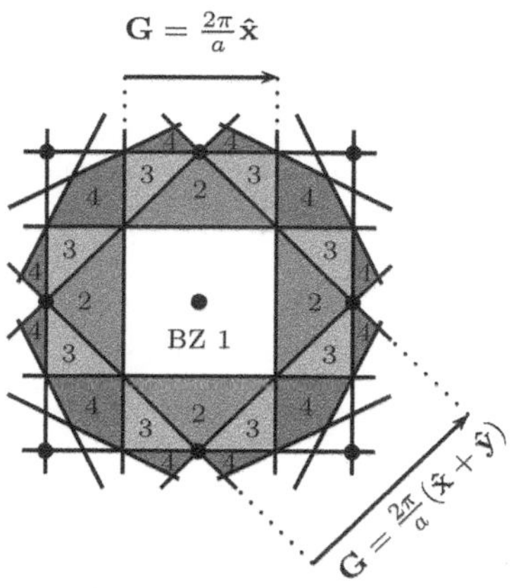

Fig. 13.5 First, second, third, and fourth Brilliouin zones of the square lattice. All of the lines drawn in this figure are perpendicular bisectors between the central point $\mathbf{0}$ and some other reciprocal lattice point. Note that zone boundaries occur in parallel pairs symmetric around the central point $\mathbf{0}$ and are separated by a reciprocal lattice vector.

Note, that as in the case of the Wigner–Seitz cell construction, the shape of the first Brillouin zone can look a bit strange, even for a relatively simple lattice (see Fig. 12.7).

The construction of the Brillouin zone is similar in three dimensions as it is in two, and is again entirely analogous to the construction of the Wigner–Seitz cell in three dimensions. For a simple cubic lattice, the first Brillouin zone is simply a cube. For fcc and bcc lattices, however, the situation is more complicated. As we mentioned in the **Aside** at the end of Section 13.1.3, the reciprocal lattice of the fcc lattice is bcc,

[17] Here's the proof for a square lattice. Let the system be N_x by N_y unit cells. With periodic boundary conditions, the value of k_x is quantized in units of $2\pi/L_x = 2\pi/(N_x a)$ and the value of k_y is quantized in units of $2\pi/L_y = 2\pi/(N_y a)$. The size of the Brillouin zone is $2\pi/a$ in each direction, so there are precisely $N_x N_y$ different values of $\mathbf{k}$ in the Brillouin zone.

and vice-versa. Thus, the Brillouin zone of the fcc lattice is the same shape as the Wigner–Seitz cell of the bcc lattice! The Brillouin zone for the fcc lattice is shown in Fig. 13.6 (compare to Fig. 12.13). Note that in Fig. 13.6, various k-points are labeled with letters. There is a complicated labeling convention that we will not discuss, but it is worth knowing that it exists. For example, we can see in the figure that the point $\mathbf{k} = \mathbf{0}$ is labeled Γ, and the point $\mathbf{k} = (2\pi/a)\hat{\mathbf{y}}$ is labeled X.

Now that we can describe the fcc Brillouin zone, we finally have a way to properly describe the physics of waves in some real crystals!

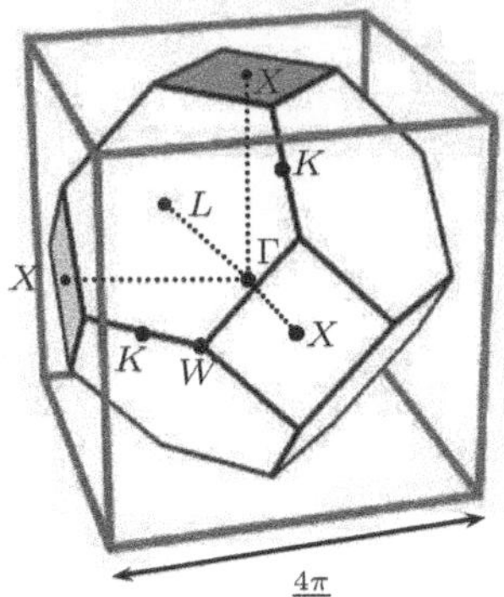

Fig. 13.6 First Brillouin zone of the fcc lattice. Note that it is the same shape as the Wigner–Seitz cell of the bcc lattice, see Fig. 12.13. Special points of the Brilliou zone are labeled with code letters such as X, K, and Γ. Note that the lattice constant of the conventional unit cell is $4\pi/a$ (see Exercise 13.1).

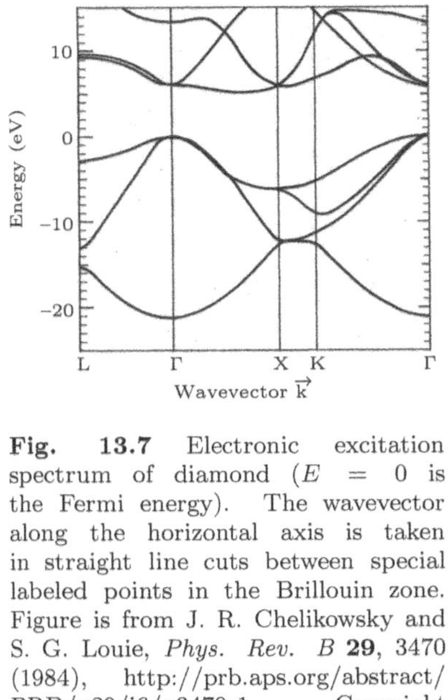

Fig. 13.7 Electronic excitation spectrum of diamond ($E = 0$ is the Fermi energy). The wavevector along the horizontal axis is taken in straight line cuts between special labeled points in the Brillouin zone. Figure is from J. R. Chelikowsky and S. G. Louie, *Phys. Rev. B* **29**, 3470 (1984), http://prb.aps.org/abstract/PRB/v29/i6/p3470_1. Copyright American Physical Society. Used by permission.

13.3 Electronic and Vibrational Waves in Crystals in Three Dimensions

In Fig. 13.7 we show the electronic band-structure (i.e., dispersion relation) of diamond, which can be described as an fcc lattice with a diatomic basis (see Fig. 12.21). As in the one-dimensional case, we can work in the reduced zone scheme where we only need to consider the first Brillouin zone. Since we are trying to display a three-dimensional spectrum (energy as a function of $\mathbf{k}$) on a one-dimensional diagram, we show several single-line cuts through reciprocal space.[18] Starting on the left of the diagram, we start at the L-point of the Brillouin zone and show $E(\mathbf{k})$ as $\mathbf{k}$ traces a straight line to the Γ point, the center of the Brillouin zone (see Fig. 13.6 for the labeling of points in the zone). Then we continue to the right and $\mathbf{k}$ traces a straight line from the Γ point to the X point. Then we make a straight line from X to K and then K back to Γ.[19] Note that the lowest band is quadratic at the center of the Brillouin zone (a dispersion $\hbar^2k^2/(2m^*)$ for some effective mass m^*).

Similarly, in Fig. 13.8, we show the phonon spectrum of diamond. There are several things to note about this figure. First of all, since diamond has a unit cell with two atoms in it (it is fcc with a basis of two atoms) there should be six modes of oscillation per k-points (three directions of motion times two atoms per unit cell). Indeed, this is what we see in the picture, at least in the central third of the picture. In the other two parts of the picture, one sees fewer modes per k-point, but this is because, due to the symmetry of the crystal along this particular direction, several excitation modes have exactly the same energy. (Note examples at the X-point where two modes come in from the right, but only one goes out to the left. This means the two modes have the same energy on the left of the X point.) Secondly, we note that at the Γ-point, $\mathbf{k} = 0$, there are exactly three modes which come down linearly to zero energy. These are the three acoustic modes- -the higher one being a longitudinal mode and the lower two being transverse. The other three modes, which are finite energy at $\mathbf{k} = 0$, are the optical modes.

[18]This type of plot, because it can look like a jumble of lines, is sometimes called a “spaghetti diagram”.

[19]In fact if one travels in a straight line from X to K and continues in a straight line, one ends up at Γ in the neighboring Brillouin zone!

Chapter Summary

- The reciprocal lattice is a lattice in k-space defined by the set of points such that $e^{i\mathbf{G}\cdot\mathbf{R}} = 1$ for all $\mathbf{R}$ in the direct lattice. Given this definition, the reciprocal lattice can be thought of as the Fourier transform of the direct lattice.
- A reciprocal lattice vector $\mathbf{G}$ defines a set of parallel equally spaced planes via $\mathbf{G}\cdot\mathbf{r} = 2\pi m$ such that every point of the direct lattice is included in one of the planes. The spacing between the planes is $d = 2\pi/|\mathbf{G}|$. If $\mathbf{G}$ is the smallest reciprocal lattice vector parallel to $\mathbf{G}$ then this set of planes is a family of lattice planes, meaning that all planes intersect points of the direct lattice.
- Miller Indices (h, k, l) are used to describe families of lattice planes, or reciprocal lattice vectors.
- The general definition of Brillouin zone is any unit cell in reciprocal space. The first Brillouin zone is the Wigner–Seitz cell around the point $\mathbf{0}$ of the reciprocal lattice. Each Brillouin zone has the same volume and contains one k-state per unit cell of the entire system. Parallel Brillouin zone boundaries are separated by reciprocal lattice vectors.

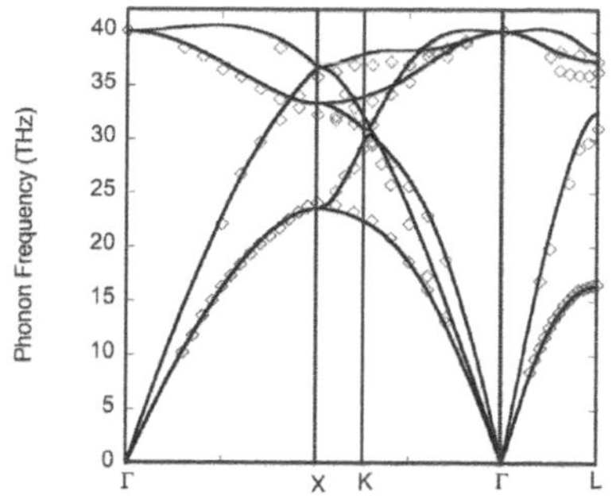

Fig. 13.8 Phonon spectrum of diamond (points are from experiment, solid line is a modern theoretical calculation). Figure is from A. Ward et al., *Phys. Rev. B* **80**, 125203 (2009), http://prb.aps.org/abstract/PRB/v80/i12/e125203, Copyright American Physical Society. Used by permission.

References

For reciprocal lattice, Miller indices and Brillouin zones, I recommend:

- Ashcroft and Mermin, chapter 5 (Again be warned of the nomenclature issue mentioned in Chapter 12, margin note 1.)
- Dove, chapter 4

Many books introduce X-ray diffraction and the reciprocal lattice at the same time. Once you have read the next chapter and studied scattering, you might go back and look at the nice introductions to reciprocal space given in the following books:

- Goodstein, sections 3.4–3.5 (very brief)
- Kittel, chapter 2
- Ibach and Luth, chapter 3
- Glazer, chapter 4

Exercises

13.1) **Reciprocal Lattice**

Show that the reciprocal lattice of a fcc (face-centered cubic) lattice is a bcc (body-centered cubic) lattice. Correspondingly, show that the reciprocal lattice of a bcc lattice is an fcc lattice. If an fcc lattice has conventional unit cell with lattice constant a, what is the lattice constant for the conventional unit cell of the reciprocal bcc lattice?

Consider now an orthorhombic face-centered lattice with conventional lattice constants a_1, a_2, a_3. What it the reciprocal lattice now?

(13.2) **Family of Planes&**

Consider the crystal shown in Exercise 12.3. Copy this figure and indicate the [210] direction and the (210) family of planes. (Why is this not a family of lattice planes?)

(13.3) **Directions and Spacings of Crystal Planes**

▷ ‡Explain briefly what is meant by the terms "crystal planes" and "Miller indices".

▷ Show that the general direction $[hkl]$ in a cubic crystal is normal to the planes with Miller indices (hkl).

▷ Is the same true in general for an orthorhombic crystal?

▷ Show that the spacing d of the (hkl) set of planes in a cubic crystal with lattice parameter a is

$$d = \frac{a}{\sqrt{h^2 + k^2 + l^2}}$$

▷ What is the generalization of this formula for an orthorhombic crystal?

(13.4) **‡Reciprocal Lattice**

(a) Define the term Reciprocal Lattice.

(b) Show that if a lattice in 3d has primitive lattice vectors $\mathbf{a_1}$, $\mathbf{a_2}$ and $\mathbf{a_3}$ then primitive lattice vectors for the reciprocal lattice can be taken as

$$\mathbf{b_1} = 2\pi \frac{\mathbf{a_2} \times \mathbf{a_3}}{\mathbf{a_1} \cdot (\mathbf{a_2} \times \mathbf{a_3})} \quad (13.13)$$

$$\mathbf{b_2} = 2\pi \frac{\mathbf{a_3} \times \mathbf{a_1}}{\mathbf{a_1} \cdot (\mathbf{a_2} \times \mathbf{a_3})} \quad (13.14)$$

$$\mathbf{b_3} = 2\pi \frac{\mathbf{a_1} \times \mathbf{a_2}}{\mathbf{a_1} \cdot (\mathbf{a_2} \times \mathbf{a_3})} \quad (13.15)$$

What is the proper formula in 2d?

(c) Define tetragonal and orthorhombic lattices. For an orthorhombic lattice, show that $|\mathbf{b_j}| = 2\pi/|\mathbf{a_j}|$. Hence, show that the length of the reciprocal lattice vector $\mathbf{G} = h\mathbf{b_1} + k\mathbf{b_2} + l\mathbf{b_3}$ is equal to $2\pi/d$, where d is the spacing of the (hkl) planes (see question 13.3)

(13.5) **More Reciprocal Lattice**

A two-dimensional rectangular crystal has a unit cell with sides $a_1 = 0.468$ nm and $a_2 = 0.342$ nm.

(a) Draw to scale a diagram of the reciprocal lattice.

▷ Label the reciprocal lattice points for indices in the range $0 \leqslant h \leqslant 3$ and $0 \leqslant k \leqslant 3$.

(b) Draw the first and second Brillouin zones using the Wigner–Seitz construction.

(13.6) **Brillouin Zones**

(a) Consider a cubic lattice with lattice constant a. Describe the first Brillouin zone. Given an arbitrary wavevector $\mathbf{k}$, write an expression for an equivalent wavevector within the first Brillouin zone (there are several possible expressions you can write).

(b) Consider a triangular lattice in two dimensions (primitive lattice vectors given by Eqs. 12.3). Find the first Brillouin zone. Given an arbitrary wavevector $\mathbf{k}$ (in two dimensions), write an expression for an equivalent wavevector within the first Brillouin zone (again there are several possible expressions you can write).

(13.7) **Number of States in the Brillouin Zone**

A specimen in the form of a cube of side L has a primitive cubic lattice whose mutually orthogonal fundamental translation vectors (primitive lattice vectors) have length a. Show that the number of different allowed $\mathbf{k}$-states within the first Brillouin zone equals the number of primitive unit cells forming the specimen. (One may assume periodic boundary conditions, although it is worth thinking about whether this still holds for hard-wall boundary conditions as well.)

(13.8) **Calculating Dispersions in $d > 1$***

(a) In Exercises 9.8 and 11.9 we discussed dispersion relations of systems in two dimensions (if you have not already solved those exercises, you should do so now).

▷ In Exercise 11.9, describe the Brillouin zone (you may assume perpendicular lattice vectors with length a_1 and a_2). Show that the tight-binding dispersion is periodic in the Brillouin zone. Show that the dispersion curve is always flat crossing a zone boundary.

▷ In Exercise 9.8, describe the Brillouin zone. Show that the phonon dispersion is periodic in the Brillouin zone. Show that the dispersion curve is always flat crossing a zone boundary.

(b) Consider a tight binding model on a three-dimensional fcc lattice where there are hopping matrix elements $-t$ from each site to each of the nearest-neighbor sites. Determine the energy spectrum $E(\mathbf{k})$ of this model. Show that near $\mathbf{k} = \mathbf{0}$ the dispersion is parabolic.

Part V

Neutron and X-Ray Diffraction

Wave Scattering by Crystals

14

In the last chapter we discussed reciprocal space, and explained how the energy dispersion of phonons and electrons is plotted within the Brillouin zone. We understand how electrons and phonons are similar to each other due to the wave-like nature of both. However, much of the same physics occurs when a crystal scatters waves (or particles[1]) that impinge upon it externally. Indeed, exposing a solid to a wave in order to probe its properties is an extremely useful thing to do. The most commonly used probe is X-rays. Another common, more modern, probe is neutrons. It can hardly be overstated how important this type of experiment is to science.

[1]Remember, in quantum mechanics there is no real difference between particles and waves! Planck and Einstein showed us that light waves are particles. Then de Broglie showed us that particles are waves!

The general setup that we will examine is shown in Fig. 14.1.

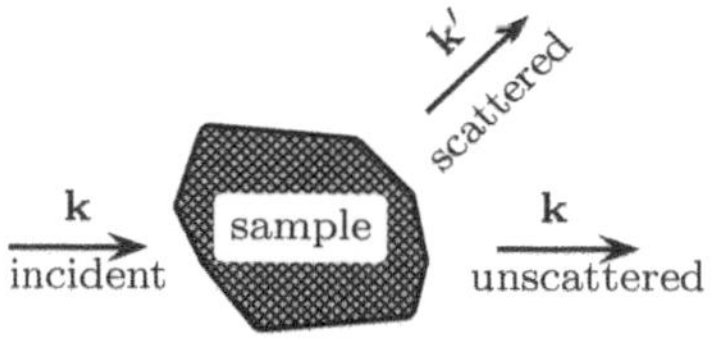

Fig. 14.1 A generic scattering experiment.

14.1 The Laue and Bragg Conditions

14.1.1 Fermi's Golden Rule Approach

If we think of the incoming wave as being a particle, then we should think of the sample as being some potential $V(\mathbf{r})$ that the particle experiences as it goes through the sample. According to Fermi's golden rule,[2] the transition rate $\Gamma(\mathbf{k}', \mathbf{k})$ per unit time for the particle scattering from $\mathbf{k}$ to $\mathbf{k}'$ is given by

$$\Gamma(\mathbf{k}', \mathbf{k}) = \frac{2\pi}{\hbar} \left|\langle \mathbf{k}'|V|\mathbf{k}\rangle\right|^2 \delta(E_{\mathbf{k}'} - E_{\mathbf{k}}).$$

[2]Fermi's golden rule should be familiar to you from quantum mechanics. Interestingly, Fermi's golden rule was actually discovered by Dirac, giving us yet another example where something is named after Fermi when Dirac really should have credit as well, or even instead. See also margin note 7 in Section 4.1.

The matrix element here

$$\langle \mathbf{k}'|V|\mathbf{k}\rangle = \int \mathbf{dr}\, \frac{e^{-i\mathbf{k}'\cdot\mathbf{r}}}{\sqrt{L^3}}\, V(\mathbf{r})\, \frac{e^{i\mathbf{k}\cdot\mathbf{r}}}{\sqrt{L^3}} = \frac{1}{L^3} \int \mathbf{dr}\, e^{-i(\mathbf{k}'-\mathbf{k})\cdot\mathbf{r}}\, V(\mathbf{r})$$

is nothing more than the Fourier transform of the potential (where L is the linear size of the sample, so the $\sqrt{L^3}$ terms just normalize the wavefunctions).

Note that these expressions are true whether or not the sample is a periodic crystal. However, if the sample is periodic the matrix element is zero unless $\mathbf{k} - \mathbf{k}'$ is a reciprocal lattice vector! To see this is true, let us write positions $\mathbf{r} = \mathbf{R} + \mathbf{x}$ where $\mathbf{R}$ is a lattice vector position and $\mathbf{x}$

is a position within the unit cell

$$\begin{aligned}\langle \mathbf{k}'|V|\mathbf{k}\rangle &= \frac{1}{L^3}\int \mathbf{dr}\, e^{-i(\mathbf{k}'-\mathbf{k})\cdot\mathbf{r}}\; V(\mathbf{r}) \\ &= \frac{1}{L^3}\sum_{\mathbf{R}}\int_{unit-cell} \mathbf{dx}\, e^{-i(\mathbf{k}'-\mathbf{k})\cdot(\mathbf{x}+\mathbf{R})}\; V(\mathbf{x}+\mathbf{R}).\end{aligned}$$

Now since the potential is assumed periodic, we have $V(\mathbf{x}+\mathbf{R}) = V(\mathbf{x})$, so this can be rewritten as

$$\langle \mathbf{k}'|V|\mathbf{k}\rangle = \frac{1}{L^3}\left[\sum_{\mathbf{R}} e^{-i(\mathbf{k}'-\mathbf{k})\cdot\mathbf{R}}\right]\left[\int_{unit-cell} \mathbf{dx}\; e^{-i(\mathbf{k}'-\mathbf{k})\cdot\mathbf{x}}\; V(\mathbf{x})\right]. \tag{14.1}$$

As we discussed in Section 13.1.3, the first term in brackets must vanish unless $\mathbf{k}' - \mathbf{k}$ is a reciprocal lattice vector.[3] This condition,

$$\mathbf{k}' - \mathbf{k} = \mathbf{G} \tag{14.2}$$

is known as the *Laue equation* (or *Laue condition*).[4,5] This condition is precisely the statement of *the conservation of crystal momentum.*[6] Note also that when the waves leave the crystal, they should have

$$|\mathbf{k}| = |\mathbf{k}'|$$

which is just the conservation of energy, which is enforced by the delta function in Fermi's golden rule. (In Section 14.4.2 we will consider more complicated scattering where energy is not conserved.)

[3]We also discussed how this first term in brackets diverges if $\mathbf{k}' - \mathbf{k}$ is a reciprocal lattice vector. This divergence is not a problem here because it gives just the number of unit cells and is canceled by the $1/L^3$ normalization factor leaving a factor of the inverse volume of the unit cell.

[6]Real momentum is conserved since the crystal itself absorbs any missing momentum. In this case, the center of mass of the crystal has absorbed momentum $\hbar(\mathbf{k}' - \mathbf{k})$. See the comment in margin note 13 in Section 9.4. Strictly speaking, when the crystal absorbs momentum, in order to conserve energy some tiny amount of energy must be lost from the scattered wave. However, in the limit that the crystal is large, this loss of energy can be neglected.

14.1.2 Diffraction Approach

It turns out that this Laue condition is nothing more than the scattering condition associated with a diffraction grating. This description of the scattering from crystals is known as the Bragg formulation of (X-ray) diffraction.[7]

Consider the configuration shown in Fig. 14.2. An incoming wave is reflected off of two adjacent layers of atoms separated by a distance d. A few things to note about this diagram. First note that the wave has been deflected by 2θ in this diagram.[8] Secondly, from simple geometry

[8]This is a *very* common source of errors on exams. The *total* deflection angle is 2θ.

[4]Max von Laue won the Nobel Prize for his work on X-ray scattering from crystals in 1914. Although von Laue never left Germany during the second world war, he remained openly opposed to the Nazi government. During the war he hid his gold Nobel medal at the Niels Bohr Institute in Denmark to prevent the Nazis from taking it. Had he been caught doing this, he may have been jailed or worse, since shipping gold out of Nazi Germany was considered a serious offense. After the occupation of Denmark in April 1940, George de Hevesy (a Nobel laureate in chemistry) decided to dissolve the medal in the solvent aqua regia to remove the evidence. He left the solution on a shelf in his lab. Although the Nazis occupied Bohr's institute and searched it very carefully, they did not find anything. After the war, the gold was recovered from solution and the Nobel Foundation presented Laue with a new medal made from the same gold.

[5]The reason this is called "Laue condition" rather than "von Laue" condition is because he was born Max Laue. In 1913 his father was elevated to the nobility and his family added the "von".

[7]William Henry Bragg and William Lawrence Bragg were a father-and-son team who won the Nobel Prize together in 1915 for their work on X-ray scattering. William L. Bragg, the driving force behind this work, was 25 years old when he won the prize. He remained the youngest Nobel laureate ever until 2014 when Malala Yousafzai won the Nobel Peace Prize at age 17.

note that the additional distance traveled by the component of the wave that reflects off of the further layer of atoms is

$$\text{extra distance} = 2d\sin\theta.$$

In order to have constructive interference, this extra distance must be equal to an integer number n of wavelengths. Thus we derive the Bragg condition for constructive interference, or what is known as *Bragg's law*

$$n\lambda = 2d\sin\theta. \tag{14.3}$$

Note that we can have diffraction from any two parallel planes of atoms such as the one shown in Fig. 14.3.

What we will see next is that this Bragg condition for constructive interference is precisely equivalent to the Laue condition!

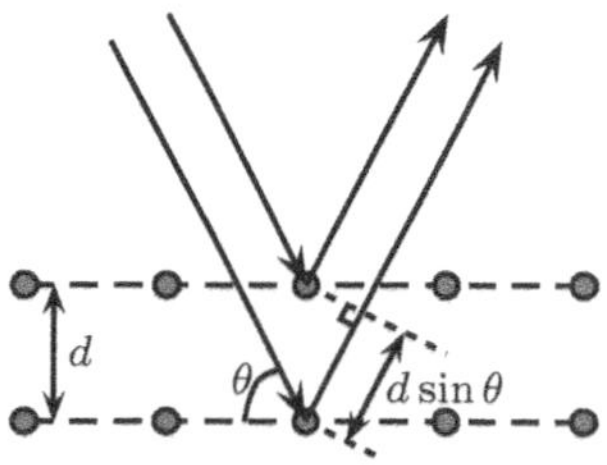

Fig. 14.2 Bragg scattering off of a plane of atoms in a crystal. The excess distance traveled by the wave striking the lower plane is $2d\sin\theta$.

14.1.3 Equivalence of Laue and Bragg conditions

Consider Fig. 14.4 (essentially the same as Fig. 14.2). Here we have shown the reciprocal lattice vector $\mathbf{G}$ which corresponds to the family of lattice planes. As we discussed in Chapter 13 the spacing between lattice planes is $d = 2\pi/|\mathbf{G}|$ (see Eqn. 13.8).

Just from geometry we have

$$\hat{\mathbf{k}}\cdot\hat{\mathbf{G}} = \sin\theta = -\hat{\mathbf{k}}'\cdot\hat{\mathbf{G}}$$

where the hats $\hat{}$ over vectors indicate unit vectors.

Suppose the Laue condition is satisfied. That is, $\mathbf{k}-\mathbf{k}' = \mathbf{G}$ with $|\mathbf{k}| = |\mathbf{k}'| = 2\pi/\lambda$ with λ the wavelength. We can rewrite the Laue equation as

$$\frac{2\pi}{\lambda}(\hat{\mathbf{k}}-\hat{\mathbf{k}}') = \mathbf{G}.$$

Now let us dot this equation with $\hat{\mathbf{G}}$ to give

$$\begin{aligned} \hat{\mathbf{G}}\cdot\frac{2\pi}{\lambda}(\hat{\mathbf{k}}-\hat{\mathbf{k}}') &= \hat{\mathbf{G}}\cdot\mathbf{G} \\ \frac{2\pi}{\lambda}(\sin\theta - \sin\theta') &= |\mathbf{G}| \\ \frac{2\pi}{|\mathbf{G}|}(2\sin\theta) &= \lambda \\ 2d\sin\theta &= \lambda \end{aligned}$$

which is the Bragg condition (in the last step we have used the relation, Eq. 13.8, between $\mathbf{G}$ and d). You may wonder why in this equation we obtained λ on the right-hand side rather than $n\lambda$ as we had in Eq. 14.3. The point here is that there if there is a reciprocal lattice vector $\mathbf{G}$, then there is also a reciprocal lattice vector $n\mathbf{G}$, and if we did the same calculation with that lattice vector we would get $n\lambda$. The plane spacing associated with $n\mathbf{G}$ does not generally correspond to a family of lattice planes (since it is not the shortest reciprocal lattice vector in the given direction) but it still allows for diffraction.

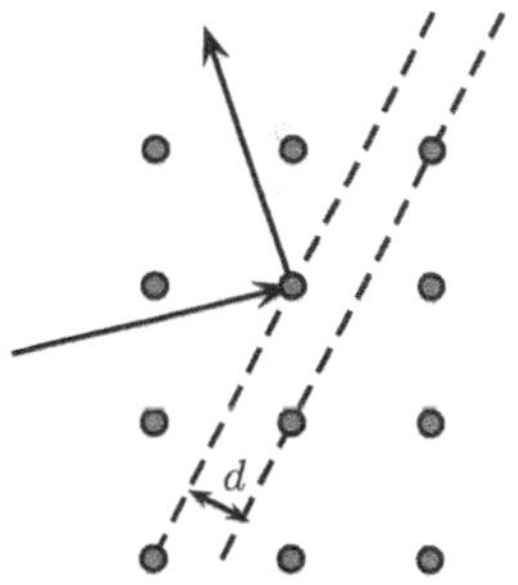

Fig. 14.3 Scattering off of the $(2\bar{1}0)$ plane of atoms.

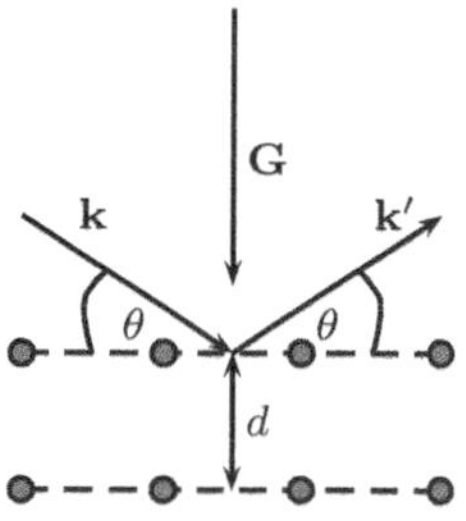

Fig. 14.4 Geometry of scattering.

Thus we conclude that the Laue condition and the Bragg condition are equivalent. It is equivalent to say that interference is constructive (as Bragg indicates) or to say that crystal momentum is conserved (as Laue indicates).

14.2 Scattering Amplitudes

If the Laue condition is satisfied, we would now like to ask how much scattering we actually get. Recall in Section 14.1.1 we started with Fermi's golden rule

$$\Gamma(\mathbf{k}', \mathbf{k}) = \frac{2\pi}{\hbar} \left|\langle \mathbf{k}'|V|\mathbf{k}\rangle\right|^2 \delta(E_{\mathbf{k}'} - E_{\mathbf{k}})$$

and we found out that if V is a periodic function, then the matrix element is given by (see Eq. 14.1)

$$\langle \mathbf{k}'|V|\mathbf{k}\rangle = \left[\frac{1}{L^3}\sum_{\mathbf{R}} e^{-i(\mathbf{k}'-\mathbf{k})\cdot\mathbf{R}}\right]\left[\int_{unit-cell} \mathbf{dx}\; e^{-i(\mathbf{k}'-\mathbf{k})\cdot\mathbf{x}}\; V(\mathbf{x})\right]. \tag{14.4}$$

The first factor in brackets gives zero unless the Laue condition is satisfied, in which case it gives a constant (due to the $1/L^3$ out front, this is now a non-divergent constant). The second term in brackets is known as the structure factor (compare to Eq. 13.6)

$$S(\mathbf{G}) = \int_{unit-cell} \mathbf{dx}\; e^{i\mathbf{G}\cdot\mathbf{x}}\; V(\mathbf{x}) \tag{14.5}$$

where we have used $\mathbf{G}$ for $(\mathbf{k}-\mathbf{k}')$, since this must be a reciprocal lattice vector or the first term in brackets vanishes.

Frequently, one writes the scattering intensity as

$$I_{(hkl)} \propto \left|S_{(hkl)}\right|^2 \tag{14.6}$$

which is shorthand for saying that $I_{(hkl)}$, the intensity of scattering off of the lattice planes defined by the reciprocal lattice vector (hkl), is proportional to the square of the structure factor at this reciprocal lattice vector. Sometimes a delta function is also written explicitly to indicate that the wavevector difference $(\mathbf{k}'-\mathbf{k})$ must be a reciprocal lattice vector.

It is usually a very good approximation to assume that the scattering potential is the sum over the scattering potentials of the individual atoms in the system,[9] so that we can write

$$V(\mathbf{x}) = \sum_{\text{atoms } j} V_j(\mathbf{x} - \mathbf{x_j})$$

where V_j is the scattering potential from atom j. The form of the function V_j will depend on what type of probe wave we are using and what type of atom j is.

We now turn to examine this structure factor more closely for our main two types of scattering probes—neutrons and X-rays.

[9] I.e., the influence of one atom on another does not affect how the atoms interact with the probe wave.

Neutrons[10]

Since neutrons are uncharged, they scatter almost exclusively from nuclei (rather than from electrons) via the nuclear forces. As a result, the scattering potential is extremely short-ranged, and can be approximated as a delta function. We thus have

$$V(\mathbf{x}) = \sum_{\text{atoms } j} f_j \, \delta(\mathbf{x} - \mathbf{x_j})$$

where $\mathbf{x_j}$ is the position of the jth atom in the unit cell. Here, f_j is known as the *form factor* or *atomic form factor*, and represents the strength of scattering from that particular nucleus. In fact, for the case of neutrons this quantity is proportional to[11] the so-called "nuclear scattering-length" b_j. Thus for neutrons we frequently write

$$V(\mathbf{x}) \sim \sum_{\text{atoms } j} b_j \, \delta(\mathbf{x} - \mathbf{x_j}).$$

Plugging this expression into Eq. 14.5, we obtain

$$S(\mathbf{G}) \sim \sum_{\text{atom } j \text{ in unit cell}} b_j \, e^{i\mathbf{G}\cdot\mathbf{x_j}} \qquad (14.7)$$

[10] Brockhouse and Shull were awarded the Nobel Prize for pioneering the use of neutron scattering experiments for understanding properties of materials. Shull's initial development of this technique began around 1946, just after the second world war, when the US atomic energy program made neutrons suddenly available. The Nobel Prize was awarded in 1994, making this one of the longest time-lags ever between a discovery and the awarding of the prize.

[11] To be precise, $f_j = 2\pi\hbar^2 b_j/m$ with m the mass of the neutron.

X-rays

X-rays scatter from the electrons in a system.[12] As a result, one can take the scattering potential $V(\mathbf{x})$ to be proportional to the electron density[13]

$$V_j(\mathbf{x} - \mathbf{x_j}) = Z_j \, g_j(\mathbf{x} - \mathbf{x_j})$$

where Z_j is the atomic number[14] of atom j (i.e., its number of electrons) and g_j is a somewhat short-ranged function (i.e., it has a few Ångstroms range—roughly the size of an atom). From this we derive

$$S(\mathbf{G}) = \sum_{\text{atom } j \text{ in unit cell}} f_j(\mathbf{G}) \, e^{i\mathbf{G}\cdot\mathbf{x_j}} \qquad (14.8)$$

where f_j, the form factor, is roughly proportional to Z_j, but has some dependence on the magnitude of the reciprocal lattice vector $\mathbf{G}$ as well (compare to Eq. 14.7). Frequently, however, we approximate f_j to be independent of $\mathbf{G}$ (which would be true if g were extremely short-ranged), although this is not strictly correct.

Aside: To be precise $f_j(\mathbf{G})$ is always just the Fourier transform of the scattering potential for atom j. We can write

$$f_j(\mathbf{G}) = \int \mathbf{dx} \, e^{i\mathbf{G}\cdot\mathbf{x}} \, V_j(\mathbf{x}) \qquad (14.9)$$

where the scattering potential $V_j(\mathbf{x})$ is just proportional to the electron density a distance $\mathbf{x}$ from the nucleus. Note that the integral here is over all of space,

[12] The coupling of photons to matter is via the usual minimal coupling $(\mathbf{p} + e\mathbf{A})^2/(2m)$. The denominator m, which is much larger for nuclei than for electrons, is why the nuclei are not important.

[13] The scattering here is essentially Thomson scattering, i.e., the scattering of light from free electrons. Here the electrons can be taken to be almost free since their binding energy to the atom is much less than the X-ray energy.

[14] If the atom occurs in an ionic solid, it may have gained or lost an electron (or two or three) in which case the number of electrons in Z_j should be appropriately modified.

not just over the unit cell (see Exercise 14.9.a). Taking the density to be a delta function results in f_j being a constant. Taking the slightly less crude approximation that the density is constant inside a sphere of radius r_0 and zero outside of this radius will result in a Fourier transform

$$f_j(\mathbf{G}) \sim 3Z_j \left(\frac{\sin(x) - x\cos(x)}{x^3} \right) \tag{14.10}$$

with $x = |\mathbf{G}r_0|$ (see Exercise 14.9.b). If the scattering angle is sufficiently small (i.e., $\mathbf{G}$ is small compared to $1/r_0$), the right-hand side is roughly Z_j with no strong dependence on $\mathbf{G}$.

Comparison of Neutrons and X-rays[15]

- For X-rays, since $f_j \sim Z_j$, the X-rays scatter very strongly from heavy atoms, and hardly at all from light atoms. This makes it very difficult to "see" light atoms like hydrogen in a solid. Further, it is hard to distinguish atoms that are very close to each other in their atomic number (since they scatter almost the same amount). Also, f_j is slightly dependent on the scattering angle.
- In comparison, for neutron scattering, the nuclear scattering length b_j varies rather erratically with atomic number (it can even be negative). In particular, hydrogen scatters fairly well, so it is easy to see. Further, one can usually distinguish atoms with similar atomic numbers rather easily.
- For neutrons, the scattering really is very short-ranged, so the form factor really is proportional to the scattering length b_j independent of $\mathbf{G}$. For X-rays there is a dependence on $\mathbf{G}$ that complicates matters.
- Neutrons also have spin. Because of this they can detect whether various electrons in the unit cell have their spins pointing up or down. The scattering of the neutrons from the electrons is much weaker than the scattering from the nuclei, but is still observable. We will return to this situation where the spin of the electron is spatially ordered in Section 20.1.2.

Electron Diffraction is Similar!

Much of the physics we learn from studying the diffraction of X-rays and neutrons from crystals can be applied just as well to other waves scattering from crystals. A particularly important technique is electron diffraction crystallography[16] which has been used very effectively to determine the structure of some very complicated biological structures.[17]

14.2.1 Simple Example

Generally, as in Eq. 14.6, we write the intensity of scattering as

$$I_{(hkl)} \propto |S_{(hkl)}|^2 \, .$$

[15] This comparison, which probably makes neutrons seem like the technique of choice, may have once been true, but is unfair to X-rays sources in the modern era. Since the development of synchrotrons (See section 14.4.3) and even more powerful free electron lasers, X-rays have been able to do some amazing things (but alas, these amazing things will not be fully covered in this book!).

[16] In fact electron diffraction can be sometimes even more powerful because $S_{(hkl)}$ can be measured directly rather than just $|S_{(hkl)}|^2$. See Hammond's book on crystallography, for example.

[17] Aaron Klug won the chemistry Nobel Prize in 1978 for developing the technique of electron crystallography and using it to deduce the structure of nucleic acid and protein complexes. However, the general idea of electron diffraction dates much further back. In 1927 Davisson and Germer, working at Bell Laboratories, demonstrated Bragg's law in the diffraction of electrons from a crystal, thus confirming de Broglie's hypothesis of the wave nature of matter. A Nobel Prize was awarded to Davisson along with George Paget Thomson, the son of J. J. Thompson, in 1937.

Assuming we have orthogonal lattice vectors, we can then generally write

$$S_{(hkl)} = \sum_{\text{atom } j \text{ in unit cell}} f_j \, e^{2\pi i (hx_j + ky_j + lz_j)} \tag{14.11}$$

where $[x_j, y_j, z_j]$ are the coordinates of atom j within the unit cell, in units of the three lattice vectors. (For X-rays, f_j may depend on $\mathbf{G}_{(hkl)}$ as well.)

Example 1: Caesium Chloride: Let us now consider the simple example of CsCl, whose unit cell is shown in Fig. 14.5. This system can be described as simple cubic with a basis given by[18]

Basis for CsCl

Cs	Position=	$[0,0,0]$
Cl	Position=	$[\frac{1}{2},\frac{1}{2},\frac{1}{2}]$.

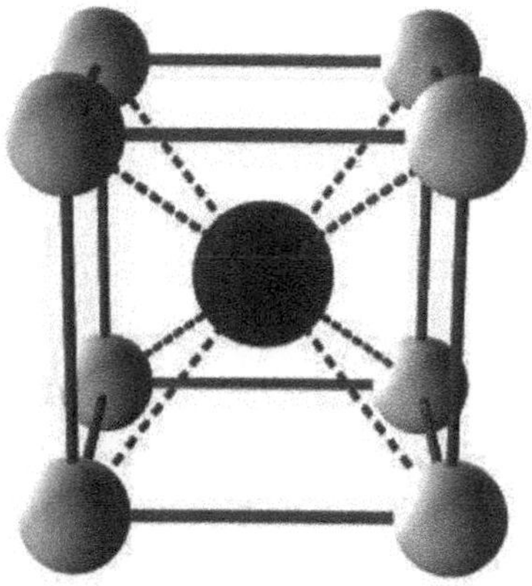

Fig. 14.5 Caesium chloride unit cell. Cs is the white corner atoms, Cl is the darker central atom. This is simple cubic with a basis. Note that bcc Cs can be thought of as just replacing the Cl with another Cs atom.

[18]Do not make the mistake of calling CsCl bcc! Bcc is a lattice where all points must be the same.

Thus the structure factor is given by

$$\begin{aligned} S_{(hkl)} &= f_{Cs} + f_{Cl} \, e^{2\pi i (h,k,l)\cdot[\frac{1}{2},\frac{1}{2},\frac{1}{2}]} \\ &= f_{Cs} + f_{Cl}(-1)^{h+k+l} \end{aligned}$$

with the f's being the appropriate form factors for the corresponding atoms. Recall that the scattered wave intensity is $I_{(hkl)} \sim |S_{(hkl)}|^2$.

14.2.2 Systematic Absences and More Examples

Example 2: Caesium bcc: Let us now consider instead a pure Cs crystal. In this case the crystal is bcc. We can think of this as simply replacing the Cl in CsCl with another Cs atom. Analogously we think of the bcc lattice as a simple cubic lattice with a basis (similar to CsCl!) which we now write as

Basis for Cs bcc
(conventional unit cell)

Cs	Position=	$[0,0,0]$
Cs	Position=	$[\frac{1}{2},\frac{1}{2},\frac{1}{2}]$.

Now the structure factor is given by

$$\begin{aligned} S_{(hkl)} &= f_{Cs} + f_{Cs} \, e^{2\pi i (h,k,l)\cdot[\frac{1}{2},\frac{1}{2},\frac{1}{2}]} \\ &= f_{Cs}\left[1 + (-1)^{h+k+l}\right] \end{aligned}$$

Crucially, note that the structure factor, and therefore the scattering intensity, vanishes for $h+k+l$ being any odd integer! This phenomenon is known as a *systematic absence.*

To understand why this absence occurs, consider the simple case of the (100) family of planes (see Fig. 13.1). This is simply a family of planes along the crystal axes with spacing a. You might expect a wave of wavelength $2a$ oriented perpendicular to these planes to scatter constructively[19]. However, if we are considering a bcc lattice, then there are additional planes of atoms half-way between the (100) planes which then cause perfect *destructive* interference. In Section 14.2.3 we will give a more geometric understanding of these absences.

[19] If two waves backscatter perpendicularly off of two successive planes, the difference in the distance traveled by the waves is twice the distance between the planes, hence the factor of 2. (See also Fig. 14.4)

Example 3: Copper fcc: Quite similarly there are systematic absences for scattering from fcc crystals as well. Recall from Eq. 12.6 that the fcc crystal can be thought of as a simple cubic lattice with a basis given by the points $[0,0,0], [\frac{1}{2},\frac{1}{2},0], [\frac{1}{2},0,\frac{1}{2}]$, and $[0,\frac{1}{2},\frac{1}{2}]$ in units of the cubic lattice constant. As a result the structure factor of fcc copper is given by (plugging into Eq. 14.11)

$$S_{(hkl)} = f_{Cu}\left[1 + e^{i\pi(h+k)} + e^{i\pi(h+l)} + e^{i\pi(k+l)}\right]. \tag{14.12}$$

It is easily shown that this expression vanishes unless h, k and l are either all odd or all even.

Summary of Systematic Absences

Systematic absences of scattering

simple cubic	all h, k, l allowed
bcc	$h+k+l$ must be even
fcc	h, k, l must be all odd or all even

Systematic absences are sometimes known as *selection rules.* Note that these selection rules do not depend on the fact that all three axes of the lattice are the same length.[20] For example, face-centered orthorhombic has the same selection rules as face-centered cubic!

[20] In Eq. 14.11, no mention is made of the lattice constant—both (h,k,l) and $[u,v,w]$ are simply written in terms of the lattice vector lengths.

It is very important to note that these absences, or selection rules, occur for any structure with the given (Bravais) lattice type. Even if a material is bcc with a basis of five different atoms per primitive unit cell, it will still show the same systematic absences as the bcc lattice we considered in Example 2 with a single atom per primitive unit cell. To see why this is true we consider yet another example.

Finally we note that the selection rules only tell you which scattering peaks *must* be absent for a given lattice. It can happen that more peaks vanish due to the details of the basis, as we shall elaborate in Eq. 14.14.

Example 4: Zinc Sulfide = fcc with a basis: As shown in Fig. 14.6, the zinc sulfide (zincblende) crystal is a an fcc lattice with a basis given by a Zn atom at $[0,0,0]$ and an S atom at $[\frac{1}{4},\frac{1}{4},\frac{1}{4}]$ (this is known as a zincblende structure). If we consider the fcc lattice to be a cubic lattice with basis given by the points $[0,0,0]$, $[\frac{1}{2},\frac{1}{2},0]$, $[\frac{1}{2},0,\frac{1}{2}]$, and $[0,\frac{1}{2},\frac{1}{2}]$, we then have the eight atoms in the conventional unit cell having positions given by the combination of the two bases, i.e.,

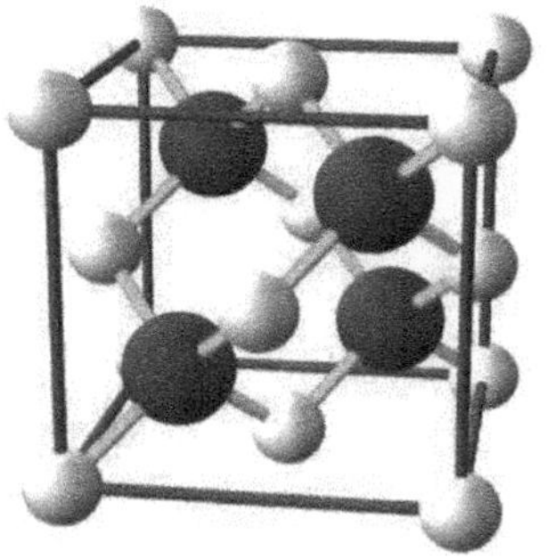

Fig. 14.6 Zinc sulfide (zincblende) conventional unit cell. This is fcc with a basis given by a Zn atom (light) at $[0,0,0]$ and a S atom (dark) at $[\frac{1}{4},\frac{1}{4},\frac{1}{4}]$.

Basis for conventional unit cell of ZnS

Zn	Positions=	$[0,0,0]$,	$[\frac{1}{2},\frac{1}{2},0]$,	$[\frac{1}{2},0,\frac{1}{2}]$,	$[0,\frac{1}{2},\frac{1}{2}]$
S	Positions=	$[\frac{1}{4},\frac{1}{4},\frac{1}{4}]$,	$[\frac{3}{4},\frac{3}{4},\frac{1}{4}]$,	$[\frac{3}{4},\frac{1}{4},\frac{3}{4}]$,	$[\frac{1}{4},\frac{3}{4},\frac{3}{4}]$.

The structure factor for ZnS is thus given by

$$\begin{aligned} S_{(hkl)} &= f_{Zn}\left[1 + e^{2\pi i (hkl)\cdot[\frac{1}{2},\frac{1}{2},0]} + \ldots\right] \\ &+ f_S\left[e^{2\pi i (hkl)\cdot[\frac{1}{4},\frac{1}{4},\frac{1}{4}]} + e^{2\pi i (hkl)\cdot[\frac{3}{4},\frac{3}{4},\frac{1}{4}]} + \ldots\right]. \end{aligned}$$

This combination of eight terms can be factored to give

$$\begin{aligned} S_{(hkl)} &= \left[1 + e^{i\pi(h+k)} + e^{i\pi(h+l)} + e^{i\pi(k+l)}\right] \\ &\times \left[f_{Zn} + f_S\, e^{2\pi i (hkl)\cdot[\frac{1}{4},\frac{1}{4},\frac{1}{4}]}\right]. \end{aligned} \tag{14.13}$$

The first term in brackets is precisely the same as the term we found for the fcc crystal in Eq. 14.12. In particular it has the same systematic absences that it vanishes unless h, k, and l are either all even or all odd. The second term gives structure associated specifically with ZnS.

Since the positions of the atoms are the positions of the underlying lattice plus the vectors in the basis, it is easy to see that the structure factor of a crystal system with a basis will always factorize into a piece which comes from the underlying lattice structure times a piece corresponding to the basis. Generalizing Eq. 14.13 we can write

$$S_{(hkl)} = S^{Lattice}_{(hkl)} \times S^{basis}_{(hkl)} \tag{14.14}$$

(where, to be precise, the form factors only occur in the latter term).

14.2.3 Geometric Interpretation of Selection Rules

The absence of certain scattering peaks has a very nice geometric interpretation. The fact that scattering does not occur at certain wavevectors for bcc and fcc lattices stems the Important Comment mentioned in Section 13.1.5. Recall that for these lattices we work with orthogonal axes for describing reciprocal space and Miller indices rather than working with the more complicated non-orthogonal primitive reciprocal lattice vectors. As a result, not all sets of Miller indices correspond to families of lattice planes.

Let us recall from Section 13.1.4 that when we give a set of Miller indices (hkl) for a (say, simple cubic) lattice, we are describing a set of parallel planes orthogonal to $\mathbf{G}_{(hkl)}$ and spaced by $2\pi/|\mathbf{G}_{(hkl)}|$. We are further guaranteed that every lattice point of the simple cubic lattice will be included in one of these planes. If we consider (010) as shown in the top of Fig. 13.2, while these planes are perfectly good lattice planes for the simple cubic lattice, they do not intersect the additional lattice point in the center of the bcc unit cell. Thus (010) is not a reciprocal lattice vector for the bcc lattice. However, as shown in the middle of Fig. 13.2, the (020) planes do intersect all of the bcc lattice points, and therefore (020) is a real reciprocal lattice vector for the bcc lattice. The general rule is that waves can always scatter by reciprocal lattice vectors, and thus the selection rule for scattering from a bcc lattice ($h + k + l$ being even) is both the condition that (hkl) is a reciprocal lattice vector and is also the condition that assures that the point in the middle of the unit cell is intersected by one of the corresponding planes. The situation is similar for the fcc lattice. The selection rule for an allowed scattering from an fcc lattice both defines which (hkl) are actual reciprocal lattice vectors and simultaneously gives a condition such that all points of the fcc lattice are included in one of the planes defined by (hkl).

14.3 Methods of Scattering Experiments

There are many methods of performing scattering experiments. In principle they are all similar—one sends in a probe wave of known wavelength (an X-ray, for example) and measures the angles at which it diffracts when it comes out. Then, using Bragg's laws (or the Laue equation) one can deduce the spacings of the lattice planes in the system.

14.3.1 Advanced Methods

Laue Method

Conceptually, perhaps the simplest method is to take a large single crystal of a material, fire waves at it (X-rays, say) from one direction, and measure the direction of the outgoing waves. However, given a single direction of the incoming wave, it is unlikely that you precisely achieve the Bragg condition for *any* set of lattice planes. In order to get more data, one can vary the wavelength of the incoming wave. This allows one to achieve the Bragg condition, at least at some wavelength.

Rotating Crystal Method

A similar technique is to rotate the crystal continuously so that at some angle of the crystal with respect to the incoming waves, one achieves the Bragg condition and measures an outcoming diffracted wave.

Both of these methods are indeed used. However, there is an important reason that they are often impossible to use. Frequently it is not possible to obtain a single crystal of a material. Growing large crystals (the beautiful ones shown in Fig. 7.1 are mined not grown) can be an enormous challenge.[21] In the case of neutron scattering, the problem is even more acute since one typically needs fairly large single crystals compared to X-rays.

[21]For example, high-temperature superconducting materials were discovered in 1986 (and resulted in a Nobel Prize the next year!). Despite a concerted world-wide effort, good single crystals of these materials were not available for 5 to 10 years.

14.3.2 Powder Diffraction

Powder diffraction, or the *Debye-Scherrer method*,[22] is the use of wave scattering on a sample which is not single crystalline, but is powdered. Because one does not need single crystals this method can be used on a much wider variety of samples.

In this case, the incoming wave can scatter off of any one of many small crystallites which may be oriented in any possible direction. In spirit this technique is similar to the rotating crystal method in that there is always some angle at which a crystal can be oriented to diffract the incoming wave. A schematic of the Debye-Scherrer setup is shown in Fig. 14.7 and sample data is shown in Fig. 14.8. Using Bragg's law, given the wavelength of the incoming wave, we can deduce the possible spacings between lattice planes.

[22]Debye is the same guy from the specific heat of solids. Paul Scherrer was Swiss but worked in Germany during the second world war, where he passed information to the famous American spy (and baseball player), Moe Berg, who had been given orders to find and shoot Heisenberg if he felt that the Germans were close to developing a bomb.

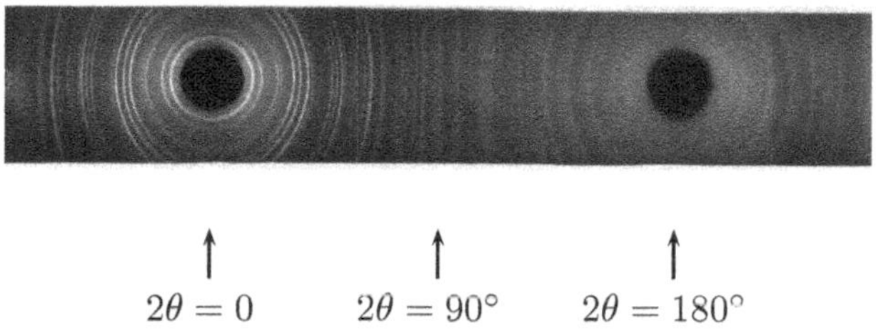

Fig. 14.8 Debye-Scherrer powder diffraction data exposed on photographic film. In modern experiments digital detectors are used.

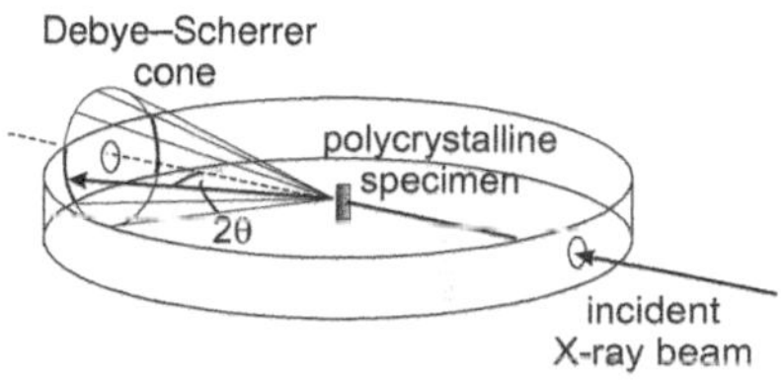

Fig. 14.7 Schematic of a Debye-Scherrer powder diffraction experiment.

A Fully Worked Example

We now present in detail a fully worked example of how to analyze a powder diffraction pattern.[23] But first, it is useful to write down a table (Table 14.1) of possible lattice planes and the selection rules that can occur for the smallest reciprocal lattice vectors.

[23]At Oxford, powder diffraction problems end up on exams every year, and unfortunately it is hard to find references that explain how to solve them.

The selection rules are those given in Section 14.2.2: simple cubic allows scattering from any plane, bcc must have $h + k + l$ be even, and fcc must have h, k, l either all odd or all even. On the table we have also added a column N which is the square magnitude of the Miller indices.

We have also added an additional column labeled "multiplicity". This quantity is important for figuring out the amplitude of scattering. The point here is that the (100) planes have some particular spacing but there

$\{hkl\}$	$N = h^2 + k^2 + l^2$	multiplicity	cubic	bcc	fcc
100	1	6	✓		
110	2	12	✓	✓	
111	3	8	✓		✓
200	4	6	✓	✓	✓
210	5	24	✓		
211	6	24	✓	✓	
220	8	12	✓	✓	✓
221	9	24	✓		
300	9	6	✓		
310	10	24	✓	✓	
311	11	24	✓		✓
222	12	8	✓	✓	✓
⋮	⋮	⋮	⋮		

Table 14.1 Selection rules for cubic, bcc, and fcc lattices.

are five other families of planes with the same spacing: $(010), (001), (\bar{1}00), (0\bar{1}0), (00\bar{1})$. (Because we mean all of these possible families of lattice planes, we use the notation $\{hkl\}$ introduced at the end of Section 13.1.5.) In the powder diffraction method, the crystal orientations are random, and here there would be six possible equivalent orientations of a crystal which will present the right angle for scattering from one of these planes, so there will be scattering intensity which is six times as large as we would otherwise calculate—this is known as the multiplicity factor. For the case of the 111 family, we would instead find eight possible equivalent planes: $(111), (11\bar{1}), (1\bar{1}1), (1\bar{1}\bar{1}), (\bar{1}11), (\bar{1}1\bar{1}), (\bar{1}\bar{1}1), (\bar{1}\bar{1}\bar{1})$. Thus, we should replace Eq. 14.6 with

$$I_{\{hkl\}} \propto M_{\{hkl\}} |S_{\{hkl\}}|^2 \tag{14.15}$$

where M is the multiplicity factor.

Calculating this intensity is straightforward for neutron scattering, but is much harder for X-ray scattering because the form factor for X-rays depends on $\mathbf{G}$. I.e, since in Eq. 14.7 the form factor (or scattering length b_j) is a constant independent of $\mathbf{G}$, it is easy to calculate the expected amplitudes of scattering based only on these constants. For the case of X-rays you need to know the functional forms of $f_j(\mathbf{G})$. At some *very* crude level of approximation it is a constant. More precisely we see in Eq. 14.10 that it is constant for small scattering angle but can vary quite a bit for large scattering angle.

Even if one knows the detailed functional form of $f_j(\mathbf{G})$, experimentally observed scattering intensities are never quite of the form predicted by Eq. 14.15. There can be several sources of corrections[24] that modify this result (these corrections are usually swept under the rug in elementary introductions to scattering, but you should at least be aware that they exist). Perhaps the most significant corrections[25] are known as

[24] Many of these corrections were first worked out by Charles Galton Darwin, the grandson of Charles Robert Darwin, the brilliant naturalist and proponent of evolution. The younger Charles was a terrific scientist in his own right. Later in life his focus turned to ideas of eugenics, predicting that the human race would eventually fail as we continue to breed unfavorable traits. (His interest in eugenics is not surprising considering that the acknowledged father of eugenics, Francis Galton, was also part of the same family.)

[25] Another important correction is due to the thermal vibrations of the crystal. Using Debye's theory of vibration, Ivar Waller derived what is now known as the Debye-Waller factor that accounts for the thermal smearing of Bragg peaks.

Lorentz corrections or Lorentz-polarization corrections. These terms, which depend on the detailed geometry of the experiment, give various prefactors (involving terms like $\cos\theta$ for example)[26] which are smooth as a function of θ.

[26] These factors are fairly flat between 80° and 140°, and can be roughly ignored. However, outside of this range these factors vary rapidly and need to be accounted for more carefully.

The Example:

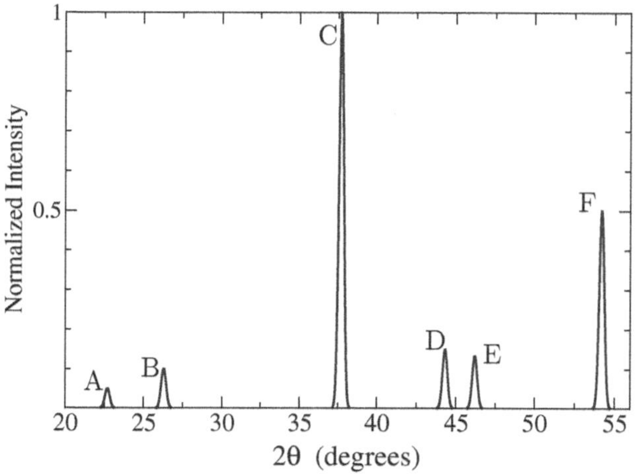

Fig. 14.9 Powder diffraction of neutrons from PrO_2. The wavelength of the neutron beam is $\lambda = .123$ nm. (One should assume that Lorentz corrections have been removed from the displayed intensities.)

Consider the powder diffraction data from PrO_2 shown in Fig. 14.9. Given the wavelength .123 nm, (and tentatively assuming a cubic lattice of some sort) we first would like to figure out the type of lattice and the lattice constant.

Note that the full deflection angle is 2θ. We will want to use Bragg's law and the expression for the spacing between planes[27]

$$d_{(hkl)} = \frac{\lambda}{2\sin\theta} = \frac{a}{\sqrt{h^2+k^2+l^2}}$$

where we have also used the expression Eq. 13.12 for the spacing between planes in a cubic lattice given the lattice constant a. Note that this then gives us

$$a^2/d^2 = h^2 + k^2 + l^2 = N$$

which is what we have labeled N in Table 14.1 of selection rules. We now make another table (Table 14.2). In the first two columns we just read the angles off of the given graph. *You should try to make the measurements of the angle from the data as carefully as possible. It makes the analysis much easier if you measure the angles right!*

In the third column of the table we calculate the distance between lattice planes for the given diffraction peak using Bragg's law. In the fourth column we have calculated the squared ratio of the lattice spacing d for the given peak to the lattice spacing for the first peak (labeled A) as a reference. We then realize that these ratios are pretty close to whole numbers divided by 3, so we try multiplying each of these quantities by 3 in the next column. If we round these numbers to integers (given in

[27] One might again wonder why we do not need the factor of n in Bragg's law as in Eq. 14.3. A factor of n will just correspond to multiplying (h, k, l) all by n, so we need not consider this case separately. See the discussion in section 14.1.3.

peak	2θ	$d = \lambda/(2\sin\theta)$	d_A^2/d^2	$3d_A^2/d^2$	$N = h^2+k^2+l^2$	$\{hkl\}$	$a = d\sqrt{h^2+k^2+l^2}$
A	22.7°	0.313 nm	1	3	3	111	.542 nm
B	26.3°	0.270 nm	1.33	3.99	4	200	.540 nm
C	37.7°	0.190 nm	2.69	8.07	8	220	.537 nm
D	44.3°	0.163 nm	3.67	11.01	11	311	.541 nm
E	46.2°	0.157 nm	3.97	11.91	12	222	.544 nm
F	54.2°	0.135 nm	5.35	16.05	16	400	.540 nm

Table 14.2 Analysis of data shown in Fig. 14.9.

the next column), we produce precisely the values of $N = h^2 + k^2 + l^2$ expected for the fcc lattice as shown in Table 14.1, and we thus conclude that we are looking at an fcc lattice.[28] The final column then calculates the lattice constant. Averaging these numbers[29] gives us a measurement of the lattice constant $a = .541 \pm .002$ nm.

The analysis thus far is equivalent to what one would do for X-ray scattering. However, with neutrons, assuming the scattering length is independent of scattering angle (which is typically a good assumption) we can go a bit further by analyzing the intensity of the scattering peaks.[30] In real data intensities are often weighted by the above-mentioned Lorentz factors. In Fig. 14.9, note that these factors have been removed so that we can expect that Eq. 14.15 holds precisely.

It turns out that the basis for the PrO_2 crystal is a Pr atom at position $[0,0,0]$ and O at $[\frac{1}{4},\frac{1}{4},\frac{1}{4}]$ and $[\frac{1}{4},\frac{1}{4},\frac{3}{4}]$. Thus, the Pr atoms form a fcc lattice and the O's fill in the holes as shown in Fig. 14.10. Given this structure, let us see what further information we might extract from the data in Fig. 14.9.

Let us start by calculating the structure factor for this crystal. Using Eq. 14.14 we have

$$\begin{aligned} S_{(hkl)} &= \left[1 + e^{i\pi(h+k)} + e^{i\pi(h+l)} + e^{i\pi(k+l)}\right] \\ &\times \left[b_{Pr} + b_O\left(e^{i(\pi/2)(h+k+l)} + e^{i(\pi/2)(h+k+3l)}\right)\right]. \end{aligned}$$

The first term in brackets is the structure factor for the fcc lattice, and it gives 4 for every allowed scattering point (when h,k,l are either all even or all odd). The second term in brackets is the structure factor for the basis.

The scattering intensity of the peaks are then given in terms of this structure factor and the peak multiplicities as shown in Eq. 14.15. We thus can write for all of our measured peaks[31]

$$I_{\{hkl\}} = CM_{\{hkl\}}\left|b_{Pr} + b_O\left(e^{i(\pi/2)(h+k+l)} + e^{i(\pi/2)(h+k+3l)}\right)\right|^2$$

where the constant C contains other constant factors (including the factor of 4^2 from the fcc structure factor). Note: We have to be a bit

[28] We emphasize that this is the general scheme for identifying a lattice type. Calculate d_A^2/d^2 and if these quantities are in the ratio of 3,4,8,11 then you have an fcc lattice. If they are in the ratios $1,2,3,4,5,6,8,\ldots$ then you have a simple cubic lattice. If they are in the ratio of $2,4,6,8,10,12,14,\ldots$ then you have a bcc lattice (see Table 14.1).

[29] Which one of these measured data points is likely to have the least error? See Exercise 14.10.

[30] For X-rays, we would need to know the form factors f_j as a function of angle to make further progress.

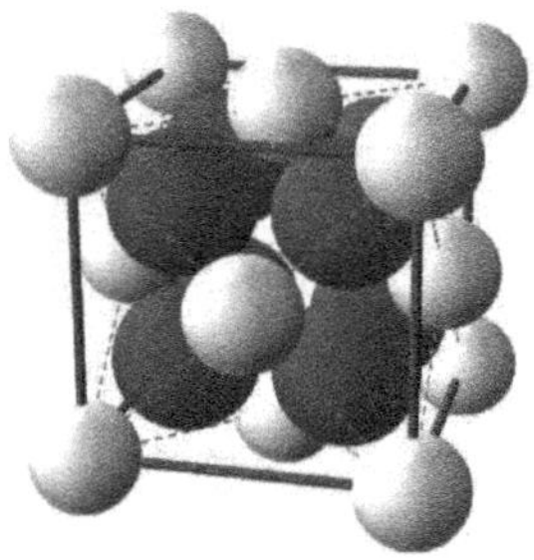

Fig. 14.10 The fluorite structure of PrO_2. This is fcc with a basis given by a white atom (Pr) at $[0,0,0]$ and dark atoms (O) at $[\frac{1}{4},\frac{1}{4},\frac{1}{4}]$ and $[\frac{1}{4},\frac{1}{4},\frac{3}{4}]$.

[31] Again assuming that smooth Lorentz correction terms have been removed from our data so that Eq. 14.15 is accurate.

careful here to make sure that the bracketed factor gives the same result for all possible (hkl) included in $\{hkl\}$, but in fact it does. Thus we can compile another table showing the predicted relative intensities of the peaks:

Scattering Intensity

peak	$\{hkl\}$	$I_{\{hkl\}}/C \propto M\|S\|^2$	Measured Intensity
A	111	$8\, b_{Pr}^2$	0.05
B	200	$6\,[b_{Pr} - 2b_O]^2$	0.1
C	220	$12\,[b_{Pr} + 2b_O]^2$	1.0
D	311	$24\, b_{Pr}^2$	0.15
E	222	$8\,[b_{Pr} - 2b_O]^2$	0.13
F	400	$6\,[b_{Pr} + 2b_O]^2$	0.5

Table 14.3 Predicted versus measured scattering intensities. Here the prediction is based purely on the scattering structure factor and the scattering multiplicity (Lorentz factors are not considered).

where the final column lists the intensities measured from the data in Fig. 14.9.

From the analytic expressions in the third column we can immediately predict that we should have

$$I_D = 3I_A \qquad I_C = 2I_F \qquad I_E = \frac{4}{3} I_B \ .$$

Examining the fourth column of this table, it is clear that these equations are properly satisfied (at least to a good approximation).

To further examine the data, we can look at the ratio I_C/I_A which in the measured data has a value of about 20. Thus we have

$$\frac{I_C}{I_A} = \frac{12[b_{Pr} + 2b_O]^2}{8\, b_{Pr}^2} = 20$$

with some algebra this can be reduced to a quadratic equation with two roots, resulting in

$$b_{Pr} = -.43\, b_O \qquad \text{or} \qquad b_{Pr} = .75\, b_O \ . \tag{14.16}$$

Further we can calculate

$$\frac{I_B}{I_A} = \frac{6[b_{Pr} - 2b_O]^2}{8\, b_{Pr}^2} = 2$$

which we can solve to give

$$b_{Pr} = .76\, b_O \qquad \text{or} \qquad b_{Pr} = -3.1\, b_O \ .$$

The former solution is consistent with Eq. 14.16, whereas the latter is not. We thus see how this neutron data can be used to experimentally determine the ratio of the nuclear scattering lengths $b_{Pr}/b_O \approx .75$. In fact, were we to look up these scattering lengths on a table we would find that this ratio is very close to correct!

14.4 Still More About Scattering

Scattering experiments such as those discussed here are *the* method for determining the microscopic structures of materials. One can use these methods (and extensions thereof) to sort out even very complicated atomic structures such as those of biological molecules.

Aside: In addition to the obvious work of von Laue and Bragg that initiated the field of X-ray diffraction (and Brockhouse and Shull for neutrons) there have been about half a dozen Nobel Prizes that have relied on, or further developed these techniques. In 1962 a chemistry Nobel Prize was awarded to Perutz and Kendrew for using X-rays to determine the structure of the biological proteins hemoglobin and myoglobin. The same year, Watson and Crick were awarded the prize in biology for determining the structure of DNA—which they did with the help of X-ray diffraction data taken by Rosalind Franklin.[32] Two years later in 1964, Dorothy Hodgkin[33] won the prize for determination of the structure of penicillin and other biological molecules. Further Nobel Prizes were given in chemistry for using X-rays to determine the structure of boranes (Lipscomb, 1976), photosynthetic proteins (Deisenhofer, Huber, Michel, 1988), and ribosomes (Ramakrishnan, Steitz, Yonath, 2009).

[32] There remains quite a controversy over the fact that Watson and Crick, at a critical juncture, were shown Franklin's data without her knowledge! Franklin may have won the prize in addition to Watson and Crick and thereby received a bit more of the appropriate credit, but she tragically died of cancer at age 37 in 1958, four years before the prize was awarded.

[33] Dorothy Hodgkin was a student and later a fellow at **Somerville College, Oxford. Yay!**

14.4.1 Variant: Scattering in Liquids and Amorphous Solids

A material need not be crystalline to scatter waves. However, for amorphous solids or liquids, instead of having delta-function peaks in the structure factor at reciprocal lattice vectors (as in Fig. 14.9), the structure factor (defined as the Fourier transform of the density) will have smooth behavior—with incipient peaks corresponding to $2\pi/d$, where d is roughly the typical distance between atoms. An example of a measured structure factor in liquid Cu is shown in Fig. 14.11. As the material gets close to its freezing point, the peaks in the structure factor will get more pronounced, becoming more like the structure of a solid where the peaks are delta functions.

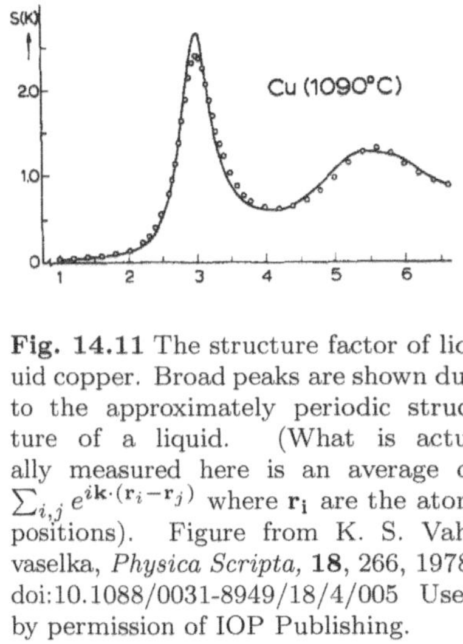

Fig. 14.11 The structure factor of liquid copper. Broad peaks are shown due to the approximately periodic structure of a liquid. (What is actually measured here is an average of $\sum_{i,j} e^{i\mathbf{k}\cdot(\mathbf{r}_i-\mathbf{r}_j)}$ where $\mathbf{r}_i$ are the atom positions). Figure from K. S. Vahvaselka, *Physica Scripta*, **18**, 266, 1978. doi:10.1088/0031-8949/18/4/005 Used by permission of IOP Publishing.

14.4.2 Variant: Inelastic Scattering

It is also possible to perform scattering experiments which are inelastic. Here, "inelastic" means that some of the energy of the incoming wave is left behind in the sample, and the energy of the outgoing wave is lower.[34] The general process is shown in Fig. 14.12. A wave is incident on the crystal with momentum $\mathbf{k}$ and energy $\epsilon(\mathbf{k})$ (For neutrons the energy is $\hbar^2\mathbf{k}^2/(2m)$, whereas for photons the energy is $\hbar c|\mathbf{k}|$). This wave transfers some of its energy and momentum to some internal excitation mode of the material—such as a phonon, or a spin or electronic excitation quanta. One measures the outgoing energy E and momentum $\mathbf{Q}$ of the wave. Since energy and crystal momentum are conserved

$$\begin{aligned} \mathbf{Q} &= \mathbf{k} - \mathbf{k}' + \mathbf{G} \\ E(\mathbf{Q}) &= \epsilon(\mathbf{k}) - \epsilon(\mathbf{k}') \end{aligned}$$

[34] At finite temperature the outgoing wave could also be at higher energy than the incoming wave, thus taking away some of the sample's thermal energy.

thus allowing one to determine the dispersion relation of the internal excitation (i.e., the relationship between $\mathbf{Q}$ and $E(\mathbf{Q})$). This technique is extremely useful for determining phonon dispersions experimentally. A reason that this technique is more difficult for X-rays than for neutrons is because it is much harder to measure small changes in the energy of X-rays than it is for neutrons (since for X-rays one needs to measure a small change in a large energy whereas for neutrons it is a small change in a small energy).

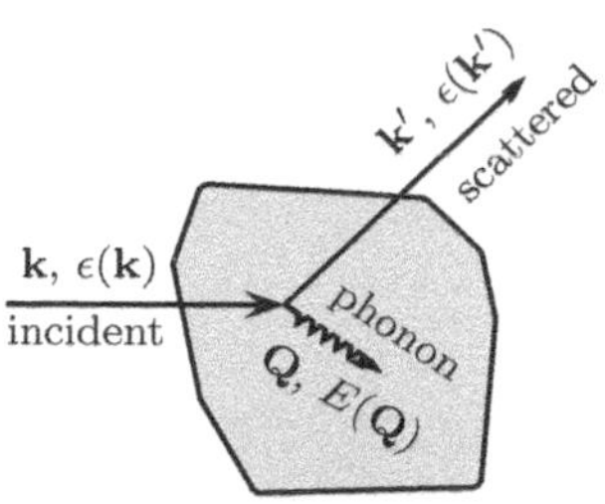

Fig. 14.12 Inelastic scattering. Energy and crystal momentum must be conserved.

14.4.3 Experimental Apparatus

Perhaps the most interesting piece of this kind of experiment is how one actually produces and measures the waves in question.

Since at the end of the day one ends up counting photons or neutrons, brighter sources (higher flux of probe particles) are always better—as it allows one to do experiments quicker, and allows one to reduce noise (since counting error on N counts is proportional to $\sqrt{N}$, meaning a fractional error that drops as $1/\sqrt{N}$). Further, with a brighter source, one can examine smaller samples more easily.

X-rays: Even small laboratories can have X-ray sources that can do very useful crystallography. A typical source accelerates electrons electrically (with 10s of keV) and smashes them into a metal target.[35] X-rays with a discrete spectrum of energies are produced when an electron is knocked out of a low atomic orbital, and an electron in a higher orbital drops down to refill the hole (this is known as X-ray fluorescence). Also, a continuous Bremsstrahlung spectrum is produced by electrons coming near the charged nuclei, but for monochromatic diffraction experiments, this is less useful. (Although one wavelength from a spectrum can always be selected—using diffraction from a known crystal!)

Much higher brightness X-ray sources are provided by huge (and hugely expensive) facilities known as synchrotron light sources—where electrons are accelerated around enormous loops (at energies in the GeV range). Then, using magnets these electrons are rapidly accelerated around corners which makes them emit X-rays extremely brightly and in a highly columnated fashion.

Detection of X-rays can be done with photographic films (the old style) but is now more frequently done with more sensitive semiconductor detectors.

Neutrons: Although it is possible to generate neutrons in a small lab, the flux of these devices is extremely small, so neutron scattering experiments are always done in large neutron source facilities. Although the first neutron sources simply used the byproduct neutrons from nuclear reactors, more modern facilities now use a technique called *spallation*,[36] where protons are accelerated into a target and neutrons are emitted. As with X-rays, neutrons can be monochromated (made into a single wavelength) by diffracting them from a known crystal. Another technique is to use time-of-flight. Since more energetic neutrons move faster, one can send a pulse of polychromatic neutrons and select only those that arrive

[35] Wilhelm Conrad Roentgen discovered X-rays using roughly this technique. In 1901 he was the first recipient of the Physics Nobel Prize for this work. In many languages, X-rays are called "Roentgen rays".

[36] The word "spallation" is generally used when some fragment is ejected from a larger body due to an impact. This word is often used in planetary science to describe the result of meteor impacts.

at a certain time in order to obtain monochromatic neutrons. On the detection side, one can again select for energy similarly. We won't say too much about neutron detection as there are many methods. Needless to say, they all involve interaction with nuclei.

Fig. 14.13 The Rutherford Appleton Laboratory in Oxfordshire, UK. (Photo used by permission of STFC). On the right, the large circular building is the DIAMOND synchrotron light source. The building on the left is the ISIS spallation neutron facility. This was the brightest neutron source on earth until August 2007, when it was surpassed by one in Oak Ridge, US. The next generation neutron source is being built in Sweden and is expected to start operating in 2019. The price tag for construction of this device is over 10^9 euros.

Chapter Summary

- Understand diffraction of waves from crystals in both the Laue and Bragg formulations (equivalent to each other).
- The structure factor (the Fourier transform of the scattering potential) in a periodic crystal has sharp peaks at allowed reciprocal lattice vectors for scattering. The scattering intensity is proportional to the square of the structure factor.
- There are systematic absences of diffraction peaks depending on the crystal structure (fcc, bcc). Know how to figure these out.
- Know how to analyze a powder diffraction pattern (very common exam question!).

References

It is hard to find references that give enough information about diffraction to suit the tastes of Oxford. These are not bad:

- Kittel, chapter 2
- Ashcroft and Mermin, chapter 6
- Dove, chapter 6 (most detailed, perhaps a bit too much information)

In addition, the following have nice, but incomplete discussions:

- Rosenberg, chapter 2
- Ibach and Luth, chapter 3
- Burns, chapter 4

Exercises

(14.1) **Reciprocal Lattice and X-ray Scattering**

Consider the lattice described in Exercise 13.5 (a two-dimensional rectangular crystal having a unit cell with sides $a_1 = 0.468$ nm and $a_2 = 0.342$ nm). A collimated beam of monochromatic X-rays with wavelength 0.166 nm is used to examine the crystal.

(a) Draw to scale a diagram of the reciprocal lattice.

(b) Calculate the magnitude of the wavevectors $\mathbf{k}$ and $\mathbf{k}'$ of the incident and reflected X-ray beams, and hence construct on your drawing the "scattering triangle" corresponding to the Laue condition $\Delta\mathbf{k} = \mathbf{G}$ for diffraction from the (210) planes (the scattering triangle includes $\mathbf{k}$, $\mathbf{k}'$ and $\Delta\mathbf{k}$).

(14.2) ‡ **X-ray scattering II**

$BaTiO_3$ has a primitive cubic lattice and a basis with atoms having fractional coordinates

Ba	[0,0,0]		
Ti	$[\frac{1}{2},\frac{1}{2},\frac{1}{2}]$		
O	$[\frac{1}{2},\frac{1}{2},0]$,	$[\frac{1}{2},0,\frac{1}{2}]$,	$[0,\frac{1}{2},\frac{1}{2}]$

▷ Sketch the unit cell.

▷ Show that the X-ray structure factor for the $(00l)$ Bragg reflections is given by

$$S_{(hkl)} = f_{Ba} + (-1)^l f_{Ti} + \left[1 + 2(-1)^l\right] f_O$$

where f_{Ba} is the atomic form factor for Ba, etc.

▷ Calculate the ratio $I_{(002)}/I_{(001)}$, where $I_{(hkl)}$ is the intensity of the X-ray diffraction from the (hkl) planes. You may assume that the atomic form factor is proportional to atomic number (Z), and neglect its dependence on the scattering vector. ($Z_{\mathrm{Ba}} = 56$, $Z_{\mathrm{Ti}} = 22$, $Z_{\mathrm{O}} = 8$.)

(14.3) ‡ **X-ray scattering and Systematic Absences**

(a) Explain what is meant by "Lattice Constant" for a cubic crystal structure.

(b) Explain why X-ray diffraction may be observed in first order from the (110) planes of a crystal with a body-centered cubic lattice, but not from the (110) planes of a crystal with a face-centered cubic lattice.

▷ Derive the general selection rules for which planes are observed in bcc and fcc lattices.

(c) Show that these selection rules hold independent of what atoms are in the primitive unit cell, so long as the lattice is bcc or fcc respectively.

(d) A collimated beam of monochromatic X-rays of wavelength 0.162 nm is incident upon a powdered sample of the cubic metal palladium. Peaks in the scattered X-ray pattern are observed at angles of 42.3°, 49.2°, 72.2°, 87.4°, and 92.3° from the direction of the incident beam.

▷ Identify the lattice type.

▷ Calculate the lattice constant and the nearest-neighbor distance.

▷ If you assume there is only a single atom in the basis does this distance agree with the known data that the density of palladium is 12023 kg m^{-3}? (Atomic mass of palladium = 106.4.)

(e) How could you improve the precision with which

the lattice constant is determined. (For one suggestion, see Exercise 14.10.)

(14.4) ‡ **Neutron Scattering**

(a) X-ray diffraction from sodium hydride (NaH) established that the Na atoms are arranged on a face-centered cubic lattice.

▷ Why is it difficult to locate the positions of the H atoms using X-rays?

The H atoms were thought to be displaced from the Na atoms either by $[\frac{1}{4}, \frac{1}{4}, \frac{1}{4}]$ or by $[\frac{1}{2}, \frac{1}{2}, \frac{1}{2}]$, to form the ZnS (zincblende) structure or NaCl (sodium chloride) structure, respectively. To distinguish these models a neutron powder diffraction measurement was performed. The intensity of the Bragg peak indexed as (111) was found to be much larger than the intensity of the peak indexed as (200).

▷ Write down expressions for the structure factors $S_{(hkl)}$ for neutron diffraction assuming NaH has

(i) the sodium chloride (NaCl) structure

(ii) the zinc blende (ZnS) structure.

▷ Hence, deduce which of the two structure models is correct for NaH. (Nuclear scattering length of Na $= 0.363 \times 10^{-5}$nm; nuclear scattering length of H $= -0.374 \times 10^{-5}$ nm.)

(b) How does one produce monochromatic neutrons for use in neutron diffraction experiments?

▷ What are the main differences between neutrons and X-rays?

▷ Explain why (inelastic) neutron scattering is well suited for observing phonons, but X-rays are not.

(14.5) **And More X-ray Scattering**

A sample of aluminum powder is put in an Debye-Scherrer X-ray diffraction device. The incident X-ray radiation is from Cu-Kα X-ray transition (this just means that the wavelength is $\lambda = .154$ nm). The following scattering angles (θ) were observed: 19.48° 22.64° 33.00° 39.68° 41.83° 50.35° 57.05° 59.42°

Given also that the atomic weight of Al is 27, and the density is 2.7 g/cm^3, use this information to calculate Avogadro's number. How far off are you? What causes the error?

(14.6) **More Neutron Scattering**

The conventional unit cell dimension for a particular bcc solid is .24nm. Two orders of diffraction are observed. What is the minimum energy of the neutrons? At what temperature would such neutrons be dominant if the distribution is Maxwell–Boltzmann.

(14.7) **Lattice and Basis**

Prove that the structure factor for any crystal (described with a lattice and a basis) is the product of the structure factor for the lattice times the structure factor for the basis (i.e., prove Eq. 14.14).

(14.8) **Cuprous Oxide and Fluorine Beta**

(a) The compound Cu_2O has a cubic conventional unit cell with the basis:

O	$[000]$; $[\frac{1}{2}, \frac{1}{2}, \frac{1}{2}]$
Cu	$[\frac{1}{4}, \frac{1}{4}, \frac{1}{4}]$; $[\frac{1}{4}, \frac{3}{4}, \frac{3}{4}]$; $[\frac{3}{4}, \frac{1}{4}, \frac{3}{4}]$; $[\frac{3}{4}, \frac{3}{4}, \frac{1}{4}]$

Sketch the conventional unit cell. What is the lattice type? Show that certain diffraction peaks depend only on the Cu form factor f_{Cu} and other reflections depend only on the O form factor f_O.

(b) Consider fluorine beta phase as described in exercise 12.5. Calculate the structure factor for this crystal. What are the selection rules?

(14.9) **Form Factors**

(a) Assume that the scattering potential can be written as the sum over the contributions of the scattering from each of the atoms in the system. Write the positions of the atoms in terms of a lattice plus a basis so that

$$V(\mathbf{x}) = \sum_{\mathbf{R},\alpha} V_\alpha(\mathbf{x} - \mathbf{R} - \mathbf{y}_\alpha)$$

where $\mathbf{R}$ are lattice points, α indexes the particles in the basis and $\mathbf{y}_\alpha$ is the position of atom α in the basis. Now use the definition of the structure factor Eq. 14.5 and derive an expression of the form of Eq. 14.8 and hence derive expression 14.9 for the form factor. (Hint: Use the fact that an integral over all space can be decomposed into a sum over integrals of individual unit cells.)

(b) Given the equation for the form factor you just derived (Eq. 14.9), assume the scattering potential from an atom is constant inside a radius a and is zero outside that radius. Derive Eq. 14.10.

(c)* Use your knowledge of the wavefunction of an electron in a hydrogen atom to calculate the X-ray form factor of hydrogen.

(14.10) **Error Analysis**

Imagine you are trying to measure the lattice constant a of some crystal using X-rays. Suppose a diffraction peak is observed at a scattering angle of 2θ. However, suppose that the value of θ is measured only within some uncertainty $\delta\theta$. What is the fractional error $\delta a/a$ in the resulting measurement of the lattice constant? How might this error be reduced? Why could it not be reduced to zero?

Part VI

Electrons in Solids

Electrons in a Periodic Potential

15

In Chapters 9 and 10 we discussed the wave nature of phonons in solids, and how crystal momentum is conserved (i.e., momentum is conserved up to reciprocal lattice vectors). Further, we found that we could describe the entire excitation spectrum within a single Brillouin zone in a reduced zone scheme. We also found in Chapter 14 that X-rays and neutrons scatter from solids by conserving crystal momentum. In this chapter we will consider the nature of electron waves in solids, and we will find that similarly crystal momentum is conserved and the entire excitation spectrum can be described within a single Brillouin zone using a reduced zone scheme.

We have seen a detailed preview of properties of electrons in periodic systems when we considered the one-dimensional tight binding model in Chapter 11, so the results of this section will be hardly surprising. However, in the current chapter we will approach the problem from a very different (and complementary) starting point. Here, we will consider electrons as free-electron waves that are only very weakly perturbed by the periodic potential from the atoms in the solid. The tight binding model is exactly the opposite limit where we consider electrons bound strongly to the atoms, and they only weakly hop from one atom to the next.

15.1 Nearly Free Electron Model

We start with completely free electrons whose Hamiltonian is

$$H_0 = \frac{\mathbf{p}^2}{2m}.$$

The corresponding energy eigenstates, the plane waves $|\mathbf{k}\rangle$, have eigenenergies given by

$$\epsilon_0(\mathbf{k}) = \frac{\hbar^2|\mathbf{k}|^2}{2m}.$$

Now consider a weak periodic potential perturbation to this Hamiltonian

$$H = H_0 + V(\mathbf{r})$$

with V periodic, meaning

$$V(\mathbf{r}) = V(\mathbf{r} + \mathbf{R})$$

where $\mathbf{R}$ is any lattice vector. The matrix elements of this potential are then just the Fourier components

$$\langle \mathbf{k}'|V|\mathbf{k}\rangle = \frac{1}{L^3}\int \mathbf{dr}\, e^{i(\mathbf{k}-\mathbf{k}')\cdot\mathbf{r}}\; V(\mathbf{r}) \equiv V_{\mathbf{k}'-\mathbf{k}} \qquad (15.1)$$

which is zero unless $\mathbf{k}' - \mathbf{k}$ is a reciprocal lattice vector (see Eq. 14.1). Thus, any plane-wave state $\mathbf{k}$ can scatter into another plane-wave state $\mathbf{k}'$ only if these two plane waves are separated by a reciprocal lattice vector.

We now apply the rules of perturbation theory. At first order in the perturbation V, we have

$$\epsilon(\mathbf{k}) = \epsilon_0(\mathbf{k}) + \langle \mathbf{k}|V|\mathbf{k}\rangle = \epsilon_0(\mathbf{k}) + V_0$$

which is just an uninteresting constant energy shift to all of the eigenstates. In fact, it is an exact statement (at any order of perturbation theory) that the only effect of V_0 is to shift the energies of all of the eigenstates by this constant.[1] Henceforth we will assume that $V_0 = 0$ for simplicity.

[1] Since the Fourier mode at $\mathbf{G} = \mathbf{0}$ is just a constant independent of position.

At second order in perturbation theory we have

$$\epsilon(\mathbf{k}) = \epsilon_0(\mathbf{k}) + \sum_{\mathbf{k}'=\mathbf{k}+\mathbf{G}}{}' \frac{|\langle \mathbf{k}'|V|\mathbf{k}\rangle|^2}{\epsilon_0(\mathbf{k}) - \epsilon_0(\mathbf{k}')} \qquad (15.2)$$

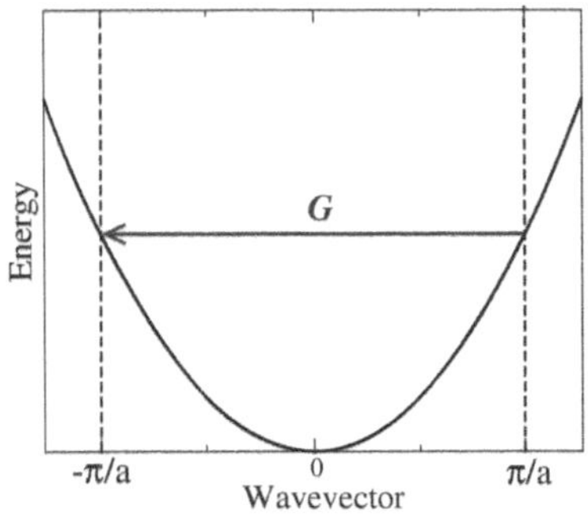

Fig. 15.1 Scattering from Brillouin zone boundary to Brillouin zone boundary. The states at the two zone boundaries are separated by a reciprocal lattice vector $\mathbf{G}$ and have the same energy. This situation leads to a divergence in perturbation theory, Eq. 15.2, because when the two energies match, the denominator is zero.

where the $'$ means that the sum is restricted to have $\mathbf{G} \neq \mathbf{0}$. In this sum, however, we have to be careful. It is possible that, for some $\mathbf{k}'$ it happens that $\epsilon_0(\mathbf{k})$ is very close to $\epsilon_0(\mathbf{k}')$ or perhaps they are even equal, in which case the corresponding term of the sum diverges and the perturbation expansion makes no sense. This is what we call a *degenerate* situation and it needs to be handled with degenerate perturbation theory, which we shall consider in a moment.

To see when this degenerate situation occurs, we look for solutions of the two equations

$$\begin{aligned} \epsilon_0(\mathbf{k}) &= \epsilon_0(\mathbf{k}') && (15.3)\\ \mathbf{k}' &= \mathbf{k}+\mathbf{G}. && (15.4) \end{aligned}$$

First, let us consider the one-dimensional case. Since $\epsilon(k) \sim k^2$, the only possible solution of Eq. 15.3 is $k' = -k$. This means the two equations are only satisfied for

$$k' = -k = \frac{n\pi}{a}$$

or precisely on the Brillouin zone boundaries (see Fig. 15.1).

In fact, this is quite general even in higher dimensions: given a point $\mathbf{k}$ on a Brillouin zone boundary, there is another point $\mathbf{k}'$ (also on a Brillouin zone boundary) such that Eqs. 15.3 and 15.4 are satisfied (see in particular Fig. 13.5 for example).[2]

[2] To see this generally, recall that a Brillouin zone boundary is a perpendicular bisector of the segment between $\mathbf{0}$ and some $\mathbf{G}$. We can write the given point $\mathbf{k} = \mathbf{G}/2 + \mathbf{k}_\perp$ where $\mathbf{k}_\perp \cdot \mathbf{G} = 0$. Then if we construct the point $\mathbf{k}' = -\mathbf{G}/2 + \mathbf{k}_\perp$, then clearly 15.4 is satisfied, $\mathbf{k}'$ is an element of the perpendicular bisector of the segment between $\mathbf{0}$ and $-\mathbf{G}$ and therefore is on a zone boundary, and $|\mathbf{k}| = |\mathbf{k}'|$ which implies that Eq. 15.3 is satisfied.

When we are very near a zone boundary, since Eq. 15.2 is divergent, we need to handle this situation with *degenerate perturbation theory*.[3] In this approach, one diagonalizes[4] the Hamiltonian within the degenerate space first (and other perturbations can be treated after this). In other words, we take states of the same energy that are connected by the matrix element and treat their mixing exactly.

[3]Hopefully you have learned this in your quantum mechanics courses already!

[4]By "diagonalizing" a matrix M we mean essentially to find the eigenvalues λ_i and normalized eigenvectors $\mathbf{v}^{(i)}$ of the matrix. Determining these is "diagonalizing" since you can then write $M = U^\dagger D U$ where $U_{ij} = v_j^{(i)*}$ and D is the diagonal matrix of eigenvalues $D_{ij} = \lambda_i \delta_{ij}$.

15.1.1 Degenerate Perturbation Theory

If two plane-wave states $|\mathbf{k}\rangle$ and $|\mathbf{k}'\rangle = |\mathbf{k}+\mathbf{G}\rangle$ are of approximately the same energy (meaning that $\mathbf{k}$ and $\mathbf{k}'$ are close to zone boundaries), then we must diagonalize the matrix elements of these states first. We have

$$\begin{aligned} \langle \mathbf{k}| H |\mathbf{k}\rangle &= \epsilon_0(\mathbf{k}) \\ \langle \mathbf{k}'| H |\mathbf{k}'\rangle &= \epsilon_0(\mathbf{k}') = \epsilon_0(\mathbf{k}+\mathbf{G}) \\ \langle \mathbf{k}| H |\mathbf{k}'\rangle &= V_{\mathbf{k}-\mathbf{k}'} = V^*_{\mathbf{G}} \\ \langle \mathbf{k}'| H |\mathbf{k}\rangle &= V_{\mathbf{k}'-\mathbf{k}} = V_{\mathbf{G}} \end{aligned} \tag{15.5}$$

where we have used the definition of $V_{\mathbf{G}}$ from Eq. 15.1, and the fact that $V_{-\mathbf{G}} = V^*_{\mathbf{G}}$ is guaranteed by the fact that $V(\mathbf{r})$ is real.

Now, within this two-dimensional space we can write any wavefunction as

$$|\Psi\rangle = \alpha|\mathbf{k}\rangle + \beta|\mathbf{k}'\rangle = \alpha|\mathbf{k}\rangle + \beta|\mathbf{k}+\mathbf{G}\rangle. \tag{15.6}$$

Using the variational principle to minimize the energy is equivalent to solving the effective Schroedinger equation[5]

[5]This should look similar to our 2 by 2 Schroedinger equation 6.9.

$$\begin{pmatrix} \epsilon_0(\mathbf{k}) & V^*_{\mathbf{G}} \\ V_{\mathbf{G}} & \epsilon_0(\mathbf{k}+\mathbf{G}) \end{pmatrix} \begin{pmatrix} \alpha \\ \beta \end{pmatrix} = E \begin{pmatrix} \alpha \\ \beta \end{pmatrix}. \tag{15.7}$$

The characteristic equation determining E is then

$$\left(\epsilon_0(\mathbf{k}) - E\right)\left(\epsilon_0(\mathbf{k}+\mathbf{G}) - E\right) - |V_{\mathbf{G}}|^2 = 0. \tag{15.8}$$

(Note that once this degenerate space is diagonalized, one could go back and treat further, non-degenerate, scattering processes in perturbation theory.)

Simple Case: k Exactly at the Zone Boundary

The simplest case we can consider is when $\mathbf{k}$ is precisely on a zone boundary (and therefore $\mathbf{k}' = \mathbf{k}+\mathbf{G}$ is also precisely on a zone boundary). In this case $\epsilon_0(\mathbf{k}) = \epsilon_0(\mathbf{k}+\mathbf{G})$ and our characteristic equation simplifies to

$$\left(\epsilon_0(\mathbf{k}) - E\right)^2 = |V_{\mathbf{G}}|^2$$

or equivalently

$$E_\pm = \epsilon_0(\mathbf{k}) \pm |V_{\mathbf{G}}|. \tag{15.9}$$

Thus we see that a gap opens up at the zone boundary. Whereas both $\mathbf{k}$ and $\mathbf{k}'$ had energy $\epsilon_0(\mathbf{k})$ in the absence of the added potential $V_{\mathbf{G}}$, when the potential is added, the two eigenstates form two linear combinations with energies split by $\pm|V_{\mathbf{G}}|$.

In One Dimension

In order to understand this better, let us focus on the one-dimensional case. Let us assume we have a potential $V(x) = \tilde{V}\cos(2\pi x/a)$ with $\tilde{V} > 0$. The Brillouin zone boundaries are at $k = \pi/a$ and $k' = -k = -\pi/a$ so that $k' - k = G = -2\pi/a$ and $\epsilon_0(k) = \epsilon_0(k')$.

Examining Eq. 15.7, we discover that the solutions (when $\epsilon_0(k) = \epsilon_0(k')$) are given by $\alpha = \pm\beta$, thus giving the eigenstates

$$|\psi_\pm\rangle = \frac{1}{\sqrt{2}}\Big(|k\rangle \pm |k'\rangle\Big) \tag{15.10}$$

corresponding to $E_\pm$ respectively. Since we can write the real space version of these $|k\rangle$ wavefunctions as[6]

$$\begin{aligned} |k\rangle &\rightarrow e^{ikx} = e^{ix\pi/a} \\ |k'\rangle &\rightarrow e^{ik'x} = e^{-ix\pi/a} \end{aligned}$$

[6] Formally what we mean here is $\langle x|k\rangle = e^{ikx}/\sqrt{L}$.

we discover that the two eigenstates are given by

$$\begin{aligned} \psi_+ &\sim e^{ix\pi/a} + e^{-ix\pi/a} &\propto \cos(x\pi/a) \\ \psi_- &\sim e^{ix\pi/a} - e^{-ix\pi/a} &\propto \sin(x\pi/a). \end{aligned}$$

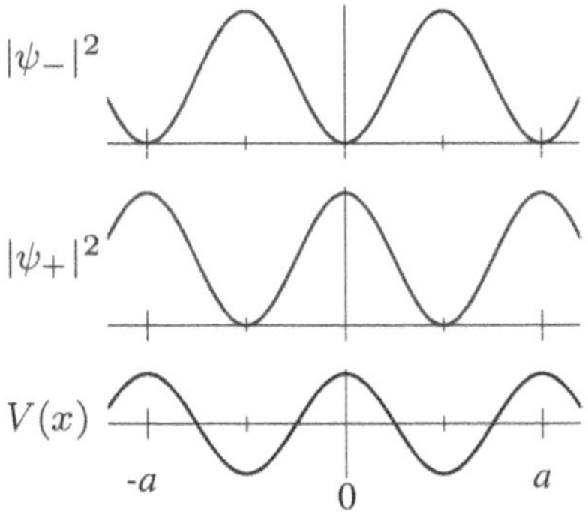

Fig. 15.2 Structure of wavefunctions at the Brillouin zone boundary. The higher-energy eigenstate ψ_+ has its density concentrated near the maxima of the potential V, whereas the lower-energy eigenstate ψ_- has its density concentrated near the minima of V.

If we then look at the densities $|\psi_\pm|^2$ associated with these two wavefunctions (see Fig. 15.2) we see that the higher energy eigenstate ψ_+ has its density concentrated mainly at the maxima of the potential V whereas the lower energy eigenstate ψ_- has its density concentrated mainly at the minima of the potential.

So the general principle is that the periodic potential scatters between the two plane waves $|\mathbf{k}\rangle$ and $|\mathbf{k}+\mathbf{G}\rangle$. If the energy of these two plane waves are the same, the mixing between them is strong, and the two plane waves can combine to form one state with higher energy (concentrated on the potential maxima) and one state with lower energy (concentrated on the potential minima).

k Not Quite on a Zone Boundary (Still in One Dimension)

It is not too hard to extend this calculation to the case where $\mathbf{k}$ is not quite on a zone boundary. For simplicity though we will stick to the one-dimensional situation. We need only solve the characteristic equation 15.8 for more general k. To do this, we expand around the zone boundaries.

Let us consider the states at the zone boundary $k = \pm n\pi/a$ which are separated by the reciprocal lattice vectors $G = \pm 2\pi n/a$. As noted in Eq. 15.9, the gap that opens up precisely at the zone boundary will be $\pm|V_G|$. Now let us consider a plane wave near this zone boundary $k = n\pi/a+\delta$ with δ being very small (and n an integer). This wavevector can scatter into $k' = -n\pi/a + \delta$ due to the periodic potential. We then have

$$\begin{aligned}\epsilon_0(n\pi/a+\delta) &= \frac{\hbar^2}{2m}\left[(n\pi/a)^2 + 2n\pi\delta/a + \delta^2\right] \\ \epsilon_0(-n\pi/a+\delta) &= \frac{\hbar^2}{2m}\left[(n\pi/a)^2 - 2n\pi\delta/a + \delta^2\right].\end{aligned}$$

The characteristic equation (Eq. 15.8) is then

$$\left(\frac{\hbar^2}{2m}\left[(n\pi/a)^2+\delta^2\right] - E + \frac{\hbar^2}{2m}2n\pi\delta/a\right) \times \left(\frac{\hbar^2}{2m}\left[(n\pi/a)^2+\delta^2\right] - E - \frac{\hbar^2}{2m}2n\pi\delta/a\right) - |V_G|^2 = 0$$

which simplifies to

$$\left(\frac{\hbar^2}{2m}\left[(n\pi/a)^2+\delta^2\right] - E\right)^2 = \left(\frac{\hbar^2}{2m}2n\pi\delta/a\right)^2 + |V_G|^2$$

or

$$E_{\pm} = \frac{\hbar^2}{2m}\left[(n\pi/a)^2+\delta^2\right] \pm \sqrt{\left(\frac{\hbar^2}{2m}2n\pi\delta/a\right)^2 + |V_G|^2}\ . \qquad (15.11)$$

Expanding the square root for small δ we obtain[7]

$$E_{\pm} = \frac{\hbar^2(n\pi/a)^2}{2m} \pm |V_G| + \frac{\hbar^2\delta^2}{2m}\left[1 \pm \frac{\hbar^2(n\pi/a)^2}{m}\frac{1}{|V_G|}\right]. \qquad (15.12)$$

Note that for small perturbation (which is what we are concerned with), the second term in the square brackets is larger than unity, so that for one of the two solutions the square bracket is negative.

Thus we see that near the band gap at the Brillouin zone boundary, the dispersion is quadratic (in δ) as shown in Fig. 15.3. In Fig. 15.4, we see (using the *repeated zone scheme*) that small gaps open at the Brillouin zone boundaries in what is otherwise a parabolic spectrum. (This plotting scheme is equivalent to the reduced zone scheme if restricted to a single zone.)

The general structure we find is thus very much like what we expected from the tight binding model we considered previously in Chapter 11. As in the tight binding picture there are *energy bands* where there are energy eigenstates, and there are gaps between bands where there are no energy eigenstates. As in the tight binding model, the spectrum is periodic in the Brillouin zone (see Fig 15.4).

In Section 11.2 we introduced the idea of the effective mass—if a dispersion is parabolic, we can describe the curvature at the bottom of

[7]The conditions of validity for this expansion is that the first term under the square root of Eqn. 15.11 is much smaller than the second, meaning that δ is very small, or we must be very close to the Brillouin zone boundary. Note that as V_G gets smaller and smaller, the expansion is valid only for k closer and closer to the zone boundary.

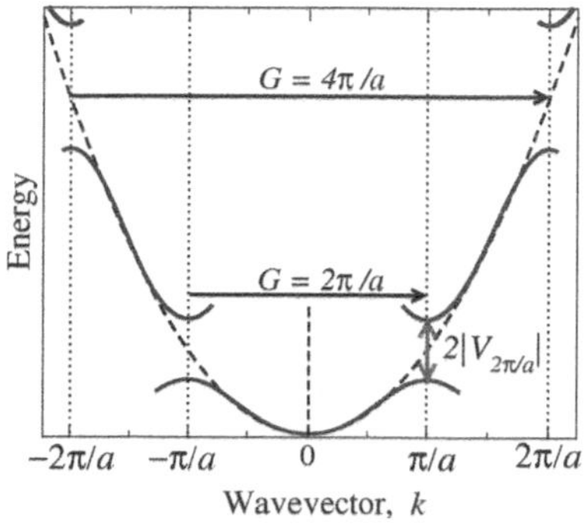

Fig. 15.3 Dispersion of a nearly free electron model. In the nearly free electron model, gaps open up at the Brillouin zone boundaries in an otherwise parabolic spectrum. Compare this to what we found for the tight binding model in Fig. 11.5.

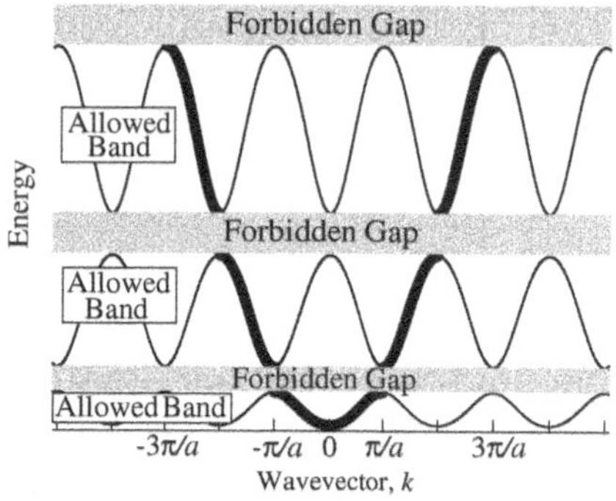

Fig. 15.4 Dispersion of a nearly free electron model. Same as Fig. 15.3, but plotted in repeated zone scheme. This is equivalent to the reduced zone scheme but the equivalent zones are repeated. Forbidden bands are marked where there are no eigenstates. The similarity to the free electron parabolic spectrum is emphasized.

the band in terms of an effective mass. In this model at every Brillouin zone boundary the dispersion is parabolic (indeed, if there is a gap, hence a local maximum and a local minimum, the dispersion must be parabolic around these extrema). Thus we can write the dispersion Eq. 15.12 (approximately) as

$$\begin{aligned} E_+(G+\delta) &= C_+ + \frac{\hbar^2\delta^2}{2m_+^*} \\ E_-(G+\delta) &= C_- - \frac{\hbar^2\delta^2}{2m_-^*} \end{aligned}$$

where C_+ and C_- are constants, and the effective masses are given by[8]

$$m_\pm^* = \frac{m}{\left|1 \pm \frac{\hbar^2(n\pi/a)^2}{m}\frac{1}{|V_G|}\right|}.$$

[8]Note that since V_G is assumed small, $1 - \frac{\hbar^2(n\pi/a)^2}{m}\frac{1}{|V_G|}$ is negative.

We will define effective mass more precisely, and explain its physics in Chapter 17. For now we just think of this as a convenient way to describe the parabolic dispersion near the Brillouin zone boundary.

Nearly Free Electrons in Two and Three Dimensions

The principles of the nearly free electron model are quite similar in two and three dimensions. In short, near the Brillouin zone boundary, a gap opens up due to scattering by a reciprocal lattice vector. States of energy slightly higher than the zone boundary intersection point are pushed up in energy, whereas states of energy slightly lower than the zone boundary intersection point are pushed down in energy. We will return to the detailed geometry of this situation in Section 16.2.

There is one more key difference between one dimension and higher dimensions. In one dimension, we found that if $\mathbf{k}$ is on a zone boundary, then there will be exactly one other $\mathbf{k}'$ such that $\mathbf{k} - \mathbf{k}' = \mathbf{G}$ is a reciprocal lattice vector and such that $\epsilon(\mathbf{k}') = \epsilon(\mathbf{k})$ (i.e., Eqs. 15.3 and 15.4 are satisfied). As described earlier in this subsection, these two plane-wave states mix with each other (see Eq. 15.6) and open up a gap. However, in higher dimensions it may occur that given $\mathbf{k}$ there may be *several* different $\mathbf{k}'$ which will satisfy these equations—i.e., many $\mathbf{k}'$ which differ from $\mathbf{k}$ by a reciprocal lattice vector and which all have the same unperturbed energy. In this case, we need to mix together all of the possible plane waves in order to discover the true eigenstates. One example of when this occurs is the two-dimensional square lattice, where the four points $(\pm\pi/a, \pm\pi/a)$ all have the same unperturbed energy and are all separated from each other by reciprocal lattice vectors.

Aside: The idea that gaps open at the Brillouin zone boundaries for nearly free waves in a periodic medium is not restricted to only electron waves. Another prominent example of very similar physics occurs for visible light. We have already seen in the previous chapter how X-ray light can scatter by reciprocal lattice vectors in a crystal. In order to arrange for *visible* light to scatter similarly we must have a material with a lattice constant that is on the order of the wavelength of the light which is roughly a fraction of a micron (which is much longer than a typical atomic length scale). Such materials can occur naturally[9] or can be made artificially by assembling small sub-micron building blocks to create what are known as *photonic crystals*. Like electron waves in solids, light in photonic crystals has a band-structure. The free light dispersion $\hbar\omega = \hbar c|\mathbf{k}|$ is modified by the periodic medium, such that gaps open at Brillouin zone boundaries entirely analogously to what happens for electrons. Such gaps in materials can be used to reflect light extremely well. When a photon strikes a material, if the photon's frequency lies within a band gap, then there are no eigenstates for that photon to enter, and it has no choice but to be fully reflected! This is just another way of saying that the Bragg condition has been satisfied, and the putative transmitted light is experiencing destructive interference.

[9]Natural examples of photonic crystals include the gemstone opal, which is a periodic array of sub-micron sized spheres of silica, and the wings of butterflies which are periodic arrays of polymer.

15.2 Bloch's Theorem

In the "nearly free electron" model we started from the perspective of plane waves that are weakly perturbed by a periodic potential. But in real materials, the scattering from atoms can be very strong so that perturbation theory may not be valid (or may not converge until very high order). How do we know that we can still describe electrons with anything remotely similar to plane waves?

In fact, by this time, after our previous experience with waves, we should know the answer in advance: the plane-wave momentum is not a conserved quantity, but the crystal momentum is. No matter how strong the periodic potential, so long as it is periodic, crystal momentum is conserved. This important fact was first discovered by Felix Bloch[10] in 1928, very shortly after the discovery of the Schroedinger equation, in what has become known as Bloch's theorem.[11]

> **Bloch's Theorem:** An electron in a periodic potential has eigenstates of the form
>
> $$\Psi^{\alpha}_{\mathbf{k}}(\mathbf{r}) = e^{i\mathbf{k}\cdot\mathbf{r}} u^{\alpha}_{\mathbf{k}}(\mathbf{r})$$
>
> where $u^{\alpha}_{\mathbf{k}}$ is periodic in the unit cell and $\mathbf{k}$ (the crystal momentum) can be chosen within the first Brillouin zone.

In reduced zone scheme there may be many states at each $\mathbf{k}$ and these are indexed by α. The periodic function u is usually known as a *Bloch function*, and Ψ is sometimes known as a *modified plane wave*. Because u is periodic, it can be rewritten as a sum over reciprocal lattice vectors

$$u^{\alpha}_{\mathbf{k}}(\mathbf{r}) = \sum_{\mathbf{G}} \tilde{u}^{\alpha}_{\mathbf{G},\mathbf{k}}\, e^{i\mathbf{G}\cdot\mathbf{r}} \,.$$

This form guarantees[12] that $u^{\alpha}_{\mathbf{k}}(\mathbf{r}) = u^{\alpha}_{\mathbf{k}}(\mathbf{r}+\mathbf{R})$ for any lattice vector $\mathbf{R}$. Therefore the full wavefunction is expressed as

$$\Psi^{\alpha}_{\mathbf{k}}(\mathbf{r}) = \sum_{\mathbf{G}} \tilde{u}^{\alpha}_{\mathbf{G},\mathbf{k}}\, e^{i(\mathbf{G}+\mathbf{k})\cdot\mathbf{r}} \,. \tag{15.13}$$

Thus an equivalent statement of Bloch's theorem is that we can write

[10]Felix Bloch later won a Nobel Prize for inventing nuclear magnetic resonance. In medicine NMR was renamed MRI (magnetic resonance imaging) when people decided the word "nuclear" sounds too much like it must be related to some sort of bomb.

[11]Bloch's theorem was actually discovered by a mathematician, Gaston Floquet in 1883, and rediscovered later by Bloch in the context of solids. This is an example of what is known as Stigler's law of eponomy: "Most things are not named after the person who first discovers them". In fact, Stigler's law was discovered by Merton.

[12]In fact, the function u is periodic in the unit cell if and only if it can be written as a sum over reciprocal lattice vectors in this way. See Exercise 15.2.

each eigenstate as being made up of a sum of plane-wave states $\mathbf{k}$ which differ by reciprocal lattice vectors $\mathbf{G}$.

Given this equivalent statement of Bloch's theorem, we now understand that the reason for Bloch's theorem is that the scattering matrix elements $\langle \mathbf{k}'|V|\mathbf{k}\rangle$ are zero unless $\mathbf{k}'$ and $\mathbf{k}$ differ by a reciprocal lattice vector. This is just the Laue condition! As a result, the Schroedinger equation is "block diagonal"[13] in the space of $\mathbf{k}$, and in any given eigenfunction only plane waves $\mathbf{k}$ that differ by some $\mathbf{G}$ can be mixed together. One way to see this more clearly is to take the Schroedinger equation

[13] No pun intended.

$$\left[\frac{\mathbf{p}^2}{2m} + V(\mathbf{r})\right] \Psi(\mathbf{r}) = E\Psi(\mathbf{r})$$

and Fourier transform it to obtain

$$\sum_{\mathbf{G}} V_{\mathbf{G}} \Psi_{\mathbf{k}-\mathbf{G}} = \left[E - \frac{\hbar^2|\mathbf{k}|^2}{2m}\right] \Psi_{\mathbf{k}}$$

where we have used the fact that $V_{\mathbf{k}-\mathbf{k}'}$ is only non-zero if $\mathbf{k} - \mathbf{k}' = \mathbf{G}$. It is then clear that for each $\mathbf{k}$ we have a Schroedinger equation for the set of $\Psi_{\mathbf{k}-\mathbf{G}}$'s and we must obtain solutions of the form of Eq. 15.13.

Although by this time it may not be surprising that electrons in a periodic potential have eigenstates labeled by crystal momenta, we should not overlook how important Bloch's theorem is. This theorem tells us that even though the potential that the electron feels from each atom is extremely strong, the electrons still behave *almost* as if they do not see the atoms at all! They still *almost* form plane-wave eigenstates, with the only modification being the periodic Bloch function u and the fact that momentum is now crystal momentum.

A quote from Felix Bloch:

> When I started to think about it, I felt that the main problem was to explain how the electrons could sneak by all the ions in a metal so as to avoid a mean free path of the order of atomic distances... By straight Fourier analysis I found to my delight that the wave differed from the plane wave of free electrons only by a periodic modulation.

Chapter Summary

- When electrons are exposed to a periodic potential, gaps arise in their dispersion relation at the Brillouin zone boundary. (The dispersion is quadratic approaching a zone boundary.)
- Thus the electronic spectrum breaks into bands, with forbidden energy gaps between the bands. In the nearly free electron model, the gaps are proportional to the periodic potential $|V_{\mathbf{G}}|$.
- Bloch's theorem guarantees that all eigenstates are some periodic function times a plane wave. In repeated zone scheme the wavevector (the *crystal momentum*) can always be taken in the first Brillouin zone.

References

Any self-respecting solid state physics book should cover this material. Here are a few good ones:

- Goodstein, section 3.6a
- Burns, sections 10.1–10.6
- Kittel, chapter 7 (Skip Kronig–Penney model. Although it is interesting, it is tedious. See Exercise 15.6.)
- Hook and Hall, section 4.1
- Ashcroft and Mermin, chapters 8–9 (not my favorite)
- Ibach and Luth, sections 7.1–7.2
- Singleton, chapters 2–3

Exercises

(15.1) ‡**Nearly Free Electron Model**

Consider an electron in a weak periodic potential in one dimension $V(x) = V(x+a)$. Write the periodic potential as

$$V(x) = \sum_G e^{iGx} V_G$$

where the sum is over the reciprocal lattice $G = 2\pi n/a$, and $V_G^* = V_{-G}$ assures that the potential $V(x)$ is real.

(a) Explain why for k near to a Brillouin zone boundary (such as k near π/a) the electron wavefunction should be taken to be

$$\psi = Ae^{ikx} + Be^{i(k+G)x} \tag{15.14}$$

where G is a reciprocal lattice vector such that $|k|$ is close to $|k+G|$.

(b) For an electron of mass m with k exactly at a zone boundary, use the above form of the wavefunction to show that the eigenenergies at this wavevector are

$$E = \frac{\hbar^2 k^2}{2m} + V_0 \pm |V_G|$$

where G is chosen so $|k| = |k+G|$.

▷ Give a qualitative explanation of why these two states are separated in energy by $2|V_G|$.

▷ Give a sketch (don't do a full calculation) of the energy as a function of k in both the extended and the reduced zone schemes.

(c) *Now consider k close to, but not exactly at, the zone boundary. Give an expression for the energy $E(k)$ correct to order $(\delta k)^2$ where δk is the wavevector difference from k to the zone boundary wavevector.

▷ Calculate the effective mass of an electron at this wavevector.

(15.2) **Periodic Functions**

Consider a lattice of points $\{\mathbf{R}\}$ and a function $\rho(\mathbf{x})$ which has the periodicity of the lattice $\rho(\mathbf{x}) = \rho(\mathbf{x}+\mathbf{R})$. Show that ρ can be written as

$$\rho(\mathbf{x}) = \sum_{\mathbf{G}} \rho_{\mathbf{G}}\, e^{i\mathbf{G}\cdot\mathbf{x}}$$

where the sum is over points $\mathbf{G}$ in the reciprocal lattice.

(15.3) **Tight Binding Bloch Wavefunctions**

Analogous to the wavefunction introduced in Chapter 11, consider a tight-binding wave ansatz of the form

$$|\psi\rangle = \sum_{\mathbf{R}} e^{i\mathbf{k}\cdot\mathbf{R}} |\mathbf{R}\rangle$$

where the sum is over the points $\mathbf{R}$ of a lattice, and $|\mathbf{R}\rangle$ is the ground-state wavefunction of an electron bound to a nucleus on site $\mathbf{R}$. In real space this ansatz can be expressed as

$$\psi(\mathbf{r}) = \sum_{\mathbf{R}} e^{i\mathbf{k}\cdot\mathbf{R}} \varphi(\mathbf{r}-\mathbf{R}).$$

Show that this wavefunction is of the form required by Bloch's theorem (i.e., show it is a modified plane wave).

(15.4) ***Nearly Free Electrons in Two Dimensions**

Consider the nearly free electron model for a square lattice with lattice constant a. Suppose the periodic potential is given by

$$\begin{aligned} V(x,y) &= 2V_{10}[\cos(2\pi x/a)+\cos(2\pi y/a)] \\ &+ 4V_{11}[\cos(2\pi x/a)\cos(2\pi y/a)] \end{aligned}$$

(a) Use the nearly free electron model to find the energies of states at wavevector $\mathbf{q} = (\pi/a, 0)$.

(b) Calculate the energies of the states at wavevector $\mathbf{q} = (\pi/a, \pi/a)$. (Hint: You should write down a 4 by 4 secular determinant, which looks difficult, but actually factors nicely. Make use of adding together rows or columns of the determinant before trying to evaluate it!)

(15.5) **Decaying Waves**

As we saw in this chapter, in one dimension, a periodic potential opens a band gap such that there are no plane-wave eigenstates between energies $\epsilon_0(G/2) - |V_G|$ and $\epsilon_0(G/2) + |V_G|$ with G a reciprocal lattice vector. However, at these forbidden energies, decaying (evanescent) waves still exist. Assume the form

$$\psi(x) = e^{ikx-\kappa x}$$

with $0 < \kappa \ll k$ and κ real. Find κ as a function of energy for $k = G/2$. For what range of V_G and E is your result valid?

(15.6) **Kronig–Penney Model*&**

Consider electrons of mass m in a so-called "delta-function comb" potential in one dimension

$$V(x) = aU\sum_n \delta(x - na)$$

(a) Argue using the Schroedinger equation that in-between delta functions, an eigenstate of energy E is always of a plane wave form $e^{iq_E x}$ with

$$q_E = \sqrt{2mE}/\hbar.$$

Using Bloch's theorem conclude that we can write an eigenstate with energy E as

$$\psi(x+a) = e^{ika}\psi(x)$$

where between delta functions ψ takes the form

$$\psi(x) = A\sin(q_E x) + B\cos(q_E x) \qquad 0 < x < a$$

and outside of this interval the function is defined by the Bloch periodicity.

(b) Using continuity of the wavefunction at $x = 0$ derive

$$B = e^{-ika}[A\sin(q_E a) + B\cos(q_E a)],$$

and using the Schroedinger equation to fix the discontinuity in slope at $x = 0$ derive

$$A - e^{-ika}k[A\cos(q_E a) - B\sin(q_E a)] = 2maUB/(q_E\hbar^2)$$

Solve these two equations to obtain

$$\cos(ka) = \cos(q_E a) + \frac{mUa}{\hbar^2 q_E}\sin(q_E a)$$

The left-hand side of this equation is always between -1 and 1, but the right-hand side is not. Conclude that there must be values of E for which there are no solutions of the Schroedinger equation—hence concluding there are gaps in the spectrum.

(c) For small values of the potential U show that this result agrees with the predictions of the nearly free electron model (i.e., determine the size of the gap at the zone boundary).

(15.7) **Zone Boundary Velocity*&**

(a) Show using the nearly-free electron model in one dimension that the group velocity of an electron goes to zero at the Brillouin zone boundary.

(b) Show the same result using a one-dimensional tight-binding model.

(c)** Extend this result to any number of dimensions. What if there is a more complicated electron hopping model, such as next-nearest neighbor hopping?

Insulator, Semiconductor, or Metal

16

In Chapter 11, when we discussed the tight-binding model in one dimension, we introduced some of the basic ideas of band structure. In Chapter 15 we found that an electron in a periodic potential shows exactly the same type of band-structure as we found for the tight-binding model. In both cases, we found that the spectrum is periodic in momentum (so all momenta can be taken to be in the first Brillouin zone, in reduced zone scheme) and that gaps open at Brillouin zone boundaries. These principles, the idea of bands and band structure, form the fundamental underpinning of our understanding of electrons in solids. In this chapter (and the next) we explore these ideas in further depth.

16.1 Energy Bands in One Dimension

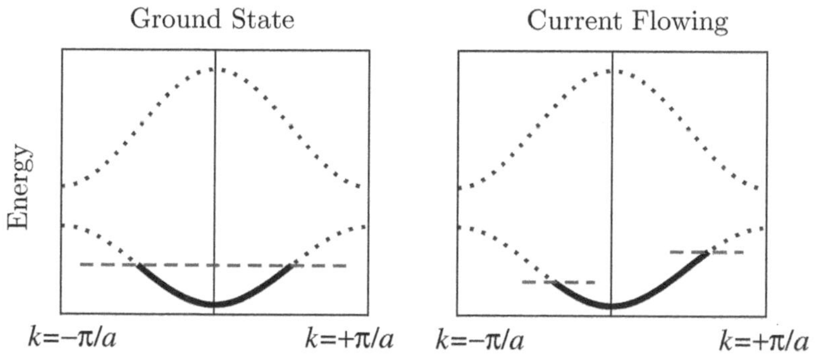

Fig. 16.1 Band diagrams of a one-dimensional monovalent chain with two orbitals per unit cell. **Left:** A band diagram with two bands is shown where each unit cell has one valence electron so that the lowest band is exactly half filled, and is therefore a metal. The filled states are thickly shaded, the chemical potential is the horizontal dashed line. **Right:** When electric field is applied, electrons accelerate, filling some of the k states to the right and emptying k-states to the left (in one dimension this can be thought of as having a different chemical potential on the left versus the right). Since there are an unequal number of left-moving versus right-moving electrons, the situation on the right represents net current flow.

As mentioned in Chapter 13, the number of k-states in a single Brillouin zone is equal to the number of primitive unit cells in the entire system. Thus, if each unit cell has exactly one free electron (i.e., is valence 1) there would be exactly enough electrons to fill the band *if there were only one spin state of the electron.* Being that there are two spin states of the electron, when each unit cell has only one valence electron, then the band is precisely half full. This is shown in the left of Fig. 16.1. Here, there is a Fermi surface where the unfilled states meet the filled states (in the figure, the Fermi energy is shown as a horizontal dashed line). When a band is partially filled, the electrons can repopulate when a small electric field is applied, allowing current to flow as shown in the right of Fig. 16.1. Thus, the partially filled band is a *metal.*

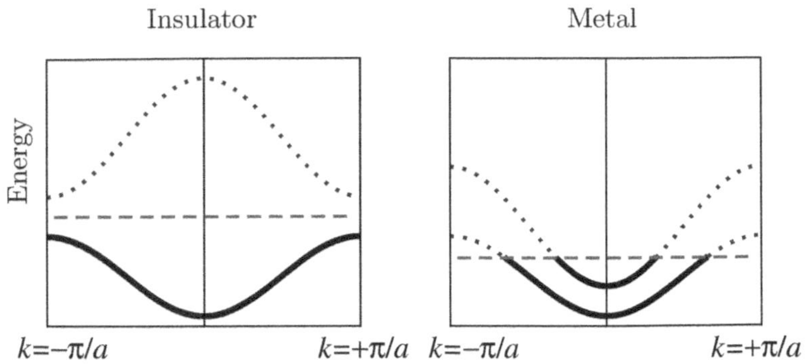

Fig. 16.2 Band diagrams of a one-dimensional divalent chain with two orbitals per unit cell. When there are two electrons per unit cell, then there are exactly enough electrons to fill the lowest band. In both pictures the chemical potential is drawn as the horizontal dashed line. **Left:** one possibility is that the lowest band (the valence band) is completely filled and there is a gap to the next band (the conduction band) in which case we get an insulator. This is a *direct* band gap as the valence band maximum and the conduction band minimum are both at the same crystal momentum (the zone boundary). **Right:** Another possibility is that the band energies overlap, in which case there are two bands, each of which is partially filled, giving a metal. If the bands were separated by more (imagine just increasing the vertical spacing between bands) we would have an insulator again, this time with an *indirect* band gap, since the valence band maximum is at the zone boundary while the conduction band minimum is at the zone center.

On the other hand, if there are two electrons per unit cell, then we have precisely enough electrons to fill one band. One possibility is shown on the left of Fig. 16.2—the entire lower band is filled and the upper band is empty, and there is a *band gap* between the two bands (note that the chemical potential is between the bands). When this is the situation, the lower (filled) band is known as the *valence band* and the upper (empty) band is known as the *conduction band*. In this situation the minimum energy excitation is created by moving an electron from the valence to the conduction band, which is non-zero energy. Because of this, at zero temperature, a sufficiently small electric perturbation will not create any excitations—the system does not respond at all to electric field.[1] Thus, systems of this type are known as (electrical) *insulators* (or more specifically *band insulators*). If the band gap is below about 4 eV, then these type of insulators are called *semiconductors*, since at room temperature a few electrons can be thermally excited into the conduction band, and these electrons then can move around freely, carrying some amount of current.

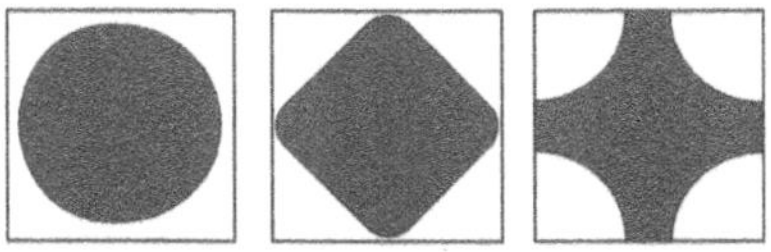

Fig. 16.3 Fermi sea of a square lattice of monovalent atoms in two dimensions as the strength of the periodic potential is varied. **Left:** In the absence of a periodic potential, the Fermi sea forms a circle whose area is precisely half that of the Brillouin zone (the black square). **Middle:** when a periodic potential is added, states closer to the zone boundary are pushed down in energy deforming the Fermi sea. **Right:** With strong periodic potential, the Fermi surface touches the Brillouin zone boundary. The Fermi surface remains continuous since the Brillouin zone is periodic. Note that the area of the Fermi sea remains fixed as the strength of the periodic potential is changed.

One might want to be aware that in the language of chemists, a band insulator is a situation where all of the electrons are tied up in bonds. For example, in diamond, carbon has valence four—meaning there are four electrons per atom in the outermost shell. In the diamond crystal, each carbon atom is covalently bonded to each of its four nearest neighbors, and each covalent bond requires two electrons. One electron is donated to each bond from each of the two atoms on either end of the bond—this completely accounts for all of the four electrons in each atom. Thus all of the electrons are tied up in bonds. This turns out to be equivalent to the statement that certain bonding bands are completely filled, and there is no mobility of electrons since there are no partially filled bands.

When there are two electrons per atom, one frequently obtains a band insulator as shown in the left of Fig. 16.2. However, another possibility is that the band energies overlap, as shown in the right of Fig. 16.2. In this case, although one has precisely the right number of electrons to fill a single band, instead one has two partially filled bands. As in Fig. 16.1 there are low-energy excitations available, and the system is metallic.

16.2 Energy Bands in Two and Three Dimensions

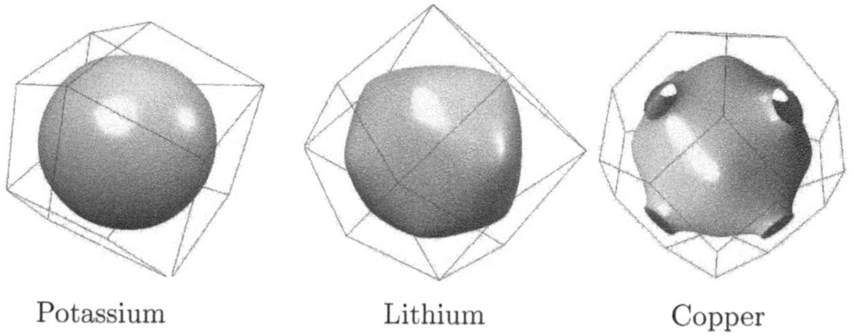

Fig. 16.4 Fermi surfaces of monovalent metals potassium (left), lithium (middle) and copper (right). **Left:** The potassium Fermi surface is almost exactly spherical, corresponding to nearly free electrons in very weak periodic potentials. **Middle:** The periodic potential for lithium is slightly stronger, and correspondingly the Fermi surface is distorted towards the zone boundaries. **Right:** The periodic potential is so strong in the case of copper that the Fermi surface intersects the Brillouin zone boundary. Compare these real Fermi surfaces to the cartoons of Fig. 16.3. The wire frames demark the first Brillouin zones, which are half filled. Potassium and lithium are bcc crystals whereas copper is fcc (thus copper has a different-shaped Brillouin zone from potassium and lithium). Figures reproduced with permission from http://www.phys.ufl.edu/fermisurface/.

It is useful to try to understand how the nearly-free electron model results in band structure in two dimensions. Let us consider a square lattice of monovalent atoms. The Brillouin zone is correspondingly square, and since there is one electron per atom, there should be enough electrons to half fill a single Brillouin zone. In the absence of a periodic potential, the Fermi sea forms a circular disc as shown in the left of Fig. 16.3. The area of this disc is precisely half the area of the zone. Now when a periodic potential is added, gaps open up at the zone boundaries. This means that states close to the zone boundary get moved down in energy—and the closer they are to the boundary, the more they get moved down. As a result, states close to the boundary get filled up preferentially at the expense of states further from the boundary. This deforms the Fermi surface[2] roughly as shown in the middle of Fig. 16.3. In either case, there are low-energy excitations possible and therefore the system is a metal.

If the periodic potential is strong enough the Fermi surface may even touch[3] the Brillouin zone boundary, as shown in the right of Fig. 16.3. Although this looks a bit strange, the Fermi surface remains perfectly continuous since the Brillouin zone is periodic in k-space. Thus if you walk off the right-hand side, you come back in the left!

The physics shown in Fig. 16.3 is quite similar to what is seen in many real materials. In Fig. 16.4 we show the Fermi surfaces for the monovalent metals potassium, lithium, and copper. Potassium is an almost ideal free electron metal, with an almost precisely spherical Fermi surface. For lithium, the Fermi surface is slightly distorted, bulging out near the zone boundaries. For copper, the Fermi surface is greatly distorted touching the Brillouin zone boundary in tubes. In all three cases, however, the Fermi seas fill exactly half the Brillouin zone volume since the metals are monovalent.

[1] The difference between filled bands and partially filled bands is dramatic. The room temperature electrical resistivity of diamond (a filled band insulator) and copper (a partially filled band metal) differ by 20 orders of magnitude!

[2] Recall that the Fermi surface is the locus of points at the Fermi energy (so all states at the Fermi surface have the same energy), separating the filled from unfilled states. Keep in mind that the area inside the Fermi surface is fixed by the total number of electrons in the system.

[3] Note that whenever a Fermi surface touches the Brillouin zone boundary, it must do so perpendicularly. This is due to the fact that the group velocity is zero at the zone boundary—i.e., the energy is quadratic as one approaches normal to the zone boundary. Since the energy is essentially not changing in the direction perpendicular to the zone boundary, the Fermi surface must intersect the zone boundary normally.

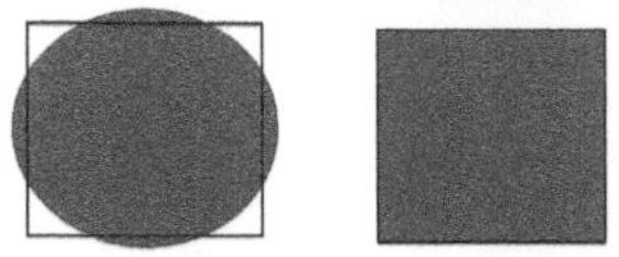

Fig. 16.5 Fermi sea of a square lattice of divalent atoms in two dimensions. **Left:** In the absence of a periodic potential, the Fermi sea forms a circle whose area is precisely that of the Brillouin zone (the black square). **Right:** When a *sufficiently strong* periodic potential is added, states inside the zone boundary are pushed down in energy so that all of these states are filled and no states outside the first Brillouin zone are filled. Since there is a gap at the zone boundary, this situation is an insulator. (Note that the area of the Fermi sea remains fixed.)

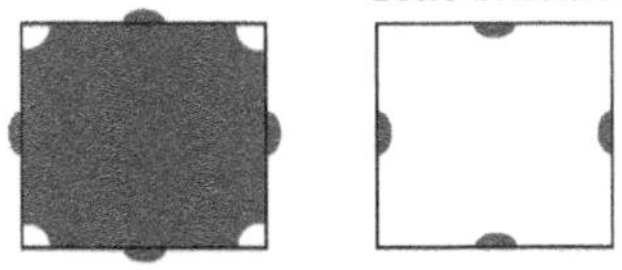

Fig. 16.6 Fermi sea of a square lattice of divalent atoms in two dimensions. **Left:** For intermediately strong periodic potential, although there are just enough electrons to completely fill the first zone, there are still some states filled in the second zone, and some states empty in the first zone, thus the system is still a metal. **Right:** The states in the second zone can be moved into the first zone by translation by a reciprocal lattice vector. This is the reduced zone scheme representation of the occupancy of the second Brillouin zone and we should think of these states as being in the second band.

Let us now consider the case of a two-dimensional square lattice of divalent atoms. In this case the number of electrons is precisely enough to fill a single zone. In the absence of a periodic potential, the Fermi surface is still circular, although it now crosses into the second Brillouin zone, as shown in the left of Fig. 16.5. Again, when a periodic potential is added a gap opens at the zone boundary—this gap opening pushes down the energy of all states within the first zone and pushes up energy of all states in the second zone. If the periodic potential is *sufficiently strong*,[4] then the states in the first zone are all lower in energy than states in the second zone. As a result, the Fermi sea will look like the right of Fig. 16.5. I.e., the entire lower band is filled, and the upper band is empty. Since there is a gap at the zone boundary, there are no low-energy excitations possible, and this system is an insulator.

It is worth considering what happens for intermediate strength of the periodic potential. Again, states outside of the first Brillouin zone are raised in energy and states inside the first Brillouin zone are lowered in energy. Therefore fewer states will be occupied in the second zone and more states occupied in the first zone. However, for intermediate strength of potential, some states will remain occupied in the second zone and some states will remain empty within the first zone as shown in Fig. 16.6. (In a reduced zone scheme we would say that there are some states filled in the second band and some empty in the first band.) This is precisely analogous to what happens in the right half of Fig. 16.2—there will still be some low-energy excitations available, and the system remains a metal. This physics is quite common in real materials. In Fig. 16.7 the Fermi surface of the divalent metal calcium is shown. As in the cartoon of Fig. 16.6 the Fermi surface intersects the Brillouin zone boundary; and although there are precisely enough electrons to completely fill the first band, instead the lowest two bands are each partially filled, and thus calcium is a metal.

We emphasize that in the case where there are many atoms per unit cell, we should count the *total valence* of all of the atoms in the unit cell put together to determine if it is possible to obtain a filled-band insulator. If the total valence of all the atoms in the unit cell is even, then for strong enough periodic potential, it is possible that some set of low-energy bands will be completely filled, there will be a gap, and the remaining bands will be empty, i.e., it will be a band insulator. However, if the periodic potential is not sufficiently strong, bands will overlap and the system will be a metal.

[4] We can estimate how strong the potential needs to be. We need to have the highest-energy state in the first Brillouin zone be lower energy than the lowest-energy state in the second zone. The highest-energy state in the first zone, in the absence of periodic potential, is in the zone corner and therefore has energy $\epsilon_{corner} = 2\hbar^2\pi^2/(2ma^2)$. The lowest-energy state in the second zone is in the middle of the zone boundary edge and in the absence of periodic potential has energy $\epsilon_{edge} = \hbar^2\pi^2/(2ma^2)$. Thus we need to open up a gap at the zone boundary which is sufficiently large that the edge becomes higher in energy than the corner. This requires roughly that $2V = \epsilon_{corner} - \epsilon_{edge}$. See Exercise 15.4 to obtain a more precise estimate.

16.3 Tight Binding

So far in this chapter we have described band structure in terms of the nearly free electron model. Similar results can be obtained starting from the opposite limit—the tight binding model introduced in Chapter 11. In this model we imagine some number of orbitals on each atom (or in each unit cell) and allow them to only weakly hop between orbitals. This spreads the eigen-energies of the atomic orbitals out into bands.

Writing down a two- (or three-)dimensional generalization of the tight binding Hamiltonian Eq. 11.4 is quite straightforward and is a good exercise to try (see for example Exercise 11.9). One only needs to allow each orbital to hop to neighbors in all available directions. The eigenvalue problem can then always be solved with a plane-wave ansatz analogous to Eq. 11.5. The solution of a tight binding model of atoms, each having a single atomic orbital, on a square lattice, is given by (compare Eq. 11.6)

$$E(k) = \epsilon_0 - 2t\cos(k_x a) - 2t\cos(k_y a). \tag{16.1}$$

Equi-energy contours for this expression are shown in Fig. 16.8. Note the similarity in the dispersion to our qualitative expectations shown in Fig. 16.3 (right) and Fig. 16.6, which were based on a nearly free electron picture.

In the above described tight binding picture, there is only a single band. However, one can make the situation more realistic by starting with several atomic orbitals per unit cell, to obtain several bands (another good exercise to try!). As mentioned in Section 6.2.2 and Chapter 11, as more and more orbitals are added to a tight binding (or LCAO) calculation, the results become increasingly accurate.

In the case where a unit cell is divalent it is crucial to determine whether bands overlap (i.e., is it insulating like the left of Fig. 16.2 or metallic type like the right of Fig. 16.2). This, of course, requires detailed knowledge of the band structure. In the tight binding picture, if the atomic orbitals start sufficiently far apart in energy, then small hopping between atoms cannot spread the bands enough to make them overlap (see Fig. 11.6). In the nearly free electron picture, the gap between bands formed at the Brillouin zone boundary is proportional to $|V_{\mathbf{G}}|$, and it is the limit of strong periodic potential that will guarantee that the bands do not overlap (see Fig. 16.5). Qualitatively these two are the same limit—very far from the idea of a freely propagating wave!

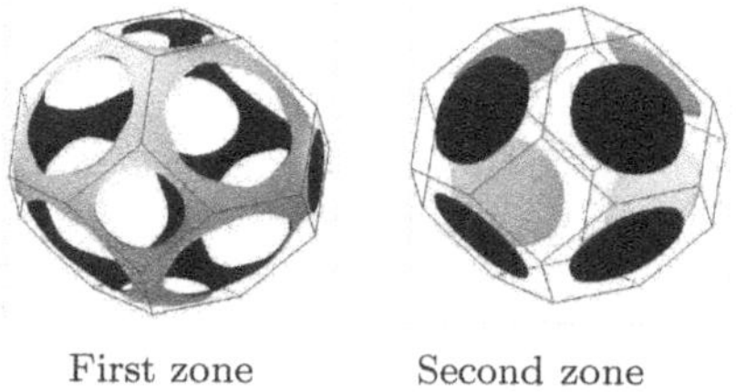

Fig. 16.7 Fermi surface of the (fcc) divalent metal calcium. **Left:** The first band is almost completely filled with electrons. The solid region is where the Fermi surface is inside the Brillouin zone boundary (the zone boundary is demarcated by the wire frame). **Right:** A few electrons fill small pockets in the second band. As in Fig. 16.6 there are just enough electrons to completely fill the lowest band, but there are states in the lowest band which are of higher energy than some states in the second band, so that a few states are empty in the first band (left), and a few states are filled in the second band (right). Figure reproduced by permission from www.physik.tu-dresden.de/~fermisur.

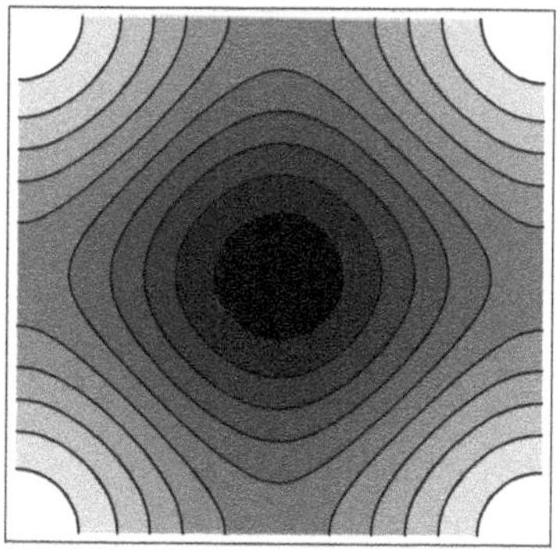

Fig. 16.8 Equi-energy contours for the dispersion of a tight binding model on a square lattice. This is a contour plot of Eq. 16.1. The first Brillouin zone is shown. Note that the contours intersect the Brillouin zone boundary normally.

16.4 Failures of the Band-Structure Picture of Metals and Insulators

The picture we have developed is that the band structure, and the filling of bands, determines whether a material is a metal or insulator (or semiconductor, meaning an insulator with a small band gap). One thing we might conclude at this point is that any system where the unit cell has a single valence electron (so the first Brillouin zone is half full) must be a metal. However, it turns out that this is not always true! The problem is that we have left out a very important effect—Coulomb interaction between electrons. We have so far completely ignored the Coulomb repulsion between electrons. Is this neglect justified at all? If we try to estimate how strong the Coulomb interaction is between electrons, (roughly $e^2/(4\pi\epsilon_0 r)$ where r is the typical distance between two electrons—i.e., the lattice constant a) we find numbers on the order of several eV. This number can be larger, or even far larger, than the Fermi energy (which is already a very large number, on the order of 10,000 K). Given this, it is hard to explain why it is at all justified to have thrown out such an important contribution. In fact, one might expect that neglecting this term would give complete nonsense! Fortunately, it turns out that in many cases it *is* justified to assume non-interacting electrons. The reason this works is actually quite subtle and was not understood until the 1950s due to the work of Lev Landau (See margin note 18 in Chapter 4 about Landau). This (rather deep) explanation, however, is beyond the scope of this book so we will not discuss it. Nonetheless, with this in mind it is perhaps not too surprising that there are cases where the non-interacting electron picture, and hence our view of band structure, fails.

Magnets

A case where the band picture of electrons fails is when the system is ferromagnetic.[5] We will discuss ferromagnetism in detail in Chapters 20–23, but in short this is where, due to interaction effects, the electron spins spontaneously align. From a kinetic energy point of view this seems unfavorable, since filling the lowest-energy eigenstates with both spin states can lower the Fermi energy compared to filling more states with only a single spin type. However, it turns out that aligning all of the spins can lower the Coulomb energy between the electrons, and thus our rules of non-interacting electron band theory no longer hold.

[5] Or antiferromagnetic or ferrimagnetic, for that matter. See Chapter 20 for definitions of these terms.

Mott Insulators

Another case where interaction physics is important is the so-called Mott insulator.[6] Consider a monovalent material. From band theory one might expect a half-filled lowest band, therefore a metal. But if one considers the limit where the electron–electron interaction is extremely strong, this is not what you get. Instead, since the electron–electron interaction is very strong, there is a huge penalty for two electrons to be on the same atom (even with opposite spins). As a result, the ground state is just one electron sitting on each atom. Since each atom has exactly one electron, no electron can move from its atom—since that would result in a double occupancy of the atom it lands on. As a result, this type of ground state is insulating. Arguably, this type of insulator—which can be thought of as more-or-less a traffic jam of electrons—is actually simpler to visualize than a band insulator! We will also discuss Mott insulators further in Sections 19.4 and particularly 23.2.

[6] Named after the English Nobel laureate, Nevill Mott. Classic examples of Mott insulators include NiO and CoO.

[7] Very weak processes can occur where, say, two photons together excite an electron.

16.5 Band Structure and Optical Properties

To the extent that electronic band structure is a good description of the properties of materials (and usually it is), one can attribute many of the optical properties of materials to this band structure.

16.5.1 Optical Properties of Insulators and Semiconductors

Band insulators cannot absorb photons which have energies less than their band-gap energy. The reason for this is that a single such photon does not have the energy to excite an electron from the valence band into the conduction band. Since the valence band is completely filled, the minimum energy excitation is of the band-gap energy—so a low-energy photon creates no excitations at all. As a result, these low-energy photons do not get absorbed by this material at all, and they simply pass right through the material.[7] Light absorption spectra are shown for three common semiconductors in Fig. 16.9. Note, for example, the strong drop in the absorption for GaAs for wavelengths greater than about .86 micron corresponding to photon energies less than the band-gap energy of 1.44 eV.

Let us now recall the properties of visible light, shown in Table 16.1. With this table in mind we see that if an insulator (or wide-bandgap semiconductor) has a band gap of greater than 3.2 eV, then it appears completely transparent since it cannot absorb any wavelength of visible light. Materials such as quartz, diamond, aluminum oxide, and so forth are insulators of this type.

Semiconductors with somewhat smaller band gaps will absorb photons with energies above the band gap (exciting electrons from the valence to the conduction band), but will be transparent to photons below this band gap. For example, cadmium sulfide (CdS) is a semiconductor with a band gap of roughly 2.6 eV, so that violet and blue light are absorbed but red and green light are transmitted. As a result this material looks reddish.[8] Semiconductors with very small band gaps (such as GaAs, Si, or Ge) look black, since they absorb all frequencies of visible light!

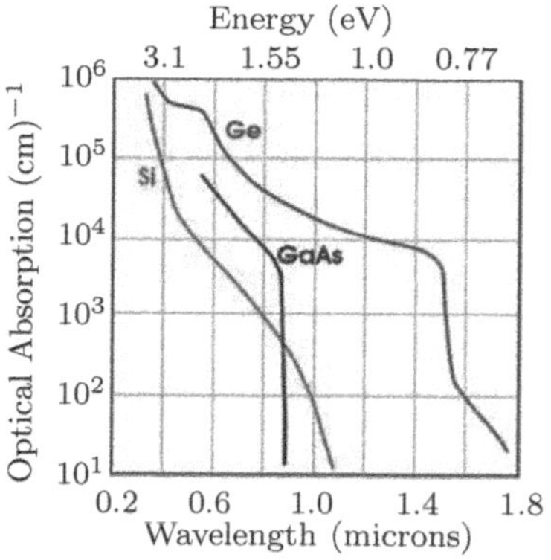

Fig. 16.9 Optical absorption of several semiconductors (Si, Ge, GaAs) as a function of photon wavelength or energy. For energies below the band-gap, the absorption is extremely small. (Note the absorption is on a log scale!) GaAs has a (direct) gap energy 1.44 eV where the absorption drops very rapidly. For Ge the drop at 0.8 eV reflects the direct band gap energy. Note however there is still some weak absorption at longer wavelengths due to the smaller energy indirect band gap. Si has a somewhat more complicated band structure with many indirect transitions and direct gap of 3.4 eV. (Figure kindly provided by David Miller, Stanford University.)

Table 16.1 Colors corresponding to photon energies.

Color	$\hbar\omega$
Infrared	< 1.65 eV
Red	~ 1.8 eV
Orange	~ 2.05 eV
Yellow	~ 2.15 eV
Green	~ 2.3 eV
Blue	~ 2.7 eV
Violet	~ 3.1 eV
Ultraviolet	> 3.2 eV

16.5.2 Direct and Indirect Transitions

While the band gap determines the minimum energy excitation that can be made in an insulator (or semiconductor), this is not the complete story in determining whether or not a photon can be absorbed by a material. It turns out to matter quite a bit at which values of **k** the maximum of the valence band and the minimum of the conduction band lies. If the value of **k** for the valence band maximum is the same as the value of **k** for the conduction band minimum, then we say that it is a *direct* band gap. If the values of **k** differ, then we say that it is an *indirect* band gap. For example, the system shown on the left of Fig. 16.2

[8]Colors of materials can be quite a bit more complicated than this simple picture, as when a color is absorbed one often needs to look at details to find out how strongly it is absorbed!

is a direct band gap, where both the valence band maximum and the conduction band minimum are at the zone boundary. In comparison, if the band shapes were as in the right of Fig. 16.2, but the band gap were large enough such that it would be an insulator (just imagine the bands separated by more), this would be an indirect band gap since the valence band maximum is at the zone boundary, but the conduction band minimum is at $\mathbf{k} = \mathbf{0}$.

One can also have both indirect and direct band gaps in the same material, as shown in Fig. 16.10. In this figure, the minimum energy excitation is the *indirect transition*—meaning an excitation of an electron across an indirect band gap, or equivalently a transition of non-zero crystal momentum[9] where the electron is excited from the top of the valence band to the bottom of the conduction band at a very different $\mathbf{k}$. While this may be the lowest energy excitation that can occur, it is very hard for this type of excitation to result from exposure of the system to light—the reason for this is energy–momentum conservation. If a photon is absorbed, the system absorbs both the energy and the momentum of the photon. But given an energy E in the eV range, the momentum of the photon $\hbar|k| = E/c$ is extremely small, because c is so large. Thus the system cannot conserve momentum while exciting an electron across an indirect band gap. Nonetheless, typically if a system like this is exposed to photons with energy greater than the indirect band gap a small number of electrons will manage to get excited—usually by some complicated process including absorption of a photon exciting an electron with simultaneous emission of a phonon[10] to arrange the conservation of energy and momentum. In comparison, if a system has a direct band gap, and is exposed to photons of energy greater than this direct band gap, then it strongly absorbs these photons while exciting electrons from the valence band to the conduction band.

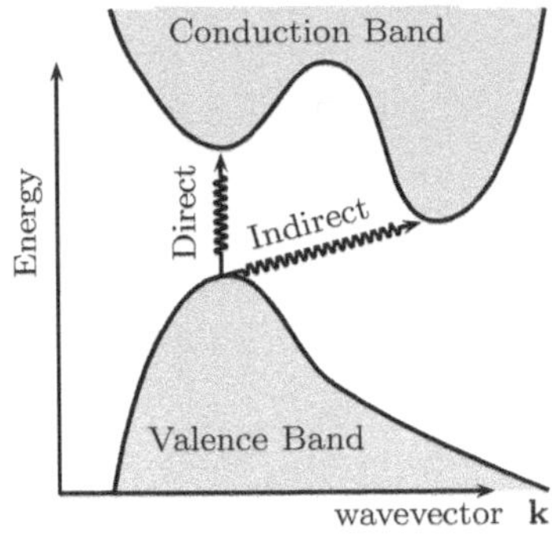

Fig. 16.10 Direct and indirect transitions. While the indirect transition is lower energy, it is hard for a photon to excite an electron across an indirect band gap because photons carry very little momentum (since the speed of light, c, is large).

[9] By "non-zero" we mean, substantially non-zero—like a fraction of the Brillouin zone.

[10] Another way to satisfy the conservation of momentum is via a "disorder assisted" process. Recall that the reason we conserve crystal momentum is because the system is perfectly periodic. If the system has some disorder, and is therefore not perfectly periodic, then crystal momentum is not perfectly conserved. Thus the greater the disorder level, the less crystal momentum needs to be conserved and the easier it is to make a transition across an indirect band gap.

In Fig. 16.9, the spectrum of Ge shows weak optical absorption for energies lower than its direct band gap of 0.8 eV—this is due to excitations across its lower-energy indirect band gap.

16.5.3 Optical Properties of Metals

The optical properties of metals are a bit more complicated than that of insulators. Since metals are very conductive, photons (which are electromagnetic) excite the electrons,[11] which then re-emit light. This re-emission (or reflection) of light is why metals look shiny. Noble metals (gold, silver, platinum) look particularly shiny because their surfaces do not form insulating oxides when exposed to air, which many metals (such as sodium) do within seconds.

[11] Note the contrast with insulators—when an electron is excited above the band gap, since the conductivity is somewhat low, the electron does not re-emit quickly, and the material mostly just absorbs the given wavelength.

Even amongst metals (ignoring possible oxide surfaces), colors vary. For example, silver looks brighter than gold and copper, which look yellow or orange-ish. This again is a result of the band structure of these materials. All of the noble metals have valence 1 meaning that a band should be half filled. However, the total energy width of the conduction band is greater for silver than it is for gold or copper (in tight-binding language t is larger for silver; see Chapter 11). This means that higher-energy electronic transitions within the band are much more possible for silver than they are for gold and copper. For copper and gold, photons with blue and violet colors are not well absorbed and re-emitted, leaving these material looking a bit more yellow and orange. For silver on the other hand, all visible colors are re-emitted well, resulting in a more perfect (or "white") mirror. While this discussion of the optical properties of metals is highly over-simplified,[12] it captures the correct essence—that the details of the band structure determine which color photons are easily absorbed and/or reflected, and this in turn determines the apparent color of the material.

[12] Really there are many bands overlapping in these materials, and the full story addresses inter- and intra-band transitions.

16.5.4 Optical Effects of Impurities

It turns out that small levels of impurities put into periodic crystals (particularly into semiconductors and insulators) can have dramatic effects on many of their optical (as well as electrical!) properties. For example, one nitrogen impurity per million carbon atoms in a diamond crystal gives the crystal a yellowish color. One boron atom per million carbon atoms give the diamond a blueish color.[13] We will discuss the physics that causes this in Section 17.2.1.

[13] Natural blue diamonds are extremely highly prized and are very expensive. Possibly the world's most famous diamond, the Hope Diamond, is of this type (it is also supposed to be cursed, but that is another story). With modern crystal growth techniques, in fact it is possible to produce man-made diamonds of "quality" better than those that are mined. Impurities can be placed in as desired to give the diamond any color you like. Due to the powerful lobby of the diamond industry, most synthetic diamonds are labeled as such—so although you might feel cheap wearing a synthetic, in fact, you probably own a better product than those that have come out of the earth! (Also you can rest with a clean conscience that the production of your diamond did not finance any wars in Africa.)

Chapter Summary

- A material is a metal if it has low-energy excitations. This happens when at least one band is partially full. (Band) insulators and semiconductors have only filled bands and empty bands and have a gap for excitations.
- A semiconductor is a (band) insulator with a small band gap.
- The valence of a material determines the number of carriers being put into the band—and hence can determine if one has a metal or insulator/semiconductor. However, if bands overlap (and frequently they do) one might not be able to fill the bands to a point where there is a gap.
- The gap between bands is determined by the strength of the periodic potential. If the periodic potential is strong enough (the atomic limit in tight binding language), bands will not overlap.
- The band picture of materials fails to account for electron–electron interaction. It cannot describe (at least without modification) interaction-driven physics such as magnetism and Mott insulators.
- Optical properties of solids depend crucially on the possible energies of electronic transitions. Photons easily create transitions with low momentum, but cannot create transitions with larger momentum easily. Optical excitations over an indirect (finite momentum) gap are therefore weak.

References

- Goodstein, section 3.6c
- Kittel, chapter 7; first section of chapter 8; first section of chapter 9
- Burns, sections 10.7, 10.10
- Hook and Hall, sections 4.2,4.3, 5.4
- Rosenberg, sections 8.9–8.19
- Singleton, chapters 3–4

Exercises

(16.1) **Metals and Insulators**

Explain the following:

(a) sodium, which has two atoms in a bcc (conventional cubic) unit cell, is a metal;

(b) calcium, which has four atoms in a fcc (conventional cubic) unit cell, is a metal;

(c) diamond, which has eight atoms in a fcc (conventional cubic) unit cell with a basis, is an electrical insulator, whereas silicon and germanium, which have similar structures, are semiconductors. (Try to think up several possible reasons!)

▷ Why is diamond transparent?

(16.2) **Fermi Surface Shapes**

(a) Consider a tight binding model of atoms on a (two-dimensional) square lattice where each atom has a single atomic orbital and electrons can hop to their four nearest neighbors with hopping amplitude t. If these atoms are monovalent, describe the shape of the Fermi surface.

(b) Now suppose the lattice is not square, but is rectangular instead with primitive lattice vectors of length a_x and a_y in the x and y directions respectively, where $a_x > a_y$. In this case, imagine that the hoppings have a value t_x in the x-direction and a value t_y in the y-direction, with $|t_y| > |t_x|$. (Why does this inequality match $a_x > a_y$?)

▷ Write an expression for the dispersion of the electronic states $\epsilon(\mathbf{k})$.

▷ Suppose again that the atoms are monovalent, what is the shape of the Fermi surface now?

(16.3) **More Fermi Surface Shapes***

Consider a divalent atom, such as Ca or Sr, that forms an fcc lattice (with a single atom basis). In the absence of a periodic potential, would the Fermi surface touch the Brillouin zone boundary? What fraction of the states in the first Brillouin zone remain empty?

(16.4) **Zone Boundary* &**

Assuming the group velocity of an electron goes to zero at the zone boundary (See exercise 15.7) show that the Fermi surface should intersect the Brillouin zone boundary perpendicularly. When does this result not hold? (Hint: Consider Exercise 16.2.a).

(16.5) **Yet More Fermi Surface Shapes* &**

Consider a tight binding model of atoms on a (two-dimensional) square lattice where each atom has a single atomic orbital as in Exercise 16.2.a. Imagine that in addition to the hopping amplitude t to the four nearest neighbor sites, there is also a hopping amplitude t' to the next nearest neighbor sites (diagonal hopping).

(a) Calculate the dispersion $\epsilon(\mathbf{k})$.

(b) Assuming the atoms are monovalent, and assuming $t = 1$eV and $t' = .1$eV, sketch the shape of the Fermi surface.

(c) Using $t = 1$eV and $t' = -.1$eV, confirm the results of Exercise 16.4.

Semiconductor Physics

17

17.1 Electrons and Holes

Suppose we start with an insulator or semiconductor and we excite one electron from the valence band to the conduction band, as shown in the left of Fig. 17.1. This excitation may be due to absorbing a photon, or it might be a thermal excitation. (For simplicity in the figure we have shown a direct band gap. For generality we have not assumed that the curvature of the two bands are the same.) When the electron has been moved up to the conduction band, there is an absence of an electron in the valence band known as a *hole*. Since a completely filled band is inert, it is very convenient to only keep track of the few holes in the valence band (assuming there are only a few) and to treat these holes as individual elementary particles. The electron can fall back into the empty state that is the hole, emitting energy (a photon, say) and "annihilating" both the electron from the conduction band and the hole from the valence band.[1] Note that while the electrical charge of an electron is negative the electrical charge of a hole (the absence of an electron) is positive—equal and opposite to that of the electron.[2]

[1]This is equivalent to pair annihilation of an electron with a positron. In fact, the analogy between electron–hole and electron–positron is fairly precise. As soon as Dirac constructed his equation (in 1928) describing the relativistic motion of electrons, and predicting positrons, it was understood that the positron could be thought of as an absence of an electron in a filled sea of states. The filled sea of electron states with a gap to exciting electron–positron pairs is the inert vacuum, which is analogous to an inert filled valence band.

[2]If this does not make intuitive sense consider the process of creating an electron–hole pair as described in Fig. 17.1. Initially (without the excited electron–hole pair) the system is charge neutral. We excite the system with a photon to create the pair, and we have not moved any additional net charge into the system. Thus if the electron is negative, the hole must be positive to preserve overall charge neutrality.

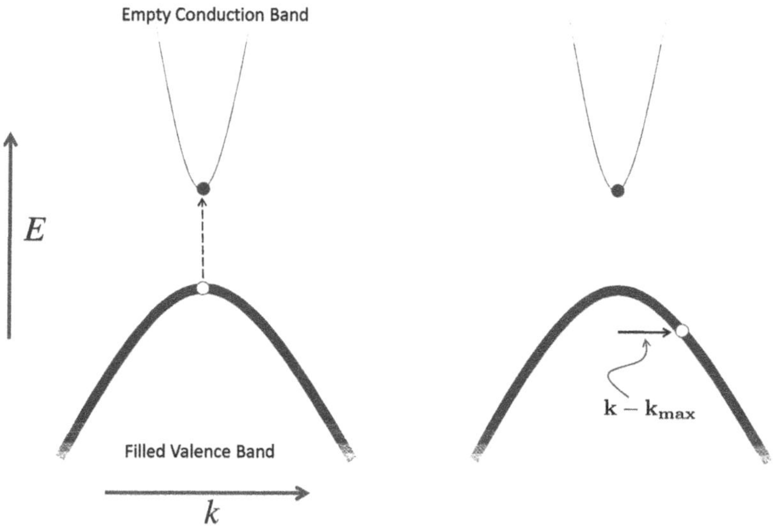

Fig. 17.1 Electrons and holes in a semiconductor. **Left:** A single hole in the valence band and a single electron in the conduction band. **Right:** Moving the hole to a momentum away from the top of the valence band costs *positive* energy—like pushing a balloon under water. As such, the effective mass of the hole is defined to be positive. The energy of the configuration on the right is greater than that on the left by $E = \hbar^2|\mathbf{k}-\mathbf{k_{max}}|^2/(2m^*)$

Effective Mass of Electrons

As mentioned in Sections 11.2 and 15.1.1, it is useful to describe the curvature at the bottom of a band in terms of an effective mass. Let us assume that near the bottom of the conduction band (assumed to be at $\mathbf{k} = \mathbf{k_{min}}$) the energy is given by[3,4,5,6]

$$E = E_{min} + \alpha|\mathbf{k} - \mathbf{k_{min}}|^2 + \ldots$$

with $\alpha > 0$, where the dots mean higher-order term in the deviation from $\mathbf{k_{min}}$. We then *define* the *effective mass* to be given by

$$\frac{\hbar^2}{m^*} = \frac{\partial^2 E}{\partial k^2} = 2\alpha \qquad (17.1)$$

at the bottom of the band (with the derivative being taken in any direction for an isotropic system). Correspondingly, the (group) velocity is given by[7]

$$\mathbf{v} = \frac{\nabla_{\mathbf{k}} E}{\hbar} = \frac{\hbar(\mathbf{k} - \mathbf{k_{min}})}{m^*} \qquad (17.2)$$

This definition is chosen to be in analogy with the free electron behavior $E = \hbar^2|\mathbf{k}|^2/(2m)$ with corresponding velocity $\mathbf{v} = \nabla_{\mathbf{k}} E/\hbar = \hbar\mathbf{k}/m$.

Effective Mass of Holes

Analogously we can define an effective mass for holes. Here things get a bit more complicated. For the top of the valence band, the energy dispersion for electrons would be

$$E = E_{\rm max} - \alpha|\mathbf{k} - \mathbf{k_{max}}|^2 + \ldots \qquad (17.3)$$

with $\alpha > 0$. The modern convention is to *define the effective mass for holes at the top of a valence band to be always positive*[8]

$$\frac{\hbar^2}{m^*_{\rm hole}} = -\frac{\partial^2 E}{\partial k^2} = 2\alpha. \qquad (17.4)$$

The convention of the effective mass being positive makes sense because the energy to boost the hole from zero velocity ($\mathbf{k} = \mathbf{k_{max}}$ at the top of the valence band) to finite velocity is positive. This energy is naturally given by

$$E_{\rm hole} = \text{constant} + \frac{\hbar^2|\mathbf{k} - \mathbf{k_{max}}|^2}{2m^*_{\rm hole}}.$$

The fact that boosting the hole away from the top of the valence band is positive energy may seem a bit counter-intuitive being that the dispersion of the hole band is an upside-down parabola. However, one should think of this as being like pushing a balloon under water. The lowest energy configuration is with the *electrons* at the lowest energy possible and the hole at the highest energy possible. So pushing the hole under the electrons costs positive energy. (This is depicted in the right-hand side of Fig. 17.1.) A good way to handle this bookkeeping is to remember

$$E(\text{absence of electron in state } \mathbf{k}) = -E(\text{electron in state } \mathbf{k}). \qquad (17.5)$$

[3] It is an important principle that near a minimum or a maximum one can always expand and get something quadratic plus higher order corrections.

[4] For simplicity we have assumed the system to be isotropic. In the more general case we would have

$$\begin{aligned} E = E_{min} &+ \alpha_x(k_x - k_x^{min})^2 \\ &+ \alpha_y(k_y - k_y^{min})^2 \\ &+ \alpha_z(k_z - k_z^{min})^2 + \ldots \end{aligned}$$

for some orthogonal set of axes (the "principal axes") x, y, z. In this case we would have an effective mass which can be different in the three different principal directions.

[5] For simplicity we also neglect the spin of the electron here. In general, spin–orbit coupling can make the dispersion depend on the spin state of the electron. Among other things, this can modify the effective electron g-factor.

[6] It often occurs that the bottom of conduction band has more than one minimum at different points $\mathbf{k}^{(n)}_{\mathbf{min}}$ in the Brillouin zone with exactly the same energy. We then say that there are multiple "valleys" in the band structure. Such a situation occurs due to the symmetry of the crystal. For example, in silicon (an fcc structure with a basis, see Fig. 12.21), six conduction band minima with the same energy occur approximately at the k-points $(\pm 5.3/a, 0, 0)$, $(0, \pm 5.3/a, 0)$, and $(0, 0, \pm 5.3/a)$.

[7] More accurately, $\mathbf{v} = \nabla_{\mathbf{k}} E(\mathbf{k})/\hbar + \mathbf{K}$ where the additional term $\mathbf{K}$ is known as the "Karplus–Luttinger" anomalous velocity and is proportional to applied electric field. This correction, resulting from subtle quantum-mechanical effects, is almost always neglected in solid state texts and rarely causes trouble (this is related to margin note 9 in Chapter 11). Only recently has research focused more on systems where such terms do matter. Proper treatment of this effect is beyond the scope of this book.

[8] Be warned: a few books define the mass of holes to be negative. This is a bit annoying but not inconsistent as long as the negative sign shows up somewhere else as well!

The momentum and velocity of a hole

There is a bit of complication with signs in keeping track of the momentum of a hole. If an electron is added to a band in a state $\mathbf{k}$ then the crystal momentum contained in the band increases by $\hbar\mathbf{k}$. Likewise, if an electron in state $\mathbf{k}$ is *removed* from an otherwise filled band, then the crystal momentum in the band must decrease by $\hbar\mathbf{k}$. Then, since a fully filled band carries no net crystal momentum the absence of an electron in state $\mathbf{k}$ should be a hole whose crystal momentum is $-\hbar\mathbf{k}$. It is thus convenient to *define* the wavevector $\mathbf{k}_{\text{hole}}$ of a hole to be the negative of the wavevector $\mathbf{k}_{\text{electron}}$ of the corresponding absent electron.[9]

[9] Other conventions are possible but this is probably the simplest.

This definition of wavevector is quite sensible when we try to calculate the group velocity of a hole. Analogous to the electron, we write the hole group velocity as the derivative of the hole energy

$$\mathbf{v}_{\text{hole}} = \frac{\nabla_{\mathbf{k}_{\text{hole}}} E_{\text{hole}}}{\hbar} \tag{17.6}$$

Now, using Eq. 17.5, and also the fact that that the wavevector of the hole is minus the wavevector of the missing electron, we get two canceling minus signs and we find that

$$\mathbf{v}_{\text{hole}} = \mathbf{v}_{\text{missing electron}}.$$

This is a rather fundamental principle. The time evolution of a quantum state is independent of whether that state is occupied with a particle or not!

Effective Mass and Equations of Motion

We have defined the effective masses above in analogy with that of free electrons, by looking at the curvature of the dispersion. An equivalent definition (equivalent at least at the top or bottom of the band) is to define the effective mass m^* as being the quantity that satisfies Newton's second law, $F = m^* a$ for the particle in question. To demonstrate this, our strategy is to imagine applying a force to an electron in the system and then equate the work done on the electron to its change in energy. Let us start with an electron in momentum state $\mathbf{k}$. Its group velocity is $\mathbf{v} = \nabla_k E(\mathbf{k})/\hbar$. If we apply a force,[10] the work done per unit time is

[10] For example, if we apply an electric field $\mathbf{E}$ and it acts on an electron of charge $-e$, the force is $\mathbf{F} = -e\mathbf{E}$.

$$dW/dt = \mathbf{F} \cdot \mathbf{v} = \mathbf{F} \cdot \nabla_k E(\mathbf{k})/\hbar.$$

On the other hand, the change in energy per unit time must also be (by the chain rule)

$$dE/dt = d\mathbf{k}/dt \cdot \nabla_k E(\mathbf{k}).$$

Setting these two expressions equal to each other we (unsurprisingly) obtain Newton's equation

$$\mathbf{F} = \hbar \frac{d\mathbf{k}}{dt} = \frac{d\mathbf{p}}{dt} \tag{17.7}$$

where we have used $\mathbf{p} = \hbar\mathbf{k}$.

If we now consider electrons near the bottom of a band, we can plug in the expression Eq. 17.2 for the velocity, and this becomes

$$\mathbf{F} = m^* \frac{d\mathbf{v}}{dt}$$

exactly as Newton would have expected. In deriving this result recall that we have assumed that we are considering an electron near the bottom of a band so that we can expand the dispersion quadratically (or similarly we assumed that holes are near the top of a band). One might wonder how we should understand electrons when they are neither near the top nor the bottom of a band. More generally Eq. 17.7 always holds, as does the fact that the group velocity is $\mathbf{v} = \nabla_k E/\hbar$. It is then sometimes convenient to define an effective mass for an electron as a function of momentum to be given by[11]

[11] For simplicity we write this in its one-dimensional form.

$$\frac{\hbar^2}{m^*(k)} = \frac{\partial^2 E}{\partial k^2}$$

which agrees with our definition (Eq. 17.1) near the bottom of band. However, near the top of a band it is the *negative* of the corresponding hole mass (note the sign in Eq. 17.4). Note also that somewhere in the middle of the band the dispersion must reach an inflection point ($\partial^2 E/\partial k^2 = 0$), whereupon the effective mass actually becomes infinite as it changes sign.

Aside: It is useful to compare the time evolution of electrons and holes near the top of bands. If we think in terms of holes (the natural thing to do near the top of a band) we have $\mathbf{F} = +e\mathbf{E}$ and the holes have a positive mass. However, if we think in terms of electrons, we have $\mathbf{F} = -e\mathbf{E}$ but the mass is negative. Either way, the acceleration of the **k**-state is the same, whether we are describing the dynamics in terms of an electron in the state or in terms of a hole in the state. As mentioned below Eq. 17.6, this equivalence is expected, since the time evolution of an eigenstate is independent of whether that eigenstate is filled with an electron or not.

17.1.1 Drude Transport: Redux

Back in Chapter 3 we studied Drude theory—a simple kinetic theory of electron motion. The main failure of Drude theory was that it did not treat the Pauli exclusion principle properly: it neglected the fact that in metals the high density of electrons makes the Fermi energy extremely high. However, in semiconductors or band insulators, when only a few electrons are in the conduction band and/or only a few holes are in the valence band, then we can consider this to be a low-density situation, and to a very good approximation, we can ignore Fermi statistics. (For example, if only a single electron is excited into the conduction band, then we can completely ignore the Pauli principle, since it is the only electron around—there is no chance that any state it wants to sit in will already be filled!) As a result, when there is a low density of conduction electrons or valence holes, it turns out that Drude theory works extremely well! We will come back to this issue later in Section 17.3, and make this statement much more precise.

At any rate, in the semiclassical picture, we can write a simple Drude transport equation (really Newton's equations!) for electrons in the conduction band

$$m_e^* \, d\mathbf{v}/dt = -e(\mathbf{E} + \mathbf{v} \times \mathbf{B}) - m_e^* \, \mathbf{v}/\tau$$

with m_e^* the electron effective mass. Here the first term on the right-hand side is the Lorentz force on the electron, and the second term is a drag force with an appropriate scattering time τ. The scattering time determines the so-called *mobility* μ which measures the ease with which the particle moves. The mobility is generally defined as the ratio of the velocity to the electric field.[12] In this Drude approach we then obtain

$$\mu = |\mathbf{v}|/|\mathbf{E}| = |e\tau/m^*|.$$

[12] Mobility is defined to be positive for both electrons and holes.

Similarly, we can write equations of motion for holes in the valence band

$$m_h^* \, d\mathbf{v}/dt = e(\mathbf{E} + \mathbf{v} \times \mathbf{B}) - m_h^* \, \mathbf{v}/\tau$$

where m_h^* is the hole effective mass. Note again that here the charge on the hole is *positive*. This should make sense—the electric field pulls on an electron in a direction opposite to the direction that it pulls on the absence of an electron!

If we think back all the way to Chapters 3 and 4, one of the physical puzzles that we could not understand is why the Hall coefficient sometimes changes sign (see Table 3.1). In some cases it looked as if the charge carrier had positive charge. Now we understand why this is true. In some materials the main charge carrier is the hole!

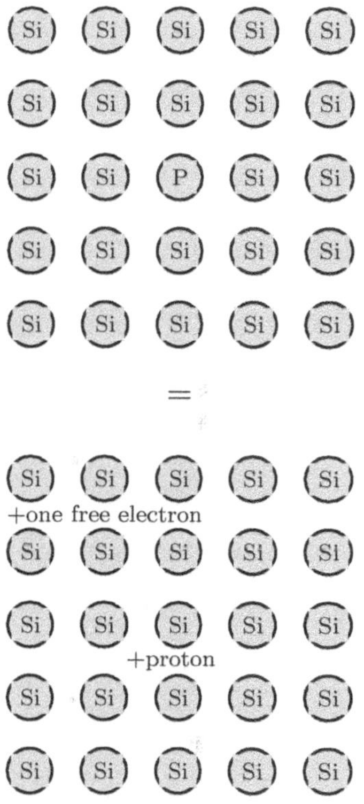

Fig. 17.2 Cartoon of doping a semiconductor. Doping Si with P adds one free electron to wander freely in the conduction band and leaves behind a positive charge on the nucleus.

17.2 Adding Electrons or Holes with Impurities: Doping

In a pure band insulator or semiconductor, if we excite electrons from the valence to the conduction band (either with photons or thermally) we can be assured that the density of electrons in the conduction band (typically called n, which stands for "negative" charges) is precisely equal to the density of holes left behind in the valence band (typically called p, which stands for "positive" charges). However, in an impure semiconductor or band insulator this is not the case.

Without impurities, a semiconductor is known as *intrinsic*. The opposite of intrinsic, the case where there are impurities present, is sometimes known as *extrinsic*.

Let us now examine the extrinsic case more carefully. Consider for example, silicon (Si), which is a semiconductor with a band gap of about 1.1 eV. Now imagine that a phosphorus (P) atom replaces one of the Si atoms in the lattice as shown on the top of Fig. 17.2. This P atom, being directly to the right of Si on the periodic table, can be thought of as nothing more than a Si atom plus an extra proton and an extra electron,[13] as shown in the bottom of Fig. 17.2. Since the valence band is already filled this additional electron must go into the conduction band. The P atom is known as a *donor* (or *electron donor*) in silicon since it donates an electron to the conduction band. It is also sometimes known as an *n-dopant*,[14] since n is the symbol for the density of electrons in the conduction band.

Analogously, we can consider aluminum, the element directly to the left of Si on the periodic table. In this case, the aluminum dopant provides one fewer electron than Si, so there will be one electron missing from the valence band. In this case Al is known as an electron *acceptor*, or equivalently as a *p-dopant*, since p is the symbol for the density of holes.[15,16]

In a more chemistry-oriented language, we can depict the donors and acceptors as shown in Fig. 17.3. In the intrinsic case, all of the electrons are tied up in covalent bonds of two electrons. With the n-dopant, there is an extra unbound electron, whereas with the p-dopant there is an extra unbound hole (one electron too few).

[13]There are extra neutrons as well, but they don't do much in this context.

[14]"Dopant" generally means a chemical inserted into an object to alter its properties. This definition is true more broadly than the field of physics (e.g. Lance Armstrong, Jerry Garcia).

[15]Yes, it is annoying that the common dopant phosphorus has the chemical symbol P, but it is not a p-dopant in Si, it is an n-dopant.

[16]More frequently than Al, boron (B) is used as a p-dopant in Si. Since B lies just above Al in the periodic table, it plays the same chemical role.

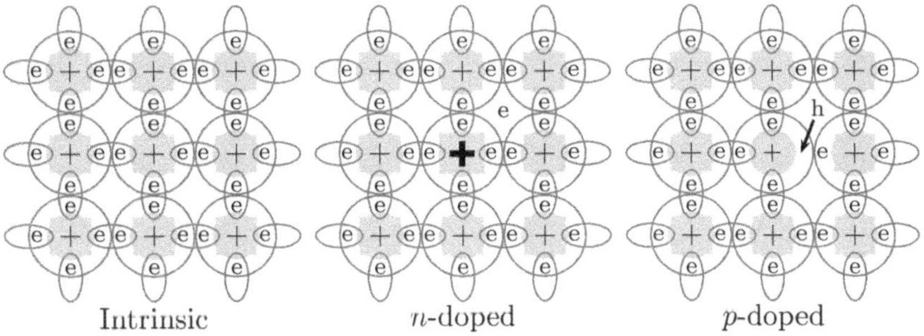

Fig.17.3 Cartoon of doping a semiconductor. **Left:** In the intrinsic case, all of the electrons are tied up in covalent bonds of two electrons. **Middle:** In the n-dopant case there is an extra unbound electron, and the dopant carries an extra nuclear charge. **Right:** In the p-dopant case there is one electron too few to complete all the bonds so there is an extra hole (denoted h) and the nuclear charge has one less positive charge than in the intrinsic case (the + sign is supposed to look slightly less large).

17.2.1 Impurity States

Let us consider even more carefully what happens when we add dopants. For definiteness let us consider adding an n-dopant such as P to a semiconductor such as Si. Once we add a single n-dopant to an otherwise intrinsic sample of Si, we get a single electron above the gap in the conduction band. This electron behaves like a free particle with mass m_e^*. However, in addition, we have a single extra positive charge $+e$ at some point in the crystal due to the P nucleus. The free electron is attracted back to this positive charge and forms a bound state that is similar to a hydrogen atom. There are two main differences between a real hydrogen atom and this bound state of an electron in the conduction band and the impurity nucleus. First of all, the electron has effective mass m_e^* which can be very different from the real (bare) mass of the electron (and is typically smaller than the bare mass of the electron). Secondly, instead of the two charges attracting each other with a potential $V = e^2/(4\pi\epsilon_0 r)$ they attract each other with a potential $V = e^2/(4\pi\epsilon_r\epsilon_0 r)$, where ϵ_r is the relative permittivity (or relative dielectric constant) of the material.

With these two small differences we can calculate the energies of the hydrogenic bound states exactly as we do for genuine hydrogen in our quantum mechanics courses.

We recall that the energy eigenstates of the hydrogen atom are given by $E_n^{H-atom} = -\text{Ry}/n^2$ where Ry is the Rydberg constant given by

$$\text{Ry} = \frac{me^4}{8\epsilon_0^2 h^2} \approx 13.6\text{eV}$$

with m the electron mass. The corresponding radius of this hydrogen atom wavefunction is $r_n \approx n^2 a_0$ with the Bohr radius given by

$$a_0 = \frac{4\pi\epsilon_0 \hbar^2}{me^2} \approx .51 \times 10^{-10}\text{m}.$$

The analogous calculation for a hydrogenic impurity state in a semiconductor gives precisely the same expression, only ϵ_0 is replaced by $\epsilon_0\epsilon_r$ and m is replaced by m_e^*. One obtains

$$\text{Ry}^{\text{eff}} = \text{Ry}\left(\frac{m_e^*}{m}\frac{1}{\epsilon_r^2}\right)$$

and

$$a_0^{\text{eff}} = a_0\left(\epsilon_r \frac{m}{m_e^*}\right).$$

Because the dielectric constant of semiconductors is typically high (on the order of 10 for most common semiconductors) and because the effective mass is frequently low (a third of m or even smaller), the effective Rydberg Ry^{eff} can be tiny compared to the real Rydberg, and the effective Bohr radius a_0^{eff} can be huge compared to the real Bohr radius.[17] For example, in silicon[18] the effective Rydberg, Ry^{eff}, is much less than .1 eV and a_0^{eff} is above 30 Ångstroms! Thus this donor impurity forms an energy eigenstate with energy just below the bottom of the conduction band (Ry^{eff} below the band bottom only). At zero temperature this eigenstate will be filled, but it takes only a small temperature to excite a bound electron out of a hydrogenic orbital and into the conduction band.

A depiction of this physics is given in Fig. 17.4 where we have plotted an energy diagram for a semiconductor with donor or acceptor impurities. Here the energy eigenstates are plotted as a function of position. Between the valence and conduction band (which are uniform in position), there are many localized hydrogen-atom-like eigenstates. The energies of these states are not all exactly the same, since each impurity atom is perturbed by other impurity atoms in its environment. If the density of impurities is high enough, electrons (or holes) can hop from one impurity to the next, forming an *impurity band.*

Note that because the effective Rydberg is very small, the impurity eigenstates are only slightly below the conduction band or above the valence band respectively. With a small temperature, these donors or acceptors can be thermally excited into the band. Thus, except at low

[17] Note that the large Bohr Radius justifies *post facto* our use of a continuum approximation for the dielectric constant ϵ_r. On small length scales, the electric field is extremely inhomogeneous due to the microscopic structure of the atoms, but on large enough length scales we can use classical electromagnetism and simply model the material as a medium with a dielectric constant.

[18] Because silicon has an anisotropic band, and therefore an anisotropic mass, the actual formula is more complicated.

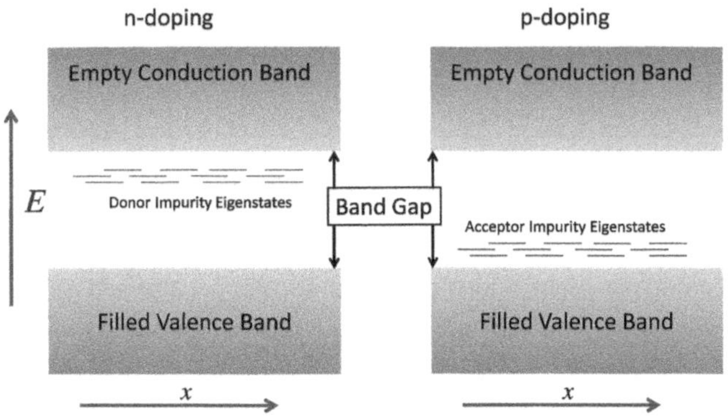

Fig. 17.4 Energy diagram of a doped semiconductor (left) with donor impurities, or (right) with acceptor impurities. The energy eigenstates of the hydrogenic orbitals tied to the impurities are not all the same because each impurity is perturbed by neighbor impurities. At low temperature, the donor impurity eigenstates are filled and the acceptor eigenstates are empty. But with increasing temperature, the electrons in the donor eigenstates are excited into the conduction band, and similarly the holes in the acceptor eigenstates are excited into the valence band.

enough temperature that the impurities bind the carrier, we can think of the impurities as simply adding carriers to the band. So the donor impurities donate free electrons to the conduction band, whereas the acceptor impurities give free holes to the valence band. However, at very low temperature these carriers get bound back to their respective nuclei so that they can no longer carry electricity—a phenomenon known as *carrier freeze out*. We will typically assume that we are at temperatures high enough (such as room temperature) such that freeze-out does not occur.

Note that in the absence of impurities, the Fermi energy (the chemical potential at zero temperature) is in the middle of the band gap. When donor impurities are added, at zero temperature, impurity states near the top of the band gap are filled. Thus the Fermi energy is moved up to the top of the band gap. On the other hand, when acceptors are added, the acceptor states near the bottom of the band gap are empty (remember it is a bound state of a hole to a nucleus!). Thus, the Fermi energy is moved down to the bottom of the band gap.

Optical Effects of Impurities (Redux)

As mentioned previously in Section 16.5.4, the presence of impurities in a material can have dramatic effects on its optical properties. There are two main optical effects of impurities. The first effect is that the impurities add charge carriers to an otherwise insulating material—turning an insulator into something that conducts at least somewhat. This obviously can have some important effects on the interaction with light. The second important effect is the introduction of new energy levels within the gap. Whereas before the introduction of impurities, the lowest energy transition that can be made is the full energy of the gap, now one can have optical transitions between impurity states, or from the bands to the impurity states.

17.3 Statistical Mechanics of Semiconductors

We now use our knowledge of statistical physics to analyze the occupation of the bands at finite temperature.

Imagine a band structure as shown in Fig. 17.5. The minimum energy of the conduction band is defined to be ϵ_c and the maximum energy of the valence band is defined to be ϵ_v. The band gap is correspondingly $E_{gap} = \epsilon_c - \epsilon_v$.

Recall from way back in Eq. 4.10 that the density of states per unit volume for free electrons (in three dimensions with two spin states) is given by

$$g(\epsilon \geqslant 0) = \frac{(2m)^{3/2}}{2\pi^2\hbar^3}\sqrt{\epsilon}.$$

The electrons in our conduction band are exactly like these free electrons, except that (a) the bottom of the band is at energy ϵ_c and (b) they have an effective mass m_e^*. Thus the density of states for these electrons near the bottom of the conduction band is given by

$$g_c(\epsilon \geqslant \epsilon_c) = \frac{(2m_e^*)^{3/2}}{2\pi^2\hbar^3}\sqrt{\epsilon - \epsilon_c}.$$

Similarly, the density of states for holes near the top of the valence band is given by

$$g_v(\epsilon \leqslant \epsilon_v) = \frac{(2m_h^*)^{3/2}}{2\pi^2\hbar^3}\sqrt{\epsilon_v - \epsilon}.$$

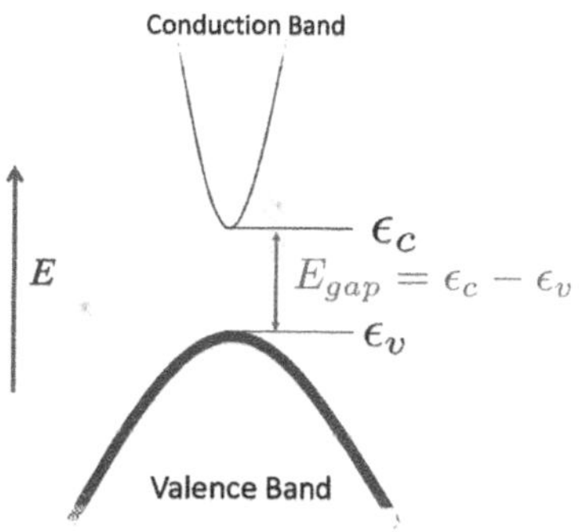

Fig. 17.5 A band diagram of a semiconductor near the top of the valence band (mostly filled) and the bottom of the conduction band (mostly empty). This diagram shows a direct band gap, but the considerations of this section apply to indirect gaps as well.

At fixed chemical potential μ the total density of electrons n in the conduction band, as a function of temperature T, is thus given by

$$n(T) = \int_{\epsilon_c}^{\infty} d\epsilon\, g_c(\epsilon)\, n_F(\beta(\epsilon - \mu)) = \int_{\epsilon_c}^{\infty} d\epsilon\, \frac{g_c(\epsilon)}{e^{\beta(\epsilon-\mu)} + 1}$$

where n_F is the Fermi occupation factor, and $\beta^{-1} = k_B T$ as usual. If the chemical potential is "well below" the conduction band (i.e., if $\beta(\epsilon - \mu) \gg 1$), then we can approximate

$$\frac{1}{e^{\beta(\epsilon-\mu)} + 1} \approx e^{-\beta(\epsilon-\mu)}.$$

In other words, Fermi statistics can be replaced by Boltzmann statistics when the temperature is low enough that the density of electrons in the band is very low. (We have already run into this principle in Section 17.1.1 when we discussed that Drude theory, a classical approach that neglects Fermi statistics, actually works very well for electrons above the band gap in semiconductors!) We thus obtain

$$\begin{aligned} n(T) &\approx \int_{\epsilon_c}^{\infty} d\epsilon g_c(\epsilon) e^{-\beta(\epsilon-\mu)} = \frac{(2m_e^*)^{3/2}}{2\pi^2\hbar^3}\int_{\epsilon_c}^{\infty} d\epsilon\,(\epsilon - \epsilon_c)^{1/2} e^{-\beta(\epsilon-\mu)} \\ &= \frac{(2m_e^*)^{3/2}}{2\pi^2\hbar^3} e^{\beta(\mu-\epsilon_c)} \int_{\epsilon_c}^{\infty} d\epsilon\,(\epsilon - \epsilon_c)^{1/2} e^{-\beta(\epsilon-\epsilon_c)}. \end{aligned}$$

The last integral is (using $y^2 = x = \epsilon - \epsilon_c$).

$$\begin{aligned}\int_0^\infty dx\, x^{1/2} e^{-\beta x} &= 2\int_0^\infty dy\, y^2 e^{-\beta y^2} = -2\frac{d}{d\beta}\int_0^\infty dy\, e^{-\beta y^2} \\ &= -\frac{d}{d\beta}\sqrt{\frac{\pi}{\beta}} = \frac{1}{2}\beta^{-3/2}\sqrt{\pi}.\end{aligned}$$

Thus we obtain the standard expression for the density of electrons in the conduction band:

$$n(T) = \frac{1}{4}\left(\frac{2m_e^* k_B T}{\pi\hbar^2}\right)^{3/2} e^{-\beta(\epsilon_c-\mu)}\,. \tag{17.8}$$

Note that this is mainly just exponential activation from the chemical potential to the bottom of the conduction band, with a prefactor which doesn't change too quickly as a function of temperature (obviously the exponential changes very quickly with temperature!).

Quite similarly, we can write the density of holes in the valence band p as a function of temperature:[19]

$$p(T) = \int_{-\infty}^{\epsilon_v} d\epsilon\, g_v(\epsilon)\left[1 - \frac{1}{e^{\beta(\epsilon-\mu)}+1}\right] = \int_{-\infty}^{\epsilon_v} d\epsilon\, \frac{g_v(\epsilon)e^{\beta(\epsilon-\mu)}}{e^{\beta(\epsilon-\mu)}+1}\,.$$

Again, if μ is substantially above the top of the valence band, we have $e^{\beta(\epsilon-\mu)} \ll 1$ so we can replace this by

$$p(T) = \int_{-\infty}^{\epsilon_v} d\epsilon\, g_v(\epsilon)e^{\beta(\epsilon-\mu)}$$

and the same type of calculation then gives

$$p(T) = \frac{1}{4}\left(\frac{2m_h^* k_B T}{\pi\hbar^2}\right)^{3/2} e^{-\beta(\mu-\epsilon_v)} \tag{17.9}$$

again showing that the holes are activated from the chemical potential *down* into the valence band (recall that pushing a hole down into the valence band costs energy!).

Law of Mass Action

A rather crucial relation is formed by combining Eqs. 17.8 and 17.9,

$$\begin{aligned}n(T)p(T) &= \frac{1}{2}\left(\frac{k_B T}{\pi\hbar^2}\right)^3 (m_e^* m_h^*)^{3/2} e^{-\beta(\epsilon_c-\epsilon_v)} \\ &= \frac{1}{2}\left(\frac{k_B T}{\pi\hbar^2}\right)^3 (m_e^* m_h^*)^{3/2} e^{-\beta E_{gap}}\end{aligned} \tag{17.10}$$

where we have used the fact that the gap energy $E_{gap} = \epsilon_c - \epsilon_v$. Eq. 17.10 is sometimes known as the *law of mass action*,[20] and it is true independent of doping of the material.

[19] If the Fermi factor n_F gives the probability that a state is occupied by an electron, then $1 - n_F$ gives the probability that the state is occupied by a hole.

[20] The nomenclature here, "law of mass action", is a reference to an analog in chemistry. In chemical reactions we may have an equilibrium between two objects A and B and their compound AB. This is frequently expressed as

$$A + B \rightleftharpoons AB$$

There is some chemical equilibrium constant K which gives the ratio of concentrations

$$K = \frac{[AB]}{[A][B]}$$

where $[X]$ is the concentration of species X. The law of mass action states that this constant K remains fixed, independent of the individual concentrations. In semiconductor physics it is quite similar, only the "reaction" is

$$e + h \rightleftharpoons 0,$$

the annihilation of an electron and a hole, so that the product of $[e] = n$ and $[h] = p$ is fixed.

Intrinsic Semiconductors

For an intrinsic (i.e., undoped) semiconductor the number of electrons excited into the conduction band must be equal to the number of holes left behind in the valence band, so $p = n$. We can then divide Eq. 17.8 by 17.9 to get

$$1 = \left(\frac{m_e^*}{m_h^*}\right)^{3/2} e^{-\beta(\epsilon_v+\epsilon_c-2\mu)} \quad .$$

Taking the log of both sides gives the useful relation

$$\mu = \frac{1}{2}(\epsilon_c + \epsilon_v) + \frac{3}{4}(k_B T)\log(m_h^*/m_e^*). \qquad (17.11)$$

Note that at zero temperature, the chemical potential is precisely mid-gap.

Using either this expression, or by using the law of mass action along with the constraint $n = p$, we can obtain an expression for the *intrinsic* density of carriers in the semiconductor

$$n_{intrinsic} = p_{intrinsic} = \sqrt{np} = \frac{1}{\sqrt{2}}\left(\frac{k_B T}{\pi\hbar^2}\right)^{3/2} (m_e^* m_h^*)^{3/4} e^{-\beta E_{gap}/2} \quad .$$

Doped Semiconductors

For doped semiconductors, the law of mass action still holds. If we further assume that the temperature is high enough so that there is no carrier freeze-out (i.e., carriers are not bound to impurities) then we have

$$n - p = (\text{density of donors}) - (\text{density of acceptors}).$$

This, along with the law of mass action, gives us two equations with two unknowns which can be solved.[21] In short, the result is that if we are at a temperature where the undoped intrinsic carrier density is much greater than the dopant density, then the dopants do not matter much, and the chemical potential is roughly midgap as in Eq. 17.11 (this is the *intrinsic* regime). On the other hand, if we are at a temperature where the intrinsic undoped density is much smaller than the dopant density, then we can think of this as a low-temperature situation where the carrier concentration is mainly set by the dopant density (this is the *extrinsic* regime). In the n-doped case, the bottom of the conduction band gets filled with the density of electrons from the donors, and the chemical potential gets shifted up towards the conduction band. Correspondingly, in the p-doped case, holes fill the top of the valence band, and the chemical potential gets shifted down towards the valence band. Note that in this case of strong doping, the majority carrier concentration is obtained just from the doping, whereas the minority carrier concentration—which might be very small—is obtained via law of mass action. The ability to add carriers of either charge to semiconductors by doping is absolutely crucial to being able to construct semiconductor devices, as we will see in the next chapter.

[21] Here is how to solve these two equations. Let

$$D = \text{doping} = n - p.$$

Let us further assume that $n > p$ so $D > 0$ (we can do the calculation again making the opposite assumption, at the end). Also let

$$I = n_{intrinsic} = p_{intrinsic}$$

so that

$$I^2 = \frac{1}{2}\left(\frac{k_B T}{\pi\hbar^2}\right)^3 (m_e^* m_h^*)^{3/2} e^{-\beta E_{gap}}$$

from the law of mass action. Using $np = I^2$, we can then construct

$$D^2 + 4I^2 = (n-p)^2 + 4np = (n+p)^2$$

So we obtain

$$n = \frac{1}{2}\left(\sqrt{D^2+4I^2} + D\right)$$
$$p = \frac{1}{2}\left(\sqrt{D^2+4I^2} - D\right).$$

As stated in the main text, if $I \gg D$ then the doping D is not important. On the other hand, if $I \ll D$ then the majority carrier density is determined by the doping only, the thermal factor I is unimportant, and the minority carrier density is fixed by the law of mass action.

Chapter Summary

- Holes are the absence of an electron in the valence band. These have positive charge (electrons have negative charge), and positive effective mass. The energy of a hole gets larger at larger momentum (away from the maximum of the band) as they get pushed down into the valence band. The positive charge of the hole as a charge carrier explains the puzzle of the sign of the Hall coefficient.
- Effective mass of electrons is determined by the curvature at the bottom of the conduction band. Effective mass of holes is determined by the curvature at top of conduction band.
- Mobility of a carrier is $\mu = |e\tau/m^*|$ in Drude theory.
- When very few electrons are excited into the conduction band, or very few holes into the valence band, Boltzmann statistics is a good approximation for Fermi statistics, and Drude theory is accurate.
- Electrons or holes can be excited thermally, or can be added to a system by *doping* and can greatly change the optical and electrical properties of the material. The law of mass action assures that the product np is fixed independent of the amount of doping (it only depends on the temperature, the effective masses, and the band gap).
- Know how to derive the law of mass action!
- At very low temperature carriers may freeze out, binding to the impurity atoms that they came from. However, because the effective Rydberg is very small, carriers are easily ionized into the bands.

References

- Ashcroft and Mermin, chapter 28. (A very good discussion of holes and their effective mass is given in chapter 12. In particular see page 225 and thereafter.)
- Rosenberg, chapter 9
- Hook and Hall, sections 5.1–5.5
- Kittel, chapter 8
- Burns, chapter 10 not including 10.17 and after
- Singleton, chapters 5–6
- Ibach and Luth, sections 12–12.5
- Sze, chapter 2

Exercises

(17.1) **Holes**

(a) In semiconductor physics, what is meant by a hole and why is it useful?

(b) An electron near the top of the valence band in a semiconductor has energy

$$E = -10^{-37}|\mathbf{k}|^2$$

where E is in Joules and k is in m^{-1}. An electron is removed from a state $\mathbf{k} = 2 \times 10^8 \mathrm{m}^{-1}\hat{x}$, where $\hat{x}$ is the unit vector in the x-direction. For a hole, calculate (and give the sign of!)

(i) the effective mass

(ii) the energy

(iii) the momentum

(iv) the velocity.

▷ If there is a density $p = 10^5 \mathrm{m}^{-3}$ of such holes all having almost exactly this same momentum, calculate the current density and its sign.

(17.2) **Law of Mass Action and Doping of Semiconductors**

(a) Assume that the band-gap energy E_g is much greater than the temperature $k_B T$. Show that in a pure semiconductor at a fixed T, the product of the number of electrons (n) and the number of holes (p) depends only on the density of states in the conduction band and the density of states in the valence band (through their effective masses), and on the band-gap energy.

▷ Derive expressions for n for p and for the product np. You may need to use the integral $\int_0^\infty dx\, x^{1/2} e^{-x} = \sqrt{\pi}/2$.

(b) The band gaps of silicon and germanium are 1.1 eV and 0.75 eV respectively. You may assume the effective masses for silicon and germanium are isotropic, roughly the same, and are roughly .5 of the bare electron mass for both electrons and holes. (Actually the effective masses are not quite the same, and furthermore the effective masses are both rather anisotropic, but we are just making a rough estimates here.)

▷ Estimate the conduction electron concentration for intrinsic (undoped) silicon at room temperature.

▷ Make a rough estimate of the maximum concentration of ionized impurities that will still allow for this "intrinsic" behavior.

▷ Estimate the conduction electron concentration for germanium at room temperature.

(c) The graph in Fig. 17.6 shows the relationship between charge-carrier concentration and temperature for a certain n-doped semiconductor.

▷ Estimate the band gap for the semiconductor and the concentration of donor ions.

▷ Describe in detail an experimental method by which these data could have been measured, and suggest possible sources of experimental error.

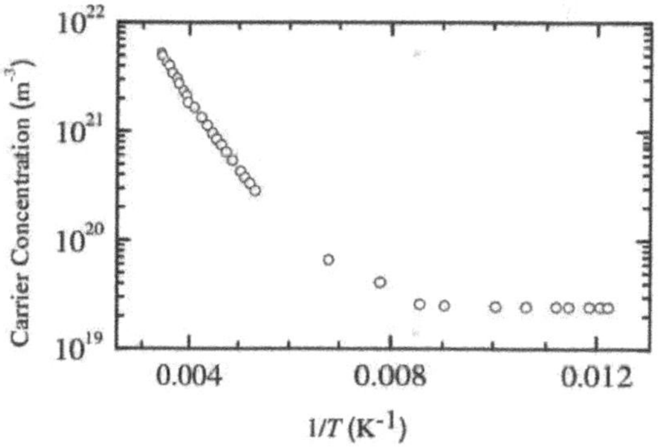

Fig. 17.6 Figure for Exercise 17.2.

(17.3) **Chemical Potential**

(a) Show that the chemical potential in an intrinsic semiconductor lies in the middle of the gap at low temperature.

(b) Explain how the chemical potential varies with temperature if the semiconductor is doped with (i) donors (ii) acceptors.

(c) A direct-gap semiconductor is doped to produce a density of 10^{23} electrons/m^3. Calculate the hole density at room temperature given that the gap is 1.0 eV, and the effective mass of carriers in the conduction and valence band are 0.25 and 0.4 electron masses respectively. Hint: use the result of Exercise 17.2.a.

(17.4) **Energy Density**

Show that the energy density of electrons in the conduction band of a semiconductor is

$$(\epsilon_c + \frac{3}{2}k_B T)\, n$$

where n is the density of these electrons and ϵ_c is the energy of the bottom of the conduction band.

(17.5) **Semiconductors**

Describe experiments to determine the following properties of a semiconductor sample: (i) sign of the majority carrier (ii) carrier concentration (assume that one carrier type is dominant) (iii) band gap (iv) effective mass (v) mobility of the majority carrier.

(17.6) **More Semiconductors**

Outline the absorption properties of a semiconductor and how these are related to the band gap. Explain the significance of the distinction between a direct and an indirect semiconductor. What region of the optical spectrum would be interesting to study for a typical semiconducting crystal?

(17.7) **Yet More Semiconductors**

Outline a model with which you could estimate the energy of electron states introduced by donor atoms into an n-type semiconductor. Write down an expression for this energy, explaining why the energy levels are very close to the conduction band edge.

(17.8) **Minimum Conductivity**

Suppose holes in a particular semiconductor have mobility μ_h and electrons in this semiconductor have mobility μ_e. The total conductivity of the semiconductor will be

$$\sigma = e\,(n\,\mu_e + p\,\mu_h)$$

with n and p the densities of electrons in the conduction band and holes in the valence band. Show that, independent of doping, the minimum conductivity that can be achieved is

$$\sigma = 2e\, n_{intrinic}\, \sqrt{\mu_e \mu_h}$$

with $n_{intrinsic}$ the intrinsic carrier density. For what value of $n - p$ is this conductivity achieved?

(17.9) **Hall Effect with Both n- and p-Dopants***

Suppose a semiconductor has a density p of holes in the valence band with mobility μ_h and a density n of electrons in the conduction band with mobility μ_n. Use Drude theory to calculate the Hall resistivity of this sample.

Semiconductor Devices

18

The development of semiconductor electronic devices no doubt changed the world. Constituting perhaps the greatest technological advance of the modern era, it is hard to overstate how much we take for granted the existence of electronics these days. We (indeed, the world!) should never lose sight of the fact that this entire industry owes its very existence to our detailed understanding of quantum condensed matter physics.

While a thorough discussion of the physics of semiconductor devices can be quite involved, it is not hard to give the general idea of how some of the elementary components work.[1] This chapter is aimed to give a brief cartoon-level description of some of the more important devices.

[1] See the references at the end of the chapter if you would like more details!

18.1 Band Structure Engineering

To make a semiconductor device one must have control over the detailed properties of materials (band gap, doping, etc.) and one must be able to assemble together semiconductors with differing such properties.

18.1.1 Designing Band Gaps

A simple example of engineering a device is given by aluminum-gallium-arsenide. GaAs is a semiconductor (zincblende structure as in Fig. 14.6) with a direct band gap (at $\mathbf{k} = 0$) of about $E_{gap}(\text{GaAs}) = 1.4$ eV. AlAs is the same structure except that the Ga has been replaced by Al and the gap[2] at $\mathbf{k} = 0$ is about 2.7 eV. One can also produce alloys (mixtures) where some fraction (x) of the Ga has been replaced by Al which we notate as $\text{Al}_x\text{Ga}_{1-x}\text{As}$. To a fairly good approximation the direct band gap just interpolates between the direct band gaps of the pure GaAs and the pure AlAs. Thus we get roughly (for $x < .4$)

$$E_{gap}(x) = (1-x)\,1.4\,\text{eV} + x\,2.7\,\text{eV}.$$

By producing this type of alloyed structure one can obtain any desired band gap in this type of material.[3] This technique of designing properties of materials (such as band gaps) by alloying *miscible* materials[4] can be applied very broadly. It is not uncommon to concoct compounds made of three, four, or even five elements in order to engineer materials with certain desired properties.

In the context of device physics one might want to build, for example, a laser out of a semiconductor. The lowest-energy transition which recombines a hole with an electron is the gap energy (this is the "lasing"

[2] AlAs is actually an indirect band-gap semiconductor, but for $x < .4$ or so $\text{Al}_x\text{Ga}_{1-x}\text{As}$ is direct band gap.

[3] By alloying the material with arbitrary x, one must accept that the system can no longer be precisely periodic but instead will be some random mixture. It turns out that as long as we are concerned with long wavelength electron waves (i.e, states near the bottom of the conduction band or the top of the valence band) this randomness is very effectively averaged out and we can roughly view the system as being a periodic crystal of As and some average of a $\text{Al}_x\text{Ga}_{1-x}$ atom. This is known as a "virtual crystal" approximation.

[4] "Miscible" means "mixable".

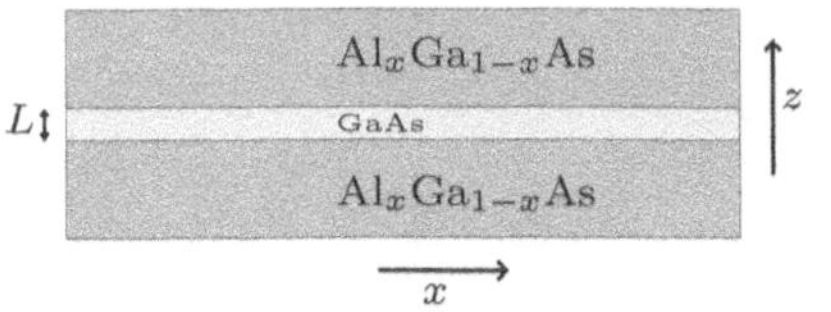

Fig. 18.1 A semiconductor heterostructure in a quantum well configuration. In the GaAs region, the conduction band is at lower energy and the valence band is at higher energy than in the AlGaAs region. Thus both electrons in the conduction band and holes in the valence band can be trapped in the GaAs region.

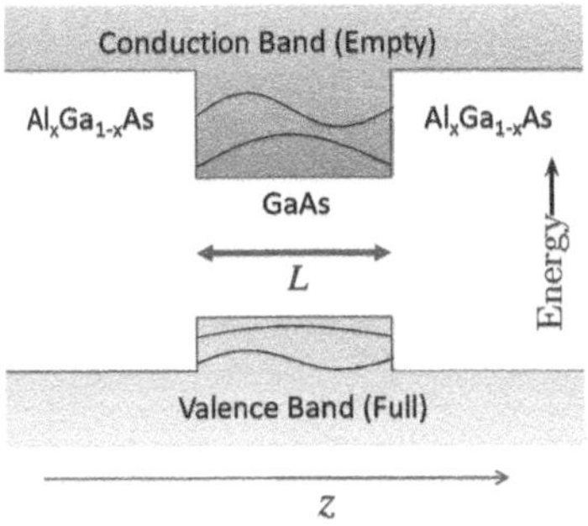

Fig. 18.2 Band diagram of a quantum well. A single electron in the conduction band can be trapped in the particle-in-a-box states in the quantum well. Similarly, a hole in the valence band can be trapped in the quantum well.

energy typically). By tuning the composition of the semiconductor, one can tune the energy of the gap and therefore the optical frequency of the laser. We will see in the rest of this chapter more examples of how band structure engineering can be very powerful for building a wide range of semiconductor devices.

18.1.2 Non-Homogeneous Band Gaps

By constructing structures where the material (or the alloying of a material) is a function of position, one can design more complex environments for electrons or holes in a system. Consider, for example, the structure shown in Fig. 18.1. Here a layer of GaAs with smaller band gap is inserted between two layers of AlGaAs[5] which has a larger band gap.[6] This structure is known as a "quantum well". In general a structure made of several varieties of semiconductor is known as a *semiconductor heterostructure*.[7] A band diagram of the quantum well structure as a function of the vertical position z is given in Fig. 18.2. The band gap is lower in the GaAs region than in the AlGaAs region. The changes in band energy can be thought of as a potential that an electron (or hole) would feel. For example, an electron in the conduction band can have a lower energy if it is in the quantum well region (the GaAs region) than it can have in the AlGaAs region. An electron in the conduction band with low energy will be trapped in this region. Just like a particle in a box, there will be discrete eigenstates of the electron's motion in the z direction, as shown in the figure. The situation is similar for holes in the valence band (recall that it requires energy to push a hole down into the valence band), so there will similarly be confined particle-in-a-box states for holes in the quantum well.

The important physics of this section is to realize that electrons in a semiconductor see the energy of the conduction band bottom (or correspondingly the holes see the energy of the valence band top) as being a potential as a function of position which they can then be trapped in!

Modulation Doping and the Two-Dimensional Electron Gas

To add electrons (or holes) to a quantum well, one typically has to include dopants to the heterostructure (n- or p-dopants respectively). A very useful trick is to put the actual dopant atoms *outside* the quantum well. Since the potential energy is lower in the well, electrons released by n-donors will fall into the well and will be trapped there.[8,9] For example, in Figs. 18.1 and 18.2, one could put the dopants in the AlGaAs region rather than in the GaAs region. This trick, known as "modulation doping",[10] allows the carriers to move freely within the well region without ever having to bump into a dopant ion. Such carriers can have enormously long mean-free paths, and this is quite useful in designing devices with very low dissipation.

[5]One frequently writes "AlGaAs" rather than "$Al_xGa_{1-x}As$" for brevity.

[6]GaAs and AlGaAs are particularly nice materials for building heterostructures because the lattice constants of GaAs and AlGaAs are extremely close. As a result, AlGaAs will attach very nicely to a GaAs surface and vice versa. If one builds heterostructures between materials with very different lattice constants, inevitably there are defects at the interface.

[7]Development of semiconductor heterostructure devices (including semiconductor lasers and heterostructure transistors) resulted in Nobel Prizes for Zhores Alferov and Herbert Kroemer in 2000.

Electrons that are trapped in a quantum well are free to travel in two dimensions, but are confined in the third direction (denoted the z-direction in Figs. 18.1 and 18.2). At low temperature, if such a confined electron does not have enough energy to jump out of the well, or even to jump to a higher particle-in-a-box state, then the electron motion is strictly two-dimensional.[11] The study of such electrons in two dimensions has turned out to be a veritable treasure trove[12] of new and exciting physics, with amazing connections to fields as diverse as string theory and quantum computation. Unfortunately, detailed study of these systems are beyond the scope of this book.[13]

18.2 *p-n* Junction

One of the simplest, yet most important, semiconductor structures is the *p-n* junction. This is nothing more than a system where a *p*-doped semiconductor is brought into direct contact with an *n*-doped semiconductor. The resulting physics is quite surprising!

To understand the *p-n* junction, let us first consider *p*-doped and *n*-doped semiconductors separately, as shown in Fig. 18.3 (compare to Fig. 17.4). Although the *n*-doped system has free negatively charged electrons and the *p*-doped system has free positively charged holes, both systems are overall electrically neutral since charged ions compensate for the charges of the mobile charge carriers. As shown in Fig. 18.3, in the *n*-doped semiconductor the chemical potential is near the very top of the band gap, whereas in the *p*-doped semiconductor the chemical potential is near the bottom of the band gap. Thus when the two materials are brought into contact the electrons in the conduction band will fall into the valence band, filling the empty hole states (as depicted by the arrow in Fig. 18.3), thus "pair-annihilating" both the electron and the hole. This pair-annihilation process amounts to a gain in energy of E_{gap} per pair annihilated (where E_{gap} is the gap energy between the bottom of the conduction band and the top of the valence band).

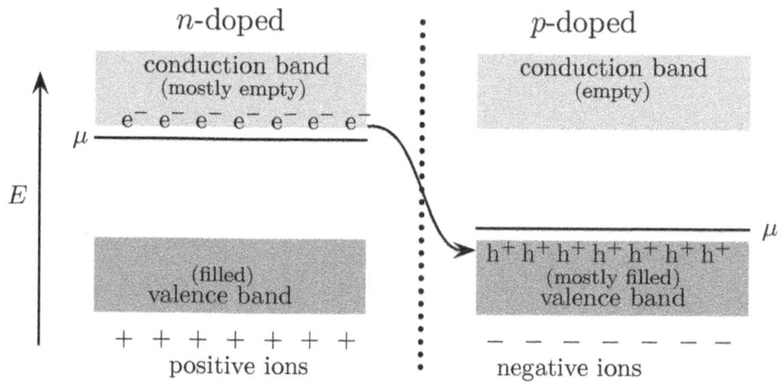

Fig. 18.3 The chemical potential for an *n*-doped semiconductor (left) is near the top of the band gap, whereas for a *p*-doped semiconductor (right) it is near the bottom of the band gap (compare to Fig. 17.4). The *n*-doped semiconductor has free electron (e^-) carriers in the mostly empty conduction band, but remains electrically neutral due to the positive ions. Similarly, the *p*-doped semiconductor has free hole (h^+) carriers in the mostly filled valence band, but remains electrically neutral due to the negative ions. When the two semiconductors are brought together the electrons want to fall down to the lower chemical potential, filling (and thus annihilating) the empty holes (as shown by the arrow).

[8] If electrons move too far from their donor ions, an electrical charge builds up (like a capacitor), so the number of electrons that fall into the well is limited by this charging energy. This is similar to the physics of the *p-n* junction which we discuss next.

[9] Similarly holes released by *p*-donors "fall-up" into the well and will be trapped!

[10] Modulation doping was invented by Horst Stormer and Ray Dingle. Stormer later won a Nobel Prize for research that was enabled by this trick! See also margin note 12. Stormer was also my boss's boss for a while when I worked at Bell Labs.

[11] This is known as a two-dimensional electron gas or "2DEG".

[12] Several Nobel Prizes have been awarded for study of two-dimensional electrons: von Klitzing in 1985 and Tsui, Stormer, and Gossard in 1998 (see margin note 3 in Chapter 1). The study of electrons in graphene, a single atomic layer of carbon, which won the prize in 2010 (Geim and Novoselov, see also margin note 8 in Chapter 19), is very closely related as well.

[13] Since this is one of my favorite topics, it may be the subject of my next book. Or maybe this will be my next next book after I write a romantic thriller about physicists in the Amazon who defeat drug smugglers.

After this process of electrons falling into holes and annihilating occurs there will be a region near the interface where there are no free carriers at all. This is known as the "depletion region" or "space charge region" (see Fig. 18.4). This region is electrically charged, since there are charged ions but no carriers to neutralize them. Thus there is a net electric field pointing from the positively charged to the negatively charged ions (i.e., the electric field points from the n-doped to the p-doped region). This electric field is very much like a capacitor—positive charge spatially separated from negative charge with an electric field in the middle. We now imagine moving an additional electron across the depletion region in order to annihilate another hole. While the annihilation process gives a gain in energy of E_{gap} the process of moving the electron across the depletion region costs an energy of $-e\Delta\phi$ where ϕ is the electrostatic potential. When the depletion region is sufficiently large (so that the charge on the capacitor is sufficiently large, and thus $\Delta\phi$ is large) it becomes no longer favorable for further electrons and holes to annihilate. Thus, the depletion region grows only to a width where these two energy scales are the same.[14]

[14] One can make a very crude approximation of the junction as a plate capacitor (it is not a plate capacitor since the charge is distributed throughout the volume of the junction, not just on two plates). For simplicity let us also assume that the acceptor doping density $n_a = p$ in the p-region is the same as the donor doping density $n_d = n$ in the n-region. The capacitance of a plate capacitor per unit area is $\epsilon_0\epsilon_r/w$ with w the depletion width, ϵ_r the relative dielectric constant, and ϵ_0 the usual vacuum permittivity. The total charge per unit area on the capacitor is $nw/2$ so the voltage across the capacitor is $\Delta\phi = Q/C = nw^2/(2\epsilon_0\epsilon_r)$. Setting $e\Delta\phi$ equal to the gap energy E_{gap} yields an approximation of w. In Exercise 18.3 you are asked to do this calculation more carefully.

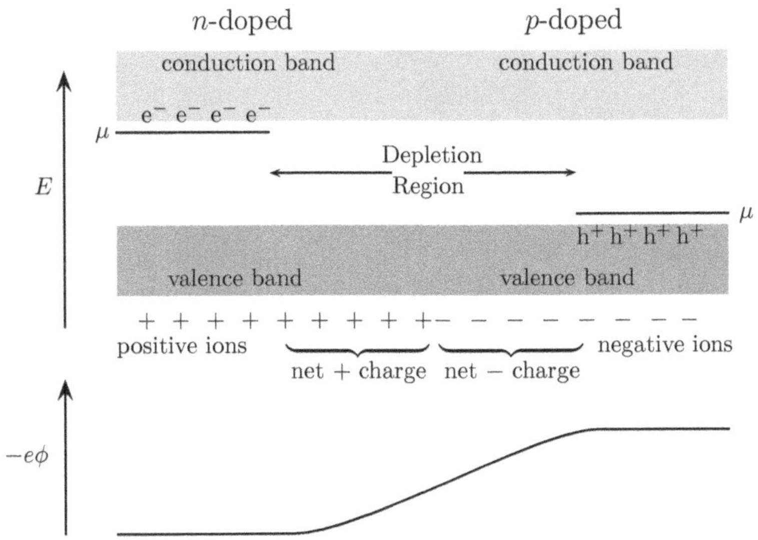

Fig. 18.4 Once electrons near the p-n interface have "fallen" into holes, thus annihilating both electron and hole, there is a depletion region near the interface where there are no free carriers. In this region, the charged ions create an electric field. The corresponding electrostatic potential $-e\phi$ is shown in the bottom of the figure. The depletion region will continue to grow until the energy for another electron to cross the depletion region (the size of the step in $-e\phi$) is larger than the gap energy which would be gained by the electron annihilating a hole.

The depiction of the p-n junction shown in Fig. 18.4 is a bit deceptive. From the top part of the figure it appears as if it would always lower the energy to move an electron across the junction to annihilate a hole (since the chemical potential of the electron on the left is plotted higher than the chemical potential of the hole on the right). What is lacking in the depiction of the top part of this figure is that it does not make apparent the electrostatic potential generated by the charges in the junction (this is shown as the plot in the lower half of the figure). It is therefore convenient to replot this figure so as to reflect the electrostatic potential as well as the (band structure) kinetic energy. This is shown in Fig. 18.5.

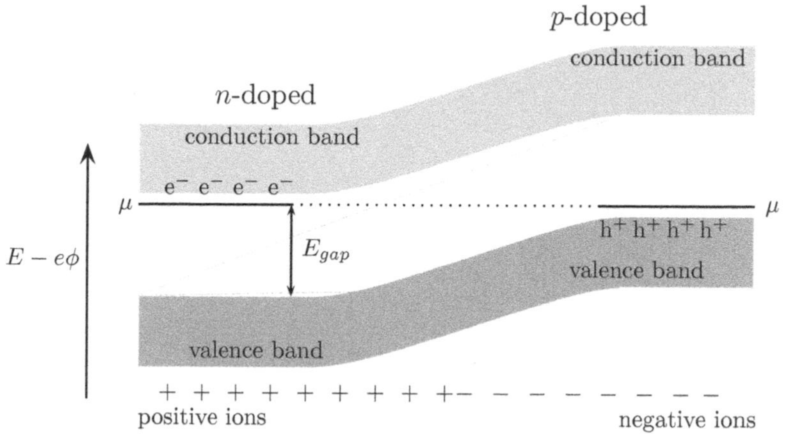

Fig. 18.5 Band diagram of an unbiaised *p-n* junction. This is precisely the same figure as Fig. 18.4, except that the electrostatic potential is added to the band energy. In this figure the equality of the (electro)chemical potential on the left versus the right indicates that there is no net driving force for electrons to flow either left or right. However, in the depletion region there is a net electric field, so if electron–hole pairs are created in this region by absorbing a photon, the electron will flow left and the hole will flow right, creating a net current. Note that the total potential voltage drop over the depletion region amounts to exactly the band gap.

Note that in this figure, the shifted chemical potentials on the two sides of the figure are now at the same level—reflecting the fact that the drop in band energy is precisely compensated by the change in electrostatic potential.[15] Thus there is no driving force for electrons to be transferred either to the right or to the left in this junction.

[15]The combination of chemical potential and electrostatic energy is usually known as electrochemical potential.

The Solar Cell

If one applies light to a semiconductor, electron–hole pairs may be excited if the energy of a photon is greater than the energy of the bandgap.[16] Consider now exposing the *p-n* junction of Fig. 18.5 to light. In most regions of the semiconductor, the created electrons and holes will quickly reannihilate. However, in the depletion region, due to the electric field in this region, electrons which are created flow off to the left (towards the *n*-doped region) and holes which are created flow off to the right (towards the *p*-doped region). In both cases, the charge current is moving to the right (negative flowing left and positive flowing right both constitute current flowing to the right). Thus a *p-n* junction spontaneously creates a current (hence a voltage, hence power) just by being exposed to light. Devices based on this principle, known as a solar cells, photovoltaics, or photodiodes, currently provide tens of billions of dollars-worth of electrical energy to the world!

[16]Optical absorption processes (and also emission processes) are very strong for direct band gaps, and are less strong for indirect band-gaps. See Section 16.5.2.

Rectification: The Diode

This *p-n* junction has the remarkable property of *rectification*: it will allow current to flow through the junction easily in one direction, but not easily (with very high resistance) in the other direction.[17] Such asymmetric devices, frequently known as *diodes*,[18] form crucial parts of many electrical circuits.

[18]"di-ode" is from Greek, meaning "two path", which refers to the fact that such devices have two different sides (the *p* and *n* sides).

To understand the rectification effect, we imagine applying some voltage to the *p-n* junction to obtain a situation as shown in Fig. 18.6.

Fig. 18.6 Band diagram of a *biased* p-n junction (compare to Fig. 18.5). In this case, the right-hand side of the diagram is bent downwards by an applied voltage ($+eV$ is negative in this figure). The four processes that can create current are labeled. In the absence of applied voltage the net current is zero. When voltage is applied, current flows—easily for eV negative, but not easily for eV positive.

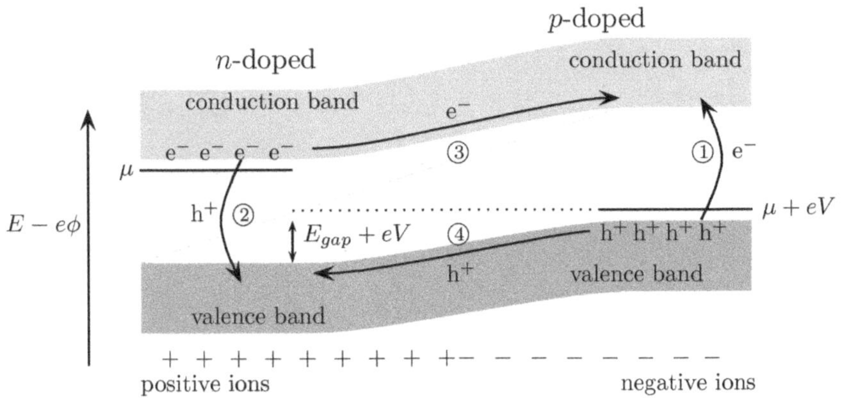

This figure is entirely analogous to Fig. 18.5 except that the potential hill (which has a total height of E_{gap} in Fig. 18.5) has been reduced by the applied voltage. Hence whereas the left and right chemical potentials line up in Fig. 18.5, here they do not align. There are four processes that can create current (labeled ①– ④ in the figure). Let us consider these one at a time to determine how much current flows.

Process ① and ② : On the right-hand side of the diagram (the p-doped side), electrons may be thermally excited into the conduction band (process ①). Some of these electrons will will flow down the slope to the left. Similarly on the left of the diagram (the n-doped side), holes may be thermally excited down into the valence band (process ②) and will flow up the slope to the right. In both cases, the number of carriers excited takes the usual activated[19] form e^{-E_{gap}/k_BT}, and in both cases the resulting charge current flows to the right (electrons flow to the left). Thus we obtain a contribution to the current of

$$I_{\text{right}} \propto e^{-E_{gap}/k_BT} \quad . \tag{18.1}$$

[19]Given the law of mass action, Eq. 17.10, $np \propto e^{-E_{gap}/k_BT}$. On each side of the system, the majority carrier density is fixed by the doping density, and it is the minority carrier density which is exponentially activated.

[20] Lest we violate the third law!

Process ③ and ④: It is also possible that electrons in the conduction band on the left-hand side of the diagram (the n-doped side) will be thermally activated to climb up the potential slope in the depletion layer (process ③) and will annihilate with holes once they arrive at the p-doped side. In the absence of applied voltage (as depicted in Fig. 18.4) the potential hill which the electrons would have to climb is precisely of height E_{gap}. Thus the amount of such current is $\propto e^{-E_{gap}/k_BT}$.

[17]The phenomenon of rectification in semiconductors was discovered by Karl Ferdinand Braun way back in 1874, but was not understood in detail until the middle of the next century. This discovery was fundamental to the development of radio technology. Braun was awarded the Nobel Prize in 1909 with Guglielmo Marconi for contributions to wireless telegraphy. Perhaps as important to modern communication, Braun also invented the cathode ray tube (CRT) which formed the display for televisions for many years until the LCD display arrived very recently. The CRT is known as a "Braun tube" in many countries.

Similarly, the holes in the valence band on the right-hand side (in the p-doped side), may be thermally activated to climb *down* the potential slope towards the n-doped side (process ④) where they annihilate with electrons. Again in the absence of an applied voltage the amount of such current is $\propto e^{-E_{gap}/k_B T}$. In both of these cases, the charge current is flowing to the left (electrons flow to the right).

When a voltage is applied to the system, the height of the potential hill is modified from E_{gap} to $E_{gap}+eV$, and correspondingly the current for these two processes (③ and ④) is modified giving

$$I_{\text{left}} \propto e^{-(E_{gap}+eV)/k_B T} \quad . \tag{18.2}$$

Note however that for processes ①and ②, voltage bias will not change the number of excited carriers so that I_{right} is independent of voltage.

Thus the total current flow in this device, is the sum of all four processes, $I_{\text{left}} + I_{\text{right}}$. While we have only kept track of the exponential factors in Eqs. 18.1 and 18.2, and not the prefactors, it is easy to argue that their prefactors must be the same. Since in the absence of applied voltage (or applied photons) there must be no net current in the system,[20] we can therefore write

$$I_{total} = J_s(T)\left(e^{-eV/k_B T} - 1\right) \tag{18.3}$$

where $J_s \propto e^{-E_{gap}/k_B T}$ is known as the saturation current. Eq. 18.3 is often known as the "diode equation" and is depicted in Fig. 18.7. Current flows easily in one direction (the so-called *forward biased* direction) whereas it flows only very poorly in the opposite direction (the *reverse biased* direction). At a cartoon level, one can think of a diode as being a circuit element whose response looks like the simplified picture in Fig. 18.8—essentially Ohmic (current proportional to voltage) in the forward biased direction and no current at all in the reverse biased direction. The reader might find it interesting to think about what sort of practical circuits can be built using diodes (see Exercise 18.5). In circuit diagrams the diode is depicted as shown in Fig. 18.9.

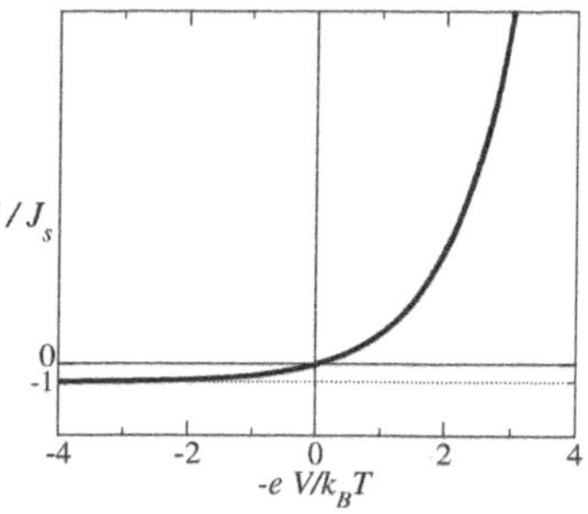

Fig. 18.7 Current–voltage relation for a diode (Eq. 18.3). Current flows easily in the forward bias direction, but not easily in the reverse bias direction. The scale of the y axis is set by the saturation current J_s, which is generally a function of temperature and other details of the device.

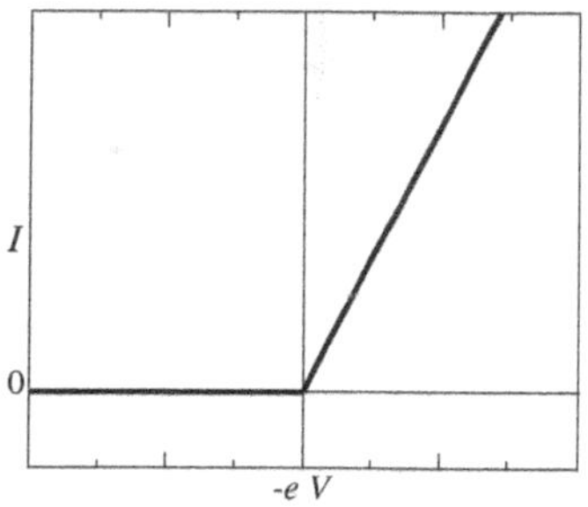

Fig. 18.8 A cartoon picture of the current–voltage relation for a diode. Roughly one imagines a diode as being Ohmic in the forward bias direction and allows no current to flow in the reverse bias direction.

Fig. 18.9 Symbol for a diode in a circuit. Current flows easily in the direction of the arrow (from the p to the n side), but not in the opposite direction. Remember that electrons flow opposite the current direction.

Light Emitting Diode

The same p-n junction structure can be used to generate light from current — the inverse process of the above mentioned solar-cell. In any region (but mainly in the depletion region) where there are both electrons in the conduction band and holes in the valence band, an electron can annihilate a hole, emitting a photon in the process.[16] When used for light generation the device is called a light-emitting diode, or LED. LED lights are rapidly displacing all other electrical room lighting sources as they are extremely power-efficient and long lasting.[21]

[21] While red LEDs have been around since the 1960s, achieving the higher frequencies required for room lighting took several decades. The development of the blue LED was considered so important that its inventors, Akasaki, Amano and Nakamura, were awarded the Nobel prize in 2014.

18.3 The Transistor

Perhaps the most important invention of the 20th century was the transistor—the simple semiconductor amplifier which forms the basis of every modern electronic circuit[22]. Every iPad, iPod, iPhone, and iBook literally contains billions of transistors. Even simple devices, like alarm clocks, TVs, or radios contain many thousands or even millions of them.[23]

[23]Potentially as important as the invention of the transistor was figuring out how to put millions of them on a single piece of silicon. The invention of the so-called "integrated circuit" earned Jack Kilby a Nobel Prize. Robert Noyce invented a similar device a few months later and founded Intel.

Although there are many types of modern transistors, by far the most common is known as the MOSFET,[24] which stands for "Metal-Oxide-Semiconductor Field Effect Transisitor". In Fig. 18.10 we show the structure of an n-MOSFET device without voltage applied to the gate. The regions around the source and drain contact are n-doped whereas the rest of the semiconductor is p-doped. Depletion layers form between the n- and p-doped regions. So in the absence of voltage applied to the gate the effective circuit is equivalent to two back-to-back p-n junctions as shown in Fig. 18.11. As a result, without voltage on the gate, current cannot flow easily between the source and drain in either direction. Note in particular the physical structure of the device which is **M**etal on top of an **O**xide insulator on top of a **S**emiconductor. These three layers comprise the MOS of the acronym MOS-FET[25] (most often in modern devices the semiconductor is silicon and the oxide is silicon dioxide).

[24]The first MOSFET was built in 1960, although a patent by Lilienfeld from 1926 (!) proposed what was essentially the same structure, although without a complete understanding of what would be required to make it work.

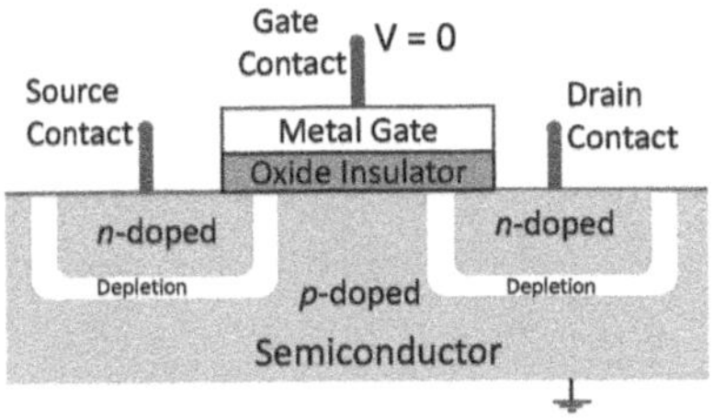

Fig. 18.10 An n-MOSFET transistor with no bias applied to the gate. Regions around the source and drain contacts are n-doped, whereas the rest of the semiconductor is p-doped. Note the depletion layers separating the n- from p-doped regions. In the absence of voltage on the gate electrode, this device is essentially two back to back diodes as depicted in Fig. 18.11 so that current cannot flow easily between source to drain in either direction. Note the **Met**al on top of **O**xide on top of **S**emiconductor which gives the device the name MOS-FET. This device has four electrical contacts including the ground (lower right).

When a positive voltage is applied to the metal gate (with respect to the ground attached to the semiconductor), the metal gate acts as one plate of a capacitor. The semiconductor forms the other plate of the capacitor, so that a positive voltage on the gate attracts negative charge to the region just under the oxide insulator. This attraction of charge is known as a "field effect", since it is the result of an electric field caused by the gate. (The term "field effect transistor" is the FET in the acronym MOSFET.) As a result, if the gate voltage is sufficiently large, larger than some particular threshold voltage $V_{threshold}$, the region under the gate becomes effectively n-like. This then means that there is a continuous channel of n-semiconductor that stretches all the way from the source to the drain and as a result the conductance between source and drain becomes very large as shown in Fig. 18.12. In this way, a (relatively small) voltage on the gate can control a large current between the source and drain, hence providing amplification of small signals.

[25]In fact a transistor can perfectly well be built using a gate which is not metal and/or an insulating layer which is not oxide. To be more inclusive of these structures as well, one occasionally hears the term "IGFET", meaning "insulated gate field effect transistor".

To determine in more detail the behavior of the n-MOSFET transistor we need to think about how electrons are attracted to the channel region. This turns out to be quite similar to the considerations we used for the p-n junction. In short, the electrostatic potential from the gate "bends" the band structure (similar to Fig. 18.6) such that very near to the gate the conduction band has been bent close to the chemical potential

[22]The invention of the transistor is usually credited to the Bell Labs team of John Bardeen, Walter Brattain, and William Shockley in 1947. Shockley, the manager of the team, although brilliant, was a very difficult person. Shockley was infuriated when he found out that Bardeen and Brattain had succeeded in making a prototype without his help. Although Shockley was (rightly) included in the Nobel Prize (due to significant improvements he made to the design) he essentially made it impossible for Bardeen and Brattain to contribute to any further development of the device. Bardeen left Bell Labs to the University of Illinois, Champagne-Urbana, where he started work on the theory of superconductivity, for which he won a second Nobel Prize (See margin note 5 in Section 6.1). Later in life Shockley became a strong proponent of eugenics, espousing opinions that were viewed as racist. He died estranged from most of his friends and family. On the more positive side of his legacy, he pioneered a nice rock climbing route in the Shawangunks of New York State which is now known as "Shockley's Ceiling".

and has started filling with electrons. Since activation of carriers into the conduction band is exponential in the energy difference between the chemical potential and the bottom of the conduction band (see Eq. 17.8) the conductivity of the channel from source to drain is exponentially sensitive to the voltage applied to the gate.

Note that we could also construct a p-MOSFET where the regions around the source and drain would be p-doped and the rest of the semiconductor would be n-doped. For such a device, conductance between source and drain would be turned on when the voltage on the gate is sufficiently *negative* so that holes are attracted to the region under the gate making a conductive p-channel between source and drain. Modern digital circuits almost always use a combination of n- and p-MOSFETS built on the same piece of silicon. This is known as Complementary MOS logic, or "CMOS". In a circuit diagram, a MOSFET is frequently drawn as shown in Fig. 18.13. Be warned, however, that there are many variations on this type of device, and they all have slightly different symbols. This particular symbol indicates that the ground contact is connected to the source electrode, which is very frequently the situation in circuits.

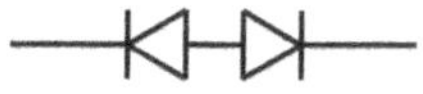

Fig. 18.11 In the absence of gate voltage, an n-MOSFET is just two p-n junctions back to back as shown here. Current cannot flow easily in either direction.

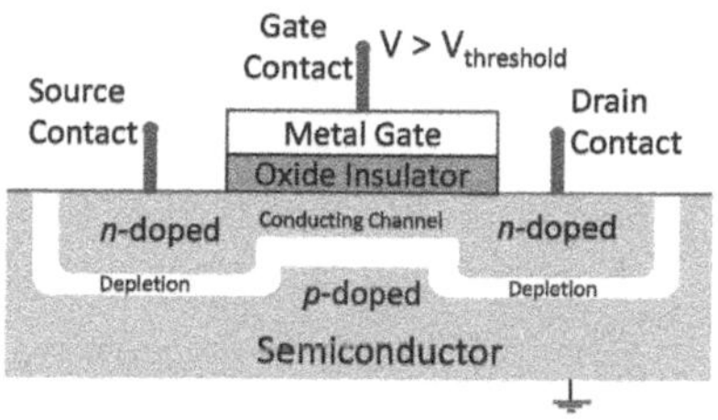

Fig. 18.12 An n-MOSFET with voltage $V > V_{threshold}$ applied to the gate. Here, electrons are attracted to the gate, making the region close to the oxide layer effectively n-doped. As a result, the n region forms a continuous conducting channel between the source and the drain and current now flows easily between these two. (Compare to Fig. 18.10.)

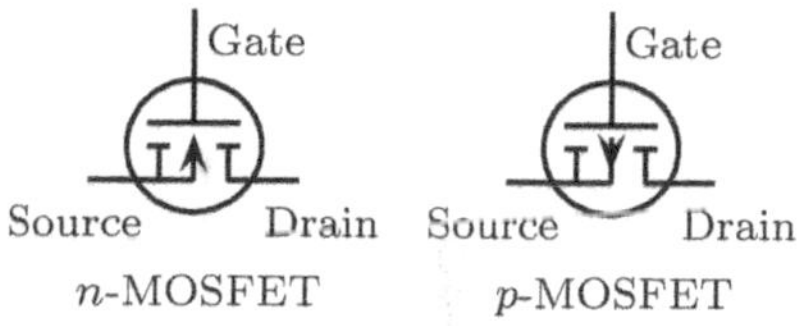

Fig. 18.13 Common symbol for n- and p-MOSFETs in circuit diagrams. Note that in this symbol it is indicated that the ground is connected to the source—which is often the case in circuits. Other (frequently similarly looking) symbols may be used if the device is used in another context.

Chapter Summary

- Alloying semiconductors can tune the band gap.
- The band gaps act as a potential for carriers so one can build particle-in-a-box potentials known as quantum wells.
- A junction between p- and n-doped semiconductor forms a depletion layer with no mobile carrier, but intrinsic electric field. This structure is the basis of the solar cell, the diode, and the LED.
- An external potential applied to a semiconductor can attract carriers to a region, greatly changing the conduction properties of the device. This "field effect" is the basis of the MOSFET.

References

There are many good references on semiconductor devices

- Hook and Hall, chapter 6 covers p-n junction fairly well
- Burns, section 10.17 covers p-n junction
- Ashcroft and Mermin, chapter 29 covers p-n junctions (This goes into just enough depth to start to be confusing.)
- Ibach and Luth, chapter 12 covers many simple semiconductor devices
- Sze (This is a good place to start if you want to learn a lot more about semiconductor devices.)

Exercises

(18.1) **Semiconductor Quantum Well**

(a) A quantum well is formed from a layer of GaAs of thickness L nm, surrounded by layers of $Ga_{1-x}Al_xAs$ (see Fig. 18.2). You may assume that the band gap of the $Ga_{1-x}Al_xAs$ is substantially larger than that of GaAs. The electron effective mass in GaAs is 0.068 m_e whereas the hole effective mass is 0.45 m_e with m_e the mass of the electron.

▷ Sketch the shape of the potential for the electrons and holes.

▷ What approximate value of L is required if the band gap of the quantum well is to be 0.1 eV larger than that of GaAs bulk material?

(b) *What might this structure be useful for?

(18.2) **Density of States for Quantum Wells**

(a) Consider a quantum well as described in the previous exercise. Calculate the density of states for electrons and holes in the quantum well. Hint: It is a 2D electron gas, but don't forget that there are several particle-in-a-box states.

(b) Consider a so-called "quantum wire" which is a one-dimensional wire of GaAs embedded in surrounding AlGaAs. (You can consider the wire cross-section to be a square with side 30nm.) Describe the density of states for electrons or holes within the quantum wire. Why might this quantum wire make a very good laser?

(18.3) ***p*-*n* Junction***

Explain the origin of the depletion layer in an abrupt *p*-*n* junction and discuss how the junction causes rectifcation to occur. Stating your assumptions, show that the total width w of the depletion layer of a *p*-*n* junction is:

$$w = w_n + w_p$$

where

$$w_n = \left(\frac{2\epsilon_r \epsilon_0 N_A \phi_0}{e N_D (N_A + N_D)} \right)^{1/2}$$

and a similar expression for w_p Here ϵ_r is the relative permittivity and N_A and N_D are the acceptor and donor densities per unit volume, while ϕ_0 is the difference in potential across the *p*-*n* junction with no applied voltage. You will have to use Poisson's equation to calculate the form of ϕ given the presence of the ion charges in the depletion region.

▷ Calculate the total depletion charge and infer how this changes when an additional voltage V is applied.

▷ What is the differential capacitance of the diode and why might it be useful to use a diode as a capacitor in an electronic circuit?

(18.4) **Single Heterojunction***

Consider an abrupt junction between an *n*-doped semiconductor with minimum conduction band energy ϵ_{c1} and an undoped semiconductor with minimum conduction band energy ϵ_{c2} where $\epsilon_{c1} > \epsilon_{c2}$. Describe qualitatively how this structure might result in a two-dimensional electron gas at the interface between the two semiconductors. Sketch the electrostatic potential as a function of position.

(18.5) **Diode Circuit**

Design a circuit using diodes (and any other simple circuit elements you need) to convert an AC (alternating current) signal into a DC (direct current) signal.

▷ *Can you use this device to design a radio reciever?

(18.6) **CMOS Circuit***

Design a circuit made of one *n*-MOSFET and one *p*-MOSFET (and some voltage sources etc.) which can act as a latch—meaning that it is stable in two possible states and can act a single bit memory (i.e., when it is turned on it stays on by itself, and when it is turned off it stays off by itself).

(18.7) **Light Emission*&**

(a) Assuming both conduction band (with effective mass m_e^*) and valence band (with effective mass m_h^*) are thermally populated, determine the intensity of light power emitted as a function of frequency. You may assume that if an electron is present with momentum $\hbar\mathbf{k}$ and spin σ in the conduction band and there is an absence of an electron with momentum $\hbar\mathbf{k}$ with spin $-\sigma$ in the valence band, then a photon is emitted in a fixed time τ.

(b) In an LED how does the population of excited electrons and holes remain constant even though annihilation is happening continuously?

Part VII

Magnetism and Mean Field Theories

Magnetic Properties of Atoms: Para- and Dia-Magnetism

19

The first question one might ask is why we are interested in magnets. While the phenomenon of magnetism was known to the ancients,[1] it has only been since the discovery of quantum mechanics that we have come to any understanding of what causes this effect.[2] It may seem like this is a relatively small corner of physics for us to focus so much attention (indeed, several chapters), but we will see that magnetism is a particularly good place to observe the effects of both statistical physics and quantum physics.[3] As we mentioned in Section 16.4, one place where the band theory of electrons fails is in trying to describe magnets. Indeed, this is precisely what makes magnets interesting! In fact, magnetism remains an extremely active area of research in physics (with many many hard and unanswered questions remaining). Much of condensed matter physics continues to use magnetism as a testing ground for understanding complex quantum and statistical physics both theoretically and in the laboratory.

We should emphasize that most magnetic phenomena are caused by the quantum-mechanical behavior of *electrons*. While nuclei do have magnetic moments, and therefore can contribute to magnetism, the magnitude of the nuclear moments is (typically) much less than that of electrons.[4]

[1] Both the Chinese and the Greeks probably knew about magnetic properties of Fe_3O_4, or magnetite (also known as loadstone when magnetized) possibly as far back as several thousand years BC (with written records existing as far back as 600 BC). One legend has it that a shepherd named Magnes, in the provence of Magnesia, had the nails in his shoes stuck to a large metallic rock, and the scientific phenomenon became named after him.

[2] Animal magnetism not withstanding... (that was a joke).

[3] There is a theorem by Niels Bohr and Hendrika van Leeuwen which shows that any treatment of statistical mechanics without quantum mechanics (i.e., classical statistical mechanics) can never produce a non-zero magnetization.

[4] To understand this, recall that the Bohr magneton, which gives the size of the magnetic moment of an electron, is given by $\mu_B = \frac{e\hbar}{2m}$ with m the electron mass. If one were to consider magnetism caused by nuclear moments, the typical moments would be smaller by a ratio of the mass of the electron to the mass of a nucleus (a factor of over 1000). Nonetheless, the magnetism of the nuclei, although small, does exist.

19.1 Basic Definitions of Types of Magnetism

Let us first make some definitions. Recall that for a small magnetic field, the magnetization of a system $\mathbf{M}$ (moment per unit volume) is typically related linearly to the applied[5] magnetic field $\mathbf{H}$ by a (magnetic) susceptibility χ. We write for small[6] fields $\mathbf{H}$,

$$\mathbf{M} = \chi \mathbf{H}. \tag{19.1}$$

[6] One can also define a more general "differential" susceptibility $d\mathbf{M}/d\mathbf{H}$ at arbitrary field.

[5] The susceptibility is defined in terms of $\mathbf{H}$. With a long rod-shaped sample oriented parallel to the applied field, $\mathbf{H}$ is the same outside and inside the sample, and is thus directly controlled by the experimentalist. The susceptibility is defined in terms of this standard configuration. However, one should remember that the field $\mathbf{B}$ that any electron in the sample experiences is related to the applied field $\mathbf{H}$ via $\mathbf{B} = \mu_0(\mathbf{H} + \mathbf{M})$.

Note that χ is dimensionless.[7] For small susceptibilities (and susceptibilities are almost always small, except in ferromagnets) there is little difference between $\mu_0\mathbf{H}$ and $\mathbf{B}$ (with μ_0 the permeability of free space), so we can also write

$$\mathbf{M} = \chi\mathbf{B}/\mu_0 \; . \tag{19.2}$$

Definition 19.1 *A* ***paramagnet*** *is a material where $\chi > 0$ (i.e., the resulting magnetization is in the same direction as the applied field).*

We have run into (Pauli) paramagnetism previously in Section 4.3. You may also be familiar with the paramagnetism of a free spin (which we will cover again in Section 19.4). Qualitatively paramagnetism occurs whenever there are magnetic moments that can be reoriented by an applied magnetic field—thus developing magnetization in the direction of the applied field.

Definition 19.2 *A* ***diamagnet*** *is a material where $\chi < 0$ (i.e., the resulting magnetization is in the opposite direction from the applied field).*

We will discuss diamagnetism more in Section 19.5. As we will see, diamagnetism is quite ubiquitous and occurs generically unless it is overwhelmed by other magnetic effects. For example, water, and almost any other biological material, is diamagnetic.[8] Qualitatively we can think of diamagnetism as being similar in spirit to Lenz's law (part of Faraday's law) that an induced current generates a field that opposes the change causing it. However, the analogy is not precise. If a magnetic field is applied to a loop of wire, current will flow to create a magnetization in the opposite direction. However, in any (non-superconducting) loop of wire, the current will eventually decay back to zero and there will be no magnetization remaining. In a diamagnet, in contrast, the magnetization remains so long as the applied magnetic field remains.

For completeness we should also define a ferromagnet—this is what we usually think of as a "magnet" (the thing that holds notes to the fridge).

Definition 19.3 *A* ***ferromagnet*** *is a material where* $\mathbf{M}$ *can be non-zero, even in the absence of any applied magnetic field.*[9,10]

It is worth already drawing the distinction between *spontaneous* and *non-spontaneous* magnetism. Magnetism is said to be *spontaneous* if it occurs even in the absence of externally applied magnetic field, as is the case for a ferromagnet. The remainder of this chapter will mainly be concerned with non-spontaneous magnetism, and we will return to spontaneous magnetism in Chapter 20.

It turns out that a lot of the physics of magnetism can be understood by just considering a single atom at a time. This will be the strategy of the current chapter—we will discuss the magnetic behavior of a single atom and only in Section 19.6 will we consider how the physics changes when we put many atoms together to form a solid. We thus start this discussion by reviewing some atomic physics that you might have learned in prior courses.[11]

[7] Many different types of units are used in studying magnetism. We will stick to SI units (for which χ is dimensionless), but be warned that other books switch between systems of units freely.

[8] It is interesting to note that a diamagnet repels the field that creates it, so it is attracted to a magnetic field minimum. Earnshaw's theorem forbids a local maximum of the $\mathbf{B}$ field in free space, but local minima can exist—and this then allows diamagnets to levitate in free space. In 1997 Andre Geim used this effect to levitate a rather confused frog. This feat earned him a so-called Ig-Nobel Prize in 2000 (Ig-Nobel Prizes are awarded for research that "cannot or should not be reproduced".) Ten years later he was awarded a real Nobel Prize for the discovery of graphene—single-layer carbon sheets. This makes him the only person so far to receive both the Ig-Nobel and the real Nobel.

[9] The definition of ferromagnetism given here is a broad definition which would also include ferrimagnets. We will discuss ferrimagnets in Section 20.1.3, and we mention that occasionally people use a more restrictive definition (also commonly used) of ferromagnetism that excludes ferrimagnets. At any rate, the broad definition given here is common.

[10] The use of the word *can* in this sentence is a bit of a weasel-wording. In Section 21.1 we will see that ferromagnets, though they have a microscopic tendency to have non-zero $\mathbf{M}$, can have zero $\mathbf{M}$ macroscopically.

[11] You *should* have learned this in prior courses. But if not, it is probably not your fault! This material is rarely taught in physics courses these days, even though it really should be. Much of this material is actually taught in chemistry courses instead!

19.2 Atomic Physics: Hund's Rules

We start with some of the fundamentals of electrons in an isolated atom (i.e., we ignore the fact that in materials atoms are not isolated, but are bound to other atoms). Recall from basic quantum mechanics that an electron in an atomic orbital can be labeled by four quantum numbers, $|n, l, l_z, \sigma_z\rangle$. In Section 5.2 we explained how the Aufbau principle and Madelung's rule determine in what order the n and l shells are filled in the periodic table. Most elements will have several filled shells and a single partially filled shell (known as the *valence* shell). To understand much of magnetism, in these cases when shells are partially filled we must first figure out which of the available (l_z) orbitals are filled in these shells and which spin states (σ_z) are filled. In particular we want to know whether these electrons will have a net magnetic moment.

For an isolated atom there is a set of rules, known as "Hund's Rules",[12] which determines how the electrons fill orbitals and what spin states they take, and hence whether the atom has a magnetic moment. Perhaps the simplest way to illustrate these rules is to consider an explicit example. Here we will again consider the atom praseodymium which we discussed earlier in Section 5.2. As mentioned there, this element in atomic form has three electrons in its outermost shell, which is an f-shell, meaning it has angular momentum $l = 3$, and therefore 7 possible values of l_z, and of course two possible values of the spin for each electron. So where in these possible orbital/spin states do we put the three electrons?

[12] Friedrich Hermann Hund was an important physicist and chemist whose work on atomic structure began in the very early days of quantum mechanics—he wrote down Hund's rules in 1925. He is also credited with being one of the inventors of molecular orbital theory which we met in Chapter 6.2.2. In fact, molecular orbital theory is sometimes known as Hund–Mulliken molecular orbital theory. Mulliken thanked Hund heavily in his Nobel Prize acceptance speech (but Hund did not share the prize). Hund died in 1997 at the age of 101. The word "Hund" means "Dog" in German.

> **Hund's First Rule** (paraphrased): Electrons try to align their spins.

Given this rule, we know that the three valence electrons in Pr will have their spins point in the same direction, thus giving us a total spin angular momentum $S = 3/2$ from the three $S = 1/2$ spins. So locally (meaning on the same atom), the three electron spins behave ferromagnetically—they all align.[13] The reason for this alignment will be discussed in Section 19.2.1, but in short, it is a result of the Coulomb interaction between electrons (and between the electrons and the nucleus)—the Coulomb energy is lower when the electron spins align.

[13] We would not call this a true ferromagnet since we are talking about a single atom here, not a macroscopic material!

We now have to decide which orbital states to put the electrons in. For this we need another rule:

> **Hund's Second Rule** (paraphrased): Electrons try to maximize their total orbital angular momentum, consistent with Hund's first rule.

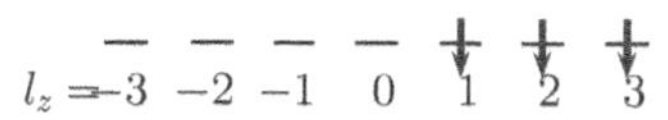

Fig. 19.1 The filling of the f shell of a Pr atom consistent with Hund's rules. We align spins and maximize L_z to maximize L.

For the case of Pr, we fill the $l_z = 3$ and $l_z = 2$ and $l_z = 1$ states to make the maximum possible total $L_z = 6$ (this gives $L = 6$, and by rotational invariance we can point $\mathbf{L}$ in any direction equally well). Thus, we fill orbitals as shown in Fig. 19.1. In the figure we have put the spins as far as possible to the right to maximize L_z (Hund's second rule) and we have aligned all the spins (Hund's first rule). Note that we could not have put two of the electrons in the same orbital, since they have to be spin-aligned and we must obey the Pauli principle. Again the rule of maximizing orbital angular momentum is driven by the physics of Coulomb interaction.

At this point we have $S = 3/2$ and $L = 6$, but we still need to think about how the spin and orbital angular momenta align with respect to each other. This brings us to the final rule:

> **Hund's Third Rule** (paraphrased): Given Hund's first and second rules, the orbital and spin angular momentum either align or antialign, so that the total angular momentum is $J = |L \pm S|$ with the sign being determined by whether the shell of orbitals is more than half filled $(+)$ or less than half filled $(-)$.

The reason for this rule is not interaction physics, but is spin–orbit coupling. The Hamiltonian will typically have a spin–orbit term $\alpha\, \mathbf{l} \cdot \boldsymbol{\sigma}$, and the sign of α determines how the spin and orbit align to minimize the energy.[14] Thus for the case of Pr, where $L = 6$ and $S = 3/2$ and the shell is less than half filled, we have total angular momentum $J = L - S = 9/2$.

One should be warned that people frequently refer to J as being the "spin" of the atom. This is a colloquial use which is very persistent but imprecise. More correctly, $\mathbf{J}$ is the total angular momentum of the electrons in the atom, whereas $\mathbf{S}$ is the spin component of $\mathbf{J}$.

[14] The fact that the sign switches at half filling does not signal a change in the sign of the underlying α (which is always positive) but rather is a feature of careful bookkeeping. So long as the shell remains less than half full, all of the spins are aligned in which case we have $\sum_i \mathbf{l}_i \cdot \boldsymbol{\sigma}_i = \mathbf{S} \cdot \mathbf{L}$ thus always favoring $\mathbf{L}$ counter-aligned with $\mathbf{S}$. When the shell is half filled $L = 0$. When we add one more spin to a half-filled shell, this spin must counter-align with the many spins that comprise the half-filled shell due to the Pauli exclusion principle. The spin–orbit coupling $\mathbf{l}_i \cdot \boldsymbol{\sigma}_i$ then makes this additional spin want to counter-align with its own orbital angular momentum $\mathbf{l}_i$, which is equal to the total orbital angular momentum $\mathbf{L}$ since the half full shell has $L = 0$. This means that the orbital angular momentum is now *aligned* with the net spin, since most of the net spin is made up of the spins comprising the half-filled shell and are counter-aligned with the spin of the electron which has been added.

19.2.1 Why Moments Align

We now return, as promised, to discuss roughly why Hund's rules work—in particular we want to know why magnetic moments (real spin moments or orbital moments) like to align with each other. This section will be only qualitative, but should give at least a rough idea of the right physics.

Let us first focus on Hund's first rule and ask why spins like to align. First of all, we emphasize that it has nothing to do with magnetic dipole interactions. While the magnetic dipoles of the spins do interact with each other, when dipole moments are on the order of the Bohr magneton, this energy scale becomes tiny—way too small to matter for anything interesting. Instead, the alignment comes from the Coulomb interaction energy. To see how this works, let us consider a wavefunction for two electrons on an atom.

Naive Argument

The overall wavefunction must be antisymmetric by Pauli's exclusion principle. We can generally write

$$\Psi(\mathbf{r}_1, \sigma_1; \mathbf{r}_2, \sigma_2) = \psi_{orbital}(\mathbf{r}_1, \mathbf{r}_2)\ \chi_{spin}(\sigma_1, \sigma_2)$$

where $\mathbf{r}_i$ are the particles' positions and σ_i are their spin. Now, if the two spins are aligned, say both are spin-up (i.e., $\chi_{spin}(\uparrow, \uparrow) = 1$ and $\chi_{spin} = 0$ for other spin configurations) then the spin wavefunction is symmetric and the spatial wavefunction $\psi_{orbital}$ must be antisymmetric. As a result we have

$$\lim_{\mathbf{r}_1 \to \mathbf{r}_2} \psi_{orbital}(\mathbf{r}_1, \mathbf{r}_2) \to 0.$$

So electrons with aligned spins cannot get close to each other, thus reducing the Coulomb energy of the system.

The argument we have just given is frequently stated in textbooks. Unfortunately, it is not the whole story.

More Correct

In fact it turns out that the crucial Coulomb interaction is that between the electron and the nucleus. Consider the case where there are two electrons and a nucleus as shown in Fig. 19.2. Note in this figure that the positive charge of the nucleus seen by one electron is screened by the negative charge of the other electron. This screening reduces the binding energy of the electrons to the nucleus. However, when the two spins are aligned, the electrons repel each other and therefore screen the nucleus less effectively. In this case, the electrons see the full charge of the nucleus and bind more strongly, thus lowering their energies.

Another way of understanding this is to realize that when the spins are not aligned, sometimes one electron gets between the other electron and the nucleus—thereby reducing the effective charge seen by the outer electron, reducing the binding energy, and increasing the total energy of the atom. However, when the electrons are spin aligned, the Pauli principle largely prevents this configuration from occurring, thereby lowering the total energy of the system.

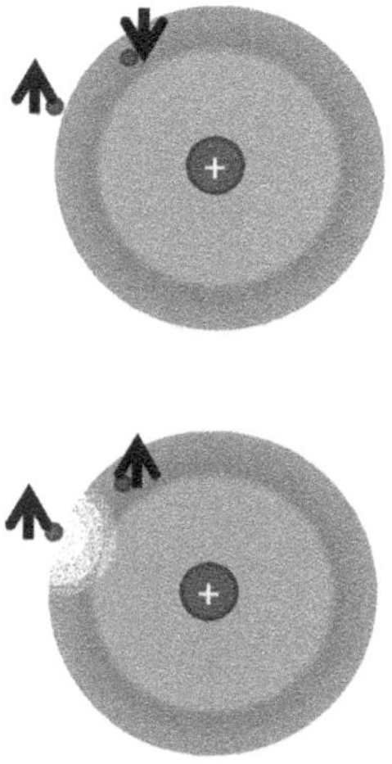

Fig. 19.2 Why aligned spins have lower energy (Hund's first rule). In this figure, the wavefunction is depicted for one of the electrons whereas the other electron (the one further left) is depicted as having fixed position. **Top:** When the two electrons have opposite spin, the effective charge of the nucleus seen by the fixed electron is reduced by the screening provided by the other electron. **Bottom:** However, when the spins are aligned, the two electrons cannot come close to each other and the fixed electron sees more of the charge of the nucleus. As such, the binding of the fixed electron to the nucleus is stronger in the case where the two electrons are spin aligned, therefore it is a lower-energy configuration.

Exchange Energy

The energy difference between having two spins aligned versus antialigned is usually known as the *exchange interaction* or *exchange energy*. The astute reader will recall that atomic physicists use the word "exchange" to refer to what we called the hopping matrix element (see margin note 13 in Section 6.2.2) which "exchanged" an electron from one orbital to another. In fact the current name is very closely related. To see the connection let us attempt a very simple calculation of the difference in energy between two electrons having their spins aligned and two electrons having their spins antialigned. Suppose we have two electrons on two different orbitals which we will call $A(\mathbf{r})$ and $B(\mathbf{r})$. We write a general wavefunction as $\psi = \psi_{spatial}\chi_{spin}$, and overall the wavefunction must be antisymmetric. If we choose the spins to be aligned (a triplet, therefore symmetric, such as $|\uparrow\uparrow\rangle$), then the spatial wavefunction must be antisymmetric, which we can write[15] as $|AB\rangle - |BA\rangle$. On the other hand, if we choose the spins to be antialigned (a singlet, therefore antisymmetric, i.e., $|\uparrow\downarrow\rangle - |\downarrow\uparrow\rangle$) then the spatial wavefunction must be symmetric $|AB\rangle + |BA\rangle$. When we add Coulomb interaction $V(\mathbf{r}_1, \mathbf{r}_2)$,

[15] Here $|AB\rangle$ means $A(\mathbf{r}_1)B(\mathbf{r}_2)$

the energy difference between the singlet and triplet can be calculated as follows:

$$\begin{aligned} E_{singlet} &= (\langle AB| + \langle BA|)\, V\, (|AB\rangle + |BA\rangle) \\ E_{triplet} &= (\langle AB| - \langle BA|)\, V\, (|AB\rangle - |BA\rangle) \\ E_{exchange} &= E_{singlet} - E_{triplet} = 4\mathrm{Re}\langle AB|V|BA\rangle. \end{aligned}$$

In the cross term $\langle AB|V|BA\rangle$ the two electrons have "exchanged" place. Hence the name.

Magnetic Interactions in Molecules and Solids

One must be somewhat careful with these types of arguments however—particularly when they are applied to molecules instead of atoms. In the case of a diatomic molecule, say H_2, we have two electrons and two nuclei. While the screening effect (Fig. 19.2) still occurs, and tries to align the electrons, it is somewhat less effective than for two electrons on a single atom—since most of the time the two electrons are near opposite nuclei anyway. Furthermore, there is a competing effect that tends to make the electrons want to antialign. As mentioned in Section 6.2.1 when we discussed covalent bonding, we can think of the two nuclei as being a square well (see Fig. 6.3), and the bonding is really a particle-in-a-box problem. There is some lowest-energy (symmetric) wavefunction in this large two-atom box, and the lowest-energy state of two electrons would be to have the two spins antialigned so that both electrons can go in the same low-energy spatial wavefunction. It can thus be quite difficult to determine whether electrons on neighboring atoms want to be aligned or antialigned. Generally either behavior is possible. (We will discuss this further in Chapter 23.)

19.3 Coupling of Electrons in Atoms to an External Field

Having discussed how electron moments (orbital or spin) can align with each other, we now turn to discuss how the electrons in atoms couple to an external magnetic field.

In the absence of a magnetic field, the Hamiltonian for an electron in an atom is of the usual form[16]

$$\mathcal{H}_0 = \frac{\mathbf{p}^2}{2m} + V(\mathbf{r})$$

where V is the electrostatic potential from the nucleus (and perhaps from the other electrons as well). Now consider adding an external magnetic field. Recall that the Hamiltonian for a charged particle in a magnetic field $\mathbf{B}$ takes the minimal coupling form[17]

$$\mathcal{H} = \frac{(\mathbf{p} + e\mathbf{A})^2}{2m} + g\mu_B \mathbf{B}\cdot\boldsymbol{\sigma} + V(\mathbf{r})$$

[16] Again, whenever we discuss magnetism it is typical to use $\mathcal{H}$ for the Hamiltonian so as not to confuse it with the magnetic field strength $H = B/\mu_0$.

[17] Recall that minimal coupling requires $\mathbf{p} \to \mathbf{p} - q\mathbf{A}$ where q is the charge of the particle. Here our particle has charge $q = -e$. The negative charge is also responsible for the fact that the electron spin magnetic moment is antialigned with its spin. Hence it is lower energy to have the spin point opposite the applied magnetic field (hence the positive sign of the so-called Zeeman term $g\mu_B\mathbf{B}\cdot\boldsymbol{\sigma}$). Blame Ben Franklin. (See margin note 15 of Section 4.3.)

where $-e$ is the charge of the particle (the electron), $\boldsymbol{\sigma}$ is the electron spin, g is the electron g-factor (approximately 2), $\mu_B = e\hbar/(2m)$ is the Bohr magneton, and $\mathbf{A}$ is the vector potential. For a uniform magnetic field, we may take $\mathbf{A} = \frac{1}{2}\mathbf{B} \times \mathbf{r}$ such that $\nabla \times \mathbf{A} = \mathbf{B}$. We then have[18]

$$\mathcal{H} = \frac{\mathbf{p}^2}{2m} + V(\mathbf{r}) + \frac{e}{2m}\mathbf{p} \cdot (\mathbf{B} \times \mathbf{r}) + \frac{e^2}{2m}\frac{1}{4}|\mathbf{B} \times \mathbf{r}|^2 + g\mu_B \mathbf{B} \cdot \boldsymbol{\sigma}. \quad (19.3)$$

The first two terms in this equation comprise the Hamiltonian $\mathcal{H}_0$ in the absence of the applied magnetic field. The next term can be rewritten

$$\frac{e}{2m}\mathbf{p} \cdot (\mathbf{B} \times \mathbf{r}) = \frac{e}{2m}\mathbf{B} \cdot (\mathbf{r} \times \mathbf{p}) = \mu_B \mathbf{B} \cdot \mathbf{l} \quad (19.4)$$

where $\hbar\mathbf{l} = \mathbf{r} \times \mathbf{p}$ is the orbital angular momentum of the electron. This can then be grouped with the so-called Zeeman term $g\mu_B\mathbf{B} \cdot \boldsymbol{\sigma}$ to give

$$\mathcal{H} = \mathcal{H}_0 + \mu_B \mathbf{B} \cdot (\mathbf{l} + g\boldsymbol{\sigma}) + \frac{e^2}{2m}\frac{1}{4}|\mathbf{B} \times \mathbf{r}|^2 . \quad (19.5)$$

The middle term on the right of this equation, known sometimes as the *paramagnetic* term, is clearly just the coupling of the external field to the total magnetic moment of the electron (both orbital moment $-\mu_B\mathbf{l}$ and spin moment $-g\mu_B\boldsymbol{\sigma}$). Note that when a $\mathbf{B}$-field is applied, these moments align with the $\mathbf{B}$-field (meaning that $\mathbf{l}$ and $\boldsymbol{\sigma}$ antialign with $\mathbf{B}$) such that the energy is lowered by the application of the field.[19] As a result a moment is created in the same direction as the applied field, and this term results in paramagnetism.

The final term of Eq. 19.5 is known as the *diamagnetic term* of the Hamiltonian, and will be responsible for the effect of diamagnetism. Since this term is quadratic in $\mathbf{B}$ it will always cause an *increase* in the total energy of the atom when the magnetic field is applied, and hence has the opposite effect from that of the above-considered paramagnetic term. Also being that this term is quadratic in B it should be expected to be less important than the paramagnetic term (which is linear in B) for small B.

These two terms of the Hamiltonian are the ones responsible for both the paramagnetic and diamagnetic response of atoms to external magnetic fields. We will treat them each in turn in the next two sections. Keep in mind that at this point we are still considering the magnetic response of a single atom!

19.4 Free Spin (Curie or Langevin) Paramagnetism

We will start by considering the effect of the paramagnetic term of Eq. 19.5. We assume that the unperturbed Hamiltonian $\mathcal{H}_0$ has been solved and we need not pay attention to this part of the Hamiltonian—we are only concerned with the reorientation of a spin $\boldsymbol{\sigma}$ and/or an orbital angular momentum $\mathbf{l}$ of an electron. At this point we also disregard the diamagnetic term of the Hamiltonian, as its effect is generally weaker than that of the paramagnetic term.

[18] Note that while p_i does not commute with r_i, it does commute with r_j for $j \neq i$, so there is no ordering problem between $\mathbf{p}$ and $\mathbf{B} \times \mathbf{r}$

[19] If the sign of the magnetic moment confuses you, it is good to remember that moment is always $-\partial F/\partial B$, and at zero temperature the free energy is just the energy.

Free Spin 1/2

As a review let us consider a simpler case that you are probably familiar with from your statistical physics course: a free spin-1/2. The Hamiltonian, you recall, of a single spin-1/2 is given by

$$\mathcal{H} = g\mu_B \mathbf{B} \cdot \boldsymbol{\sigma} \tag{19.6}$$

with g the g-factor of the spin which we set to be 2, and $\mu_B = e\hbar/(2m)$ is the Bohr magneton. We can think of this as being a simplified version of the paramagnetic term of Eq. 19.5, for a single free electron where we ignore the orbital moment. The eigenstates of $\mathbf{B} \cdot \boldsymbol{\sigma}$ are $\pm B/2$ so we have a partition function

$$Z = e^{-\beta\mu_B B} + e^{\beta\mu_B B} \tag{19.7}$$

and a corresponding free energy $F = -k_B T \log Z$, giving us a magnetic moment (per spin) of

$$\text{moment} = -\frac{\partial F}{\partial B} = \mu_B \tanh(\beta\mu_B B). \tag{19.8}$$

If we have many such atoms together in a volume, we can define the magnetization M to be the magnetic moment per unit volume. Then, at small field (expanding the tanh for small argument) we obtain a susceptibility of

$$\chi = \lim_{H\to 0} \frac{\partial M}{\partial H} = \frac{n\mu_0\mu_B^2}{k_B T} \tag{19.9}$$

where n is the number of spins per unit volume (and we have used $B \approx \mu_0 H$ with μ_0 the permeability of free space). Expression 19.9 is known as the "Curie law"[20] susceptibility (actually any susceptibility of the form $\chi \sim C/(k_B T)$ for any constant C is known as Curie law), and paramagnetism involving free spins like this is often called Curie paramagnetism or Langevin[21] paramagnetism.

[20]Named after Pierre Curie. Pierre's work on magnetism was well before he married his mega-brilliant wife Marie Skłodowska-Curie. She won one physics Nobel with Pierre, and then another one in chemistry after he died. (See margin note 5 in Section 6.1.) Half-way between the two prizes, Pierre was killed when he was run over by a horse-drawn vehicle while crossing the street. (Be careful!)

[21]Paul Langevin was Pierre Curie's student. He is well known for many important scientific discoveries. He is also well known for creating quite the scandal by having an affair with Marie Curie a few years after her husband's death (Langevin was married at the time). Although the affair quickly ended, ironically, the grandson of Langevin married the granddaughter of Curie and they had a son—all three of them are physicists.

Free Spin J

The actual paramagnetic term in the Hamiltonian will typically be more complicated than our simple spin-1/2 model, Eq. 19.6. Instead, examining Eq. 19.5 and generalizing to multiple electrons in an atom, we expect to need to consider a Hamiltonian of the form

$$\mathcal{H} = \mu_B \mathbf{B} \cdot (\mathbf{L} + g\mathbf{S}) \tag{19.10}$$

where $\mathbf{L}$ and $\mathbf{S}$ are the orbital and spin components of all of the electrons in the atom put together. Recall that Hund's rules tell us the value of L, S, and J. The form of Eq. 19.10 looks a bit inconvenient, since Hund's third rule tells us not about $\mathbf{L} + g\mathbf{S}$ but rather tells us about $\mathbf{J} = \mathbf{L} + \mathbf{S}$. Fortunately, for the type of matrix elements we are concerned with (reorientations of $\mathbf{J}$ without changing the value of J, S, or L which are dictated by Hund's rules) the Hamiltonian Eq. 19.10 turns out to be precisely equivalent to

$$\mathcal{H} = \tilde{g}\mu_B \mathbf{B} \cdot \mathbf{J} \tag{19.11}$$

where $\tilde{g}$ is an effective g-factor (known as the Landé g-factor) given by[22]

$$\tilde{g} = \frac{1}{2}(g+1) + \frac{1}{2}(g-1)\left[\frac{S(S+1) - L(L+1)}{J(J+1)}\right]$$

From our new Hamiltonian, it is easy enough to construct the partition function

$$Z = \sum_{J_z=-J}^{J} e^{-\beta \tilde{g}\mu_B B J_z} \ . \tag{19.12}$$

Analogous to the spin-1/2 case above one can differentiate to obtain the moment as a function of temperature. If one considers a density n of these atoms, one can then determine the magnetization and the susceptibility (see Exercise 19.7). The result, of the Curie form, is that the susceptibility per unit volume is given by (compare Eq. 19.9)

$$\chi = \frac{n\mu_0(\tilde{g}\mu_B)^2}{3}\frac{J(J+1)}{k_B T} \ .$$

Note that the Curie law susceptibility *diverges* at low temperature.[23] If this term is non-zero (i.e., if J is non-zero) then the Curie paramagnetism is dominant compared to any other type of paramagnetism or diamagnetism.[24]

Aside: From Eqs. 19.7 or 19.12 we notice that the partition function of a free spin is only a function of the dimensionless ratio $\mu_B B/(k_B T)$. From this we can derive that the entropy S is also a function only of the same dimensionless ratio. Let us imagine that we have a system of free spins at magnetic field B and temperature T, and we thermally isolate it from the environment. If we adiabatically reduce B, then since S must stay fixed, the temperature must drop proportionally to the reduction in B. This is the principle of the adiabatic demagnetization refrigerator.[25,26]

[23] The current calculation is a finite temperature thermodynamic calculation resulting in divergent susceptibility at zero temperature. In the next few sections we will study Larmor and Landau diamagnetism as well as Pauli and van Vleck paramagnetism. All of these calculations will be zero temperature quantum calculations and will always give much smaller finite susceptibilities.

[24] Not including superconductivity.

[25] Very-low-temperature adiabatic demagnetization refrigerators usually rely on using *nuclear* moments rather than electronic moments. The reason for this is that the (required) approximation of spins being independent holds down to much lower temperature for nuclei, which are typically quite decoupled from their neighbors. Achieving nuclear temperatures below 1μK is possible with this technique.

[26] The idea of adiabatic demagnetization was thought up by Debye.

19.5 Larmor Diamagnetism

Since Curie paramagnetism is dominant whenever $J \neq 0$, the only time we can possibly observe diamagnetism is if an atom has $J = 0$. A classic

[22] The derivation of this formula (although a bit off-topic) is not difficult. We are concerned in determining matrix elements of $\mathbf{B} \cdot (\mathbf{L} + g\mathbf{S})$ between different J_z states. To do this we write

$$\mathbf{B} \cdot (\mathbf{L} + g\mathbf{S}) = \mathbf{B} \cdot \mathbf{J}\left[\frac{\mathbf{L}\cdot\mathbf{J}}{|\mathbf{J}|^2} + g\frac{\mathbf{S}\cdot\mathbf{J}}{|\mathbf{J}|^2}\right].$$

The final bracket turns out to be just a number, which we evaluate by rewriting it as

$$\left[\frac{|\mathbf{J}|^2 + |\mathbf{L}|^2 - |\mathbf{J}-\mathbf{L}|^2}{2|\mathbf{J}|^2}\right] + g\left[\frac{|\mathbf{J}|^2 + |\mathbf{S}|^2 - |\mathbf{J}-\mathbf{S}|^2}{2|\mathbf{J}|^2}\right]$$

Finally replacing $\mathbf{J} - \mathbf{L} = \mathbf{S}$ and $\mathbf{J} - \mathbf{S} = \mathbf{L}$, then substituting in $|\mathbf{J}|^2 = J(J+1)$ and $|\mathbf{S}|^2 = S(S+1)$ and $|\mathbf{L}|^2 = L(L+1)$, with a small bit of algebra gives the desired result.

situation in which this occurs is for atoms with filled shell configurations, like the noble gases[27] where $L = S = J = 0$. Another possibility is that $J = 0$ even though $L = S$ is non-zero (one can use Hund's rules to show that this occurs if a shell has one electron fewer than being half filled). In either case, the paramagnetic term of Eq. 19.5 has zero expectation and the term can be mostly ignored.[28] We thus need to consider the effect of the final term in Eq. 19.5, the diamagnetic term.

[27] Molecules with filled shells of *molecular* orbitals, such as N_2, are very similar.

If we imagine that **B** is applied in the $\hat{z}$ direction, the expectation of the diamagnetic term of the Hamiltonian (Eq. 19.5) can be written as

$$\delta E = \frac{e^2}{8m}\langle |\mathbf{B}\times\mathbf{r}|^2\rangle = \frac{e^2B^2}{8m}\langle x^2+y^2\rangle.$$

Using the fact that the atom is rotationally symmetric, we can write

$$\langle x^2+y^2\rangle = \frac{2}{3}\langle x^2+y^2+z^2\rangle = \frac{2}{3}\langle r^2\rangle,$$

so we have

$$\delta E = \frac{e^2B^2}{12m}\langle r^2\rangle.$$

Thus the magnetic moment per electron is[29]

$$\text{moment} = -\frac{dE}{dB} = -\left[\frac{e^2}{6m}\langle r^2\rangle\right]B.$$

[29] Here the magnetic moment is proportional to the area $\langle r^2\rangle$ enclosed by the orbit of the electron. Note that the magnetization we get for a current loop is also proportional to the area of the loop. This is in accordance with our understanding of diamagnetism as being vaguely similar to Lenz's law.

Assuming that there is a density ρ of such electrons in a system, we can then write the susceptibility as

$$\chi = -\frac{\rho e^2\mu_0\langle r^2\rangle}{6m}. \tag{19.13}$$

This result, Eq. 19.13, is known as *Larmor diamagnetism*.[30] For most atoms, $\langle r^2\rangle$ is on the order of a few Bohr radii squared. In fact, the same expression can sometimes be applied for large conductive molecules if the electrons can freely travel the length of the molecule—by taking $\langle r^2\rangle$ to be the radius squared of the molecule instead of that of the atom.

[30] Joseph Larmor was a rather important physicist in the late 1800s. Among other things, he published the Lorentz transformations for time dilation and length contraction two years before Lorentz, and seven years before Einstein. However, he insisted on the aether, and rejected relativity at least until 1927 (maybe longer).

19.6 Atoms in Solids

Up to this point, we have always been considering the magnetism (paramagnetism or diamagnetism) of a single isolated atom. Although the

[28] Actually, to be more precise, even though $\mathbf{J} = \mathbf{L} + \mathbf{S}$ may be zero, the paramagnetic term in Eq. 19.5 may be important in second-order perturbation theory if **L** and **S** are individually nonzero. At second order, the energy of the system will be corrected by a term proportional to

$$\delta E_0 \sim + \sum_{p>0} \frac{|\langle p|\mathbf{B}\cdot(\mathbf{L}+g\mathbf{S})|0\rangle|^2}{E_0 - E_p}$$

and the matrix element in the numerator is generally non-zero if the state $|p\rangle$ has the same L and S as the ground state but a different J. (Recall that our effective Hamiltonian, Eq. 19.11, is valid only within a space of fixed J). Since this energy *decreases* with increasing B, this term is paramagnetic. This type of paramagnetism is known as *van Vleck paramagnetism* after the Nobel laureate J. H. van Vleck, who was a professor at Balliol College Oxford in 1961–62 but spent most of his later professional life at Harvard.

atomic picture gives a very good idea of how magnetism occurs, the situation in solids can be somewhat different. As we have discussed in Chapters 15 and 16 when atoms are put together the electronic band structure defines the physics of the material—we cannot usually think of atoms as being isolated from each other. We thus must think a bit more carefully about how our atomic calculations may or may not apply to real materials.

19.6.1 Pauli Paramagnetism in Metals

Recall that in Section 4.3 we calculated the susceptibility of the free Fermi gas. We found

$$\chi_{Pauli} = \mu_0 \mu_B^2 g(E_F) \tag{19.14}$$

with $g(E_F)$ the density of states at the Fermi surface. We might expect that such an expression would hold for metals with non-trivial band structure—with the only change being that the density of states would need to be modified. Indeed, such an expression holds fairly well for simple metals such as Li or Na.

Note that the susceptibility, per spin, of a Fermi gas (Eq. 19.14) is smaller than the susceptibility of a free spin (Eq. 19.9) by roughly a factor of $k_B T/E_F$ (this can be proven using Eq. 4.11 for a free electron gas). We should be familiar with this idea, that due to the Pauli exclusion principle, only the small fraction of spins near the Fermi surface can be flipped over, therefore giving a small susceptibility.

19.6.2 Diamagnetism in Solids

Our calculation of Larmor diamagnetism (Section 19.5) was applied to isolated atoms each having $J = L = S = 0$, such as noble gas atoms. At low temperature, noble gas atoms form very weakly bonded crystals and the same calculation continues to apply (with the exception of the case of helium which does not crystalize but rather forms a superfluid at low temperature[31]). To apply Eq. 19.13 to a noble gas crystal, one simply sets the density of electrons ρ to be equal to the density of atoms n times the number of electrons per atom (the atomic number) Z. Thus for noble gas atoms we obtain

$$\chi_{Larmor} = -\frac{Zne^2\mu_0\langle r^2\rangle}{6m} \tag{19.15}$$

where $\langle r^2\rangle$ is set by the atomic radius.

In fact, for any material, the diamagnetic term of the Hamiltonian (the coupling of the orbital motion to the magnetic field) will result in some amount of diamagnetism. To account for the diamagnetism of electrons in core orbitals, Eq. 19.15 is usually fairly accurate. For the conduction electrons in a metal, however, a much more complicated calculation gives the so-called Landau diamagnetism (see margin note 14 of Chapter 4)

$$\chi_{Landau} = -\tfrac{1}{3}\chi_{Pauli}$$

[31] Alas, superfluidity is beyond the scope of this book. It is extremely interesting and I encourage you to learn more about it!

which combines with the Pauli paramagnetism to reduce the total paramagnetism of the conduction electrons by 1/3.

If one considers, for example, a metal like copper, one might be tempted to conclude that it should be a paramagnet, due to the above-described Pauli paramagnetism (corrected by the Landau effect). However, copper is actually a diamagnet! The reason for this is that the core electrons in copper have enough Larmor diamagnetism to overwhelm the Pauli paramagnetism of the conduction electrons! In fact, Larmor diamagnetism is often strong enough to overwhelm Pauli paramagnetism in metals (this is particularly true in heavy elements where there are many core electrons that can contribute to the diamagnetism). Note however, if there are free spins in the material, then Curie paramagnetism occurs which is always stronger than any diamagnetism.[24]

19.6.3 Curie Paramagnetism in Solids

Where to find free spins?

As discussed in section 19.4, Curie paramagnetism describes the reorientation of free spins in an atom. We might ask how a "free spin" can occur in a solid? Our understanding of electrons in solids so far describes electrons as being either in full bands, in which case they cannot be flipped over at all, or in partially full bands, in which case the calculation of the Pauli susceptibility in Section 4.3 is valid—albeit possibly with a modified density of states at the Fermi surface to reflect the details of the band structure (and with the Landau correction). So how is it that we can have a free spin?

Let us think back to the description of Mott insulators in Section 16.4. In these materials, the Coulomb interaction between electrons is strong enough that no two electrons can double occupy the same site of the lattice. As a result, having one electron per site results in a "traffic jam" of electrons where no electron can hop to any other site. When this sort of Mott insulator forms, there is exactly one electron per site, which can be either spin-up or spin-down. Thus we have a free spin on each site and we should expect Curie paramagnetism![32]

[32]This picture of a Mott insulator resulting in independent free spins will be examined more closely in Chapter 23. Very weakly, some amount of (virtual) hopping can always occur and this will change the behavior at low enough temperatures.

More generally we might expect that we could have some number N valence electrons per atom, which fill orbitals to form free spins as dictated by Hund's rules. Again, if the Coulomb interaction is sufficiently strong that electrons cannot hop to neighboring sites, then the system will be Mott-insulating and we can think of the spins as being free.

Modifications of Free Spin Picture

Given that we have found free spins in a material, we can ask whether there are substantial differences between a free spin in an isolated atom and a free spin in a material.

One possible modification is that the number of electrons on an atom become modified in a material. For example, we found in Section 5.2 that praseodymium (Pr) has three free electrons in its valence (4f) shell which form a total angular momentum of $J = 9/2$ (as we found in

Section 19.2). However, in many compounds Pr exists as a +3 ion. In this case it turns out that both of the 6s electrons are donated as well as a single f electron. This leaves the Pr atom with two electrons in its f shell, thus resulting in a $J = 4$ angular momentum instead (you should be able to check this with Hund's rules!).

Another possibility is that the atoms are no longer in a rotationally symmetric environment; they see the potential due to neighboring atoms, the so-called "crystal field". In this case orbital angular momentum is not conserved and the degeneracy of states all having the same L^2 is broken, a phenomenon known as *crystal field splitting*.

As a (very) cartoon picture of this physics, we can imagine a crystal which is highly tetragonal (see Fig. 12.11) where the lattice constant in one direction is quite different from the constant in the other two. We might imagine that an atom that is living inside such an elongated box would have a lower energy if its orbital angular momentum pointed along the long axes (say, the z-axis), rather than in some other direction. In this case, we might imagine that $L_z = +L$ and $L_z = -L$ might be lower energy than any of the other possible values of L.

Another thing that may happen due to crystal field splitting is that the orbital angular momentum may be pinned to have zero expectation (for example, if the ground state is a superposition of $L_z = +L$ and $L_z = -L$). In this case, the orbital angular momentum decouples from the problem completely (a phenomenon known as *quenching* of the orbital angular momentum), and the only magnetically active degrees of freedom are the spins—only **S** can be reoriented by a magnetic field. This is precisely what happens for most transition metals.[33]

[33]The 3d shell of transition metals is shielded from the environment only by the 4s electrons, whereas for rare earths the 4f shell is shielded by 6s and 5p. Thus the transition metals are much more sensitive to crystal field perturbations than the rare earths.

The most important message to take home from this section is that atoms can have many different effective values of J, and one needs to know the microscopic details of the system before deciding which spin and orbital degrees of freedom are active. Remember that whenever there is a magnetic moment that can be reoriented by a magnetic field the material will be paramagnetic.

One final word of caution. Throughout most of this chapter we have treated the magnetism of atoms one atom at a time. In the next chapters we will consider what happens when atoms magnetically couple to their neighbors. Our approximation of treating atoms one at a time is typically a good approximation when the temperature is much larger than any coupling strength between atoms. However, at low enough temperatures, even a very weak coupling will become important, as we shall see in some detail in Chapter 22.

Chapter Summary

- Susceptibility $\chi = dM/dH$ is positive for paramagnets and negative for diamagnets.
- Sources of paramagnetism: (a) Pauli paramagnetism of free electron gas (see Section 4.3) (b) Free spin paramagnetism—know how to do the simple stat-mech exercise of calculating the paramagnetism of a free spin.
- The magnitude of the free spin is determined by Hund's rules. The bonding of the atom, or environment of this atom (crystal field) can modify this result.
- Larmor diamagnetism can occur when atoms have $\mathbf{J} = \mathbf{0}$, therefore not having strong paramagnetism. This comes from the diamagnetic term of the Hamiltonian in first-order perturbation theory. The diamagnetism per electron is proportional to the radius of the orbital squared.

References

- Ibach and Luth, section 8.1
- Hook and Hall, chapter 7
- Ashcroft and Mermin, chapter 31 plus 32 for discussion of exchange
- Kittel, chapter 11
- Blundell, chapter 2 plus 4.2 for exchange
- Burns, chapter 15A
- Goodstein, sections 5.4a–c (doesn't cover diamagnetism)
- Rosenberg, chapter 11 (doesn't cover diamagnetism)
- Pauling, for more on Hund's rules and relation to chemistry
- Hook and Hall, appendix D for discussion of exchange

Exercises

(19.1) ‡ **Atomic Physics and Magnetism**

(a) Explain qualitatively why some atoms are paramagnetic and others are diamagnetic with reference to the electronic structure of these materials.

(b) Use Hund's rules and the Aufbau principle to determine L, S, and J for the following isolated atoms:

(i) Sulfur (S) atomic number = 16
(ii) Vanadium (V), atomic number = 23
(iii) Zirconium (Zr), atomic number = 40
(iv) Dysprosium (Dy), atomic number = 66

(19.2) **More Atomic Physics**

(a) In solid erbium (atomic number=68), one electron from each atom forms a delocalized band so each Er atom has eleven f electrons on it. Calculate the Landé g-factor for the eleven electrons (the localized moment) on the Er atom.

(b)& Europeum (atomic number =63), in materials such as EuO, EuS, EuSe, EuTe, forms a localized moment with all seven of its f electrons. Calculate the Landé g-factor for the localized moment on the Eu atom.

(19.3) **Hund's Rules***

Suppose an atomic shell of an atom has angular momentum l ($l = 0$ means an s-shell, $l = 1$ means a p-shell etc, with an l-shell having $2l + 1$ orbital states and two spin states per orbital). Suppose this shell is filled with n electrons. Derive a general formula for S, L, and J as a function of l and n based on Hund's rules.

(19.4) ‡ **Para and Diamagnetism**

Manganese (Mn, atomic number=25) forms an atomic vapor at 2000K with vapor pressure 10^5 Pa. You can consider this vapor to be an ideal gas.

(a) Determine L, S, and J for an isolated manganese atom. Determine the paramagnetic contribution to the (Curie) susceptibility of this gas at 2000K.

(b) In addition to the Curie susceptibility, the manganese atom will also have some diamagnetic susceptibility due to its filled core orbitals. Determine the Larmor diamagnetism of the gas at 2000K. You may assume the atomic radius of an Mn atom is one Ångstrom.

Make sure you know the derivations of all the formulas you use!

(19.5) ‡**Diamagnetism**

(a) Argon is a noble gas with atomic number 18 and atomic radius of about .188 nm. At low temperature it forms an fcc crystal. Estimate the magnetic susceptibility of solid argon.

(b) The wavefunction of an electron bound to an impurity in n-type silicon is hydrogenic in form. Estimate the impurity contribution to the diamagnetic susceptibility of a Si crystal containing a density of 10^{20} m^{-3} donors given that the electron effective mass $m^* = 0.4m_e$ and the relative permittivity is $\epsilon_r = 12$. How does this value compare to the diamagnetism of the underlying silicon atoms? Si has atomic number 14, atomic weight 28.09, and density 2.33 g/cm^3.

(19.6) ‡**Paramagnetism**

Consider a gas of monovalent atoms with spin $S = 1/2$ (and $L = 0$, so g factor is 2) in a magnetic field B. The gas has density n.

(a) Calculate the magnetization as a function of B and T. Determine the susceptibility.

(b) Calculate the contribution to the specific heat of this gas due to the spins. Sketch this contribution as a function of $\mu_B B/k_B T$.

(19.7) **Spin J Paramagnet***

Given the Hamiltonian for a system of non-interacting spin-J atoms

$$\mathcal{H} = \tilde{g}\mu_B \mathbf{B} \cdot \mathbf{J}$$

(a)* Determine the magnetization as a function of B and T.

(b) Show that the susceptibility is given by

$$\chi = \frac{n\mu_0(\tilde{g}\mu_B)^2}{3} \frac{J(J+1)}{k_B T}$$

where n is the density of spins. (You can do this part of the exercise without having a complete closed-form expression for part a!)

(19.8) **Hund's Rules and g-factors&**

A particular insulating material has no magnetic elements except for a density ρ of Gd^{+3} ions (which have 7 electrons in a partially filled f-shell). A second material is identical to the first except that instead of Gd^{+3} ions, it has Dy^{+3} ions (which have 9 electrons in a partially filled f-shell).

(a) When these materials are put in a large magenetic field, the magnetic moments of the ions all align. Calculate the ratio of the saturation magnetic moments of the two materials.

(b) At small magnetic field, calculate the ratio of magnetic susceptibilities of the two materials.

Consider a similar experiment where one material has Fe^{+3} ions (which have 5 electrons in a partially filled d-shell) and the other has Mn^{+3} ions (which have 4 electrons in a partially filled d-shell).

(c) Show that the magnetic susceptibility of the Fe^{+3} containing material is roughly 1.45 times as big as that of the Mn^{+3} containing material. Hint: If this seems wrong, read margin note 33 and try again!

Spontaneous Magnetic Order: Ferro-, Antiferro-, and Ferri-Magnetism

20

At the end of Section 19.2.1 we commented that applying Hund's rules to molecules and solids can be unreliable, since spins on neighboring atoms could favor either having their spins aligned or having their spins antialigned, depending on which of several effects is stronger. (In Chapter 23 we will show a detailed models of how either behavior might occur.) In the current chapter we will simply *assume* that there is an interaction between neighboring spins (a so-called *exchange* interaction, see the discussion at the end of Section 19.2.1) and we will explore how the interaction between neighboring spins can align many spins on a macroscopic scale.

We first assume that we have an insulator, i.e., electrons do not hop from atom to atom.[1] We then write a model Hamiltonian as

$$\mathcal{H} = -\frac{1}{2}\sum_{i,j} J_{ij}\mathbf{S}_i \cdot \mathbf{S}_j + \sum_i g\mu_B \mathbf{B}\cdot\mathbf{S}_i \qquad (20.1)$$

where $\mathbf{S}_i$ is the spin[2] on atom[3] i and $\mathbf{B}$ is the magnetic field experienced by the spins.[4] Here $J_{ij}\mathbf{S}_i \cdot \mathbf{S}_j$ is the interaction energy[5] between spin i and spin j. Note that we have included a factor of $1/2$ out front to avoid overcounting, since the sum actually counts both J_{ij} and J_{ji} (which are equal to each other).

If $J_{ij} > 0$ it is lower energy when spins i and j are aligned, whereas if $J_{ij} < 0$ it is lower energy when the spins are antialigned. This energy difference is usually called the *exchange energy* as described in Section 19.2.1. Correspondingly J_{ij} is called the exchange constant.

The coupling between spins typically drops rapidly as the distance between the spins increases. A good model to use is one where only nearest-neighbor spins interact with each other. Frequently one writes (neglecting the magnetic field $\mathbf{B}$)

$$\mathcal{H} = -\frac{1}{2}\sum_{i,j\; neighbors} J_{ij}\,\mathbf{S}_i \cdot \mathbf{S}_j \quad .$$

One can use brackets $\langle i,j\rangle$ to indicate that i and j are neighbors:

$$\mathcal{H} = -\frac{1}{2}\sum_{\langle i,j\rangle} J_{ij}\,\mathbf{S}_i \cdot \mathbf{S}_j \quad .$$

[1] This might be the situation if we have a Mott insulator, as described in Sections 16.4 and 19.6.3, where strong interaction prevents electron hopping. In Chapter 23 we will also consider the more complicated, but important, case where the electrons in a ferromagnet are mobile.

[2] When one discusses simplified models of magnetism, very frequently one writes angular momentum as $\mathbf{S}$ without regards as to whether it is really $\mathbf{S}$, or $\mathbf{L}$ or $\mathbf{J}$. It is also conventional to call this variable the "spin" even if it is actually from orbital angular momentum in a real material.

[3] A moment associated with a specific atom is known as a *local moment.*

[4] Once again the plus sign in the final term assumes that we are talking about electronic moments (see margin note 15 of Section 4.3).

[5] *Warning:* Many references use Heisenberg's original convention that the interaction energy should be defined as $2J_{ij}\mathbf{S}_i \cdot \mathbf{S}_j$ rather than $J_{ij}\mathbf{S}_i \cdot \mathbf{S}_j$. However, more modern researchers use the latter, as we have here. This matches up with the convention used for the Ising model in Eq. 20.5, where the convention $2J$ is never used. At any rate, if someone on the street tells you J, you should ask whether they intend a factor of 2 or not.

In a uniform system where each spin is coupled to its neighbors with the same strength, we can drop the indices from J_{ij} (since they all have the same value) and obtain the so-called *Heisenberg Hamiltonian*

$$\mathcal{H} = -\frac{1}{2}\sum_{\langle i,j\rangle} J\,\mathbf{S}_i \cdot \mathbf{S}_j \quad . \tag{20.2}$$

20.1 (Spontaneous) Magnetic Order

As in the case of a ferromagnet, it is possible that even in the absence of any applied magnetic field, magnetism—or ordering of magnetic moments—may occur. This type of phenomenon is known as *spontaneous* magnetic order (since it occurs without application of any field). It is a subtle statistical mechanical question as to when magnetic interaction in a Hamiltonian actually results in spontaneous magnetic order. At our level of analysis we will assume that systems can always find ground states which "satisfy" the magnetic Hamiltonian. In Chapter 22 we will consider how temperature might destroy this magnetic ordering.

20.1.1 Ferromagnets

If $J > 0$ then neighboring spins want to be aligned. In this case the ground state occurs when all spins align together developing a macroscopic magnetic moment—this is what we call a *ferromagnet*,[6] and is depicted on the left of Fig. 20.1. The study of ferromagnetism will occupy us for most of the remainder of this book.

[6] A classic example of a Heisenberg ferromagnet is EuO, which is ferromagnetic below about 70K. This material has very simple NaCl structure where the moment $\mathbf{S}$ is on the Eu. Here Eu has seven electrons in an f shell, so by Hund's rule it has $J = S = 7/2$.

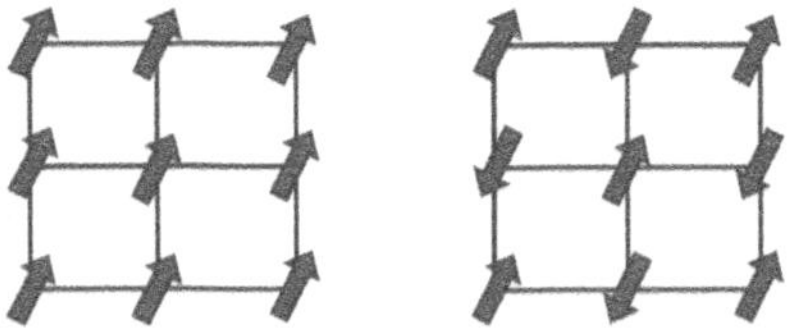

Fig. 20.1 Magnetic Spin Orderings. **Left:** Ferromagnet—all spins aligned (at least over some macroscopic regions) giving finite magnetization. **Right:** Antiferromagnet—Neighboring spins antialigned, but periodic. This so-called Néel state has zero net magnetization.

20.1.2 Antiferromagnets

On the other hand, if $J < 0$, neighboring spins want to point in opposite directions, and the most natural ordered arrangement is a periodic situation where alternating spins point in opposite directions, as shown on the right of Fig. 20.1—this is known as an *antiferromagnet*. Such an antiferromagnet has zero net magnetization but yet is magnetically ordered. This type of antiperiodic ground state is sometimes known as a *Néel state* after Louis Néel who proposed in the 1930s that such states exist.[7,8] We should be cautioned that our picture of spins pointing in directions is a classical picture, and is not quite right quantum mechanically (see Exercise 20.1). Particularly when the spin is small (like spin-1/2) the effects of quantum mechanics are strong and classical intuition can fail us. However, for spins larger than 1/2, the classical picture is still fairly good.

[7] Néel won a Nobel Prize for this work in 1970. The great Lev Landau (see margin note 18 in Chapter 4 about Landau) also proposed antiferromagnetism at about the same time as Néel. However, soon thereafter, Landau started to doubt that antiferromagnets actually existed in nature (his reasoning was based on the fact that two spins will form a quantum-mechanical singlet rather than having one always spin-up and one always spin-down). It was almost fifteen years before scattering experiments convinced Landau that antiferromagnets are indeed real. For one of these fifteen years, Landau was in jail, having compared Stalin to the Nazis. He was very lucky not to have been executed.

[8] Some examples of antiferromagnets include NiO and MnO (both having NaCl structure, with the spins on Ni and Mn respectively). A very important class of antiferromagnets are materials such as $LaCuO_2$, which when doped give high-temperature superconductors.

Detecting Antiferromagnetism with Diffraction

Being that antiferromagnets have zero net magnetization, how do we know they exist? What is their signature in the macroscopic world? It is possible (using the techniques of Section 22.2.2, see in particular Exercise 22.5) to find signatures of antiferromagnetism by examining the susceptibility as a function of temperature.[9] However, this method is somewhat indirect. A more direct approach is to examine the spin configuration using diffraction of neutrons. As mentioned in Section 14.2, neutrons are sensitive to the spin direction of the object they scatter from. If we fix the spin polarization of an incoming neutron, it will scatter differently from the two different possible spin states of atoms in an antiferromagnet. The neutrons then see that the unit cell in this antiferromagnet is actually of size $2a$, where a is the distance between atoms (i.e., the distance between two atoms with the same spin is $2a$). Thus when the spins align antiferromagnetically, the neutrons will develop scattering peaks at reciprocal wavevectors $G = 2\pi/(2a)$ which would not exist if all the atoms were aligned the same way. This type of neutron diffraction experiment are definitive in showing that antiferromagnetic order exists.[10]

[9] In fact it was this type of experiment that Néel was analyzing when he realized that antiferromagnets exist!

[10] These are the experiments that won the Nobel Prize for Clifford Shull. See margin note 10 from Chapter 14.

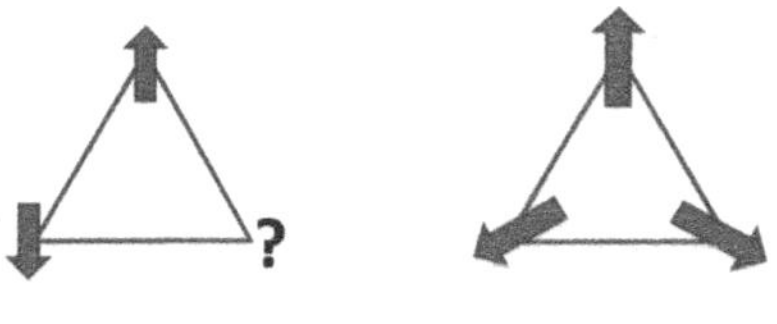

Fig. 20.2 Cartoon of a triangular antiferromagnet. **Left:** An antiferromagnetic interaction on a triangular lattice is frustrated—not all spins can be antialigned with all of their neighbors. **Right:** The ground state of antiferromagnetic interaction on a triangle for classical spins (large S) is the state on the right, where spins are at 120° to their neighbor.

Frustrated Antiferromagnets

On certain lattices, for certain interactions, there is no ground state that fully "satisfies" the interaction for all spins. For example, on a triangular lattice if there is an antiferromagnetic interaction, there is no way that all the spins can point in the opposite direction from their neighbors. As shown in the left of Fig. 20.2 on a triangle, once two of the spins are aligned opposite each other, independent of which direction the spin on the last site points, it cannot be antiparallel to both of its neighbors. It turns out that (assuming the spins are classical variables) the ground state of the antiferromagnetic Heisenberg Hamiltonian on a triangle is the configuration shown on the right of Fig. 20.2. While each bond is not quite optimally antialigned, the overall energy is optimal for this Hamiltonian (at least if the spins are classical).

20.1.3 Ferrimagnets

Once one starts to look for magnetic structure in materials, one can find many other interesting possibilities. One very common possibility is where you have a unit cell with more than one variety of atom, where the atoms have differing moments, and although the ordering is antiferromagnetic (neighboring spins point in opposite direction) there is still a net magnetic moment. An example of this is shown in Fig. 20.3. Here, the smaller moments point opposite the larger moments. This type of configuration, where one has antiferromagnetic order, yet a net magnetization due to differing spin species, is known as *ferrimagnetism*. In fact, many of the most common magnets, such as magnetite (Fe_3O_4) are ferrimagnetic. Sometimes people speak of ferrimagnets as being a subset of ferromagnets (since they have non-zero net magnetic moment in zero field) whereas other people think the word "ferromagnet" excludes ferrimagnets.[11]

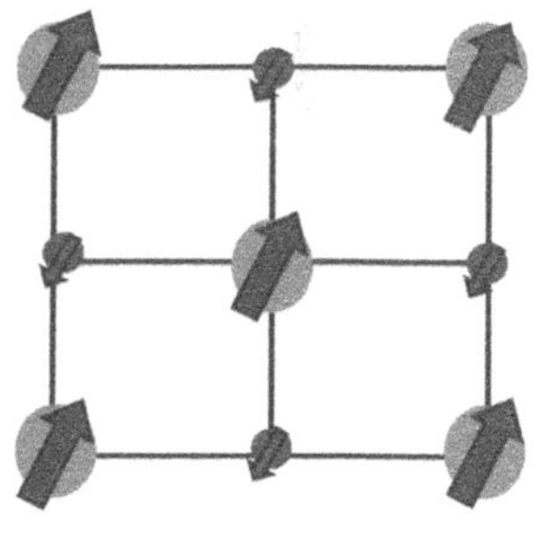

Fig. 20.3 Cartoon of a ferrimagnet. Ordering is antiferromagnetic, but because the different spin species have different moments, there is a net magnetization.

[11] The fact that the scientific community cannot come to agreement on so many definitions does make life difficult sometimes.

20.2 Breaking Symmetry

In any of these ordered states, we have not yet addressed the question of which direction the spins will actually point. Strictly speaking, the Hamiltonian Eq. 20.2 is rotationally symmetric—the magnetization can point in any direction and the energy will be the same! In a real system, however, this is rarely the case: due to the asymmetric environment the atom feels within the lattice, there will be some directions that the spins would rather point than others (this physics was also discussed in Section 19.6). Thus to be more accurate we might need to add an additional term to the Heisenberg Hamiltonian. One possibility is to write[12] (again dropping any external magnetic field)

$$\mathcal{H} = -\frac{1}{2}\sum_{\langle i,j\rangle} J\mathbf{S}_i \cdot \mathbf{S}_j - \kappa \sum_i (S_i^z)^2 \quad . \tag{20.3}$$

The κ term here favors the spin to be pointing in the $+\hat{z}$ direction or the $-\hat{z}$ direction, but not in any other direction (you could imagine this being appropriate for a tetragonal crystal elongated in the $\hat{z}$ direction). This energy from the κ term is sometimes known as the *anisotropy energy* since it favors certain directions over others. Another possible Hamiltonian is[13]

$$\mathcal{H} = -\frac{1}{2}\sum_{\langle i,j\rangle} J\mathbf{S}_i \cdot \mathbf{S}_j - \tilde{\kappa} \sum_i [(S_i^x)^4 + (S_i^y)^4 + (S_i^z)^4] \tag{20.4}$$

which favors the spin pointing along any of the orthogonal axis directions—but not towards any in-between angle.

In some cases (as we discussed in Section 19.6) the coefficient κ may be substantial. In other cases it may be very small. However, since the pure Heisenberg Hamiltonian Eq. 20.2 does not prefer any particular direction, even if the anisotropy (κ) term is extremely small, it will determine the direction of the magnetization in the ground state (in the absence of any external B field). We say that this term "breaks the symmetry". Of course, there may be some symmetry remaining. For example, in Eq. 20.3, if the interaction is ferromagnetic, the ground-state magnetization may be all spins pointing in the $+\hat{z}$ direction, or equally favorably, all spins pointing in the $-\hat{z}$ direction.

[12] For small values of the spin quantum number, these added terms may be trivial. For example, for spin 1/2, we have $(S^x)^2 = (S^y)^2 = (S^z)^2 = 1/4$. However, as S becomes larger, the spin becomes more like a classical vector and such anisotropy (κ) terms will favor the spin pointing in the corresponding directions.

[13] The latter term here has fourth powers since if we were to use $(S^x)^2 + (S^y)^2 + (S^z)^2$ this would just give us S^2 which would be a constant independent of the direction the spin points.

20.2.1 Ising Model

If the anisotropy (κ) term is extremely large, then this term can fundamentally change the Hamiltonian. For example, let us take a spin-S Heisenberg model. Adding the κ term in 20.3 with a large coefficient, forces the spin to be either $S_z = +S$ or $S_z = -S$ with all other values of S_z having a much larger energy. In this case, a new effective model may be written

$$\mathcal{H} = -\frac{1}{2}\sum_{\langle i,j\rangle} J\sigma_i\sigma_j + g\mu_B B \sum_i \sigma_i \tag{20.5}$$

[14] "Ising" is properly pronounced "Ee-sing" or "Ee-zing". In the United States it is habitually mispronounced "Eye-sing". The Ising model was actually invented by Wilhelm Lenz (another example of Stigler's law, see margin note 11 in Section 15.2). Ising was the graduate student who worked on this model for his graduate dissertation, but soon left research to work at a teaching college in the United States, where people inevitably called him "Eye-sing".

where $\sigma_i = \pm S$ only (and we have re-introduced the magnetic field B). This model is known as the *Ising model*[14] and is an extremely important model in statistical physics.[15]

[15]The Ising model is frequently referred to as the "hydrogen atom" of statistical mechanics since it is extremely simple, yet it shows many of the most important features of complex statistical mechanical systems. The one-dimensional version of the model was solved by Ising in 1925, and the two-dimensional version of the model was solved by Onsager in 1944 (a chemistry Nobel laureate, who was amusingly fired by my alma-mater, Brown University, in 1933). Onsager's achievement was viewed as so important that Wolfgang Pauli wrote after world war two that "nothing much of interest has happened [in physics during the war] except for Onsager's exact solution of the two-dimensional Ising model". (Perhaps Pauli was spoiled by the years of amazing progress in physics between the wars). Just in the last few years certain aspects of the three-dimensional Ising model appear to have been calculated exactly for the first time. If you are brave, you might try solving the one-dimensional Ising model (see Exercises 20.5 and 20.6).

Chapter Summary

- Ferromagnets: spins align. Antiferromagnets: spins antialign with neighbors so there is no net magnetization. Ferrimagnets: spins antialign with neighbors, *but* alternating spins are different magnitude so there is a net magnitization anyway. Microscopic spins stuctures of this sort can be observed with neutrons.
- Useful model Hamiltonians include the Heisenberg $-J\mathbf{S_i}\cdot\mathbf{S_j}$ Hamiltonian for isotropic spins, and the Ising $-JS_i^zS_j^z$ Hamiltonian for spins that prefer to align along only one axis.
- Spins generally do not equally favor all directions (as the Heisenberg model suggest). Anisotropy terms that favor spins along particular axes may be weak or strong. Even if they are weak, they will pick a direction among otherwise equally likely directions.

References

- Blundell, sections 5.1–5.3 (Very nice discussion, but covers mean field theory at the same time which we will cover in Chapter 22.)
- Burns, sections 15.4–15.8 (same comment)
- Hook and Hall, chapter 8 (same comment)
- Kittel, chapter 12
- Blundell, chapter 4 (mechanisms of getting spin-spin interactions)

Exercises

(20.1) **Ferromagnetic vs Antiferromagnetic States**

Consider the Heisenberg Hamiltonian

$$\mathcal{H} = -\frac{1}{2}\sum_{\langle i,j\rangle} J\,\mathbf{S}_i\cdot\mathbf{S}_j + \sum_i g\mu_B\mathbf{B}\cdot\mathbf{S}_i \qquad (20.6)$$

and for this exercise set $\mathbf{B}=0$.

(a) For $J>0$, i.e., for the case of a ferromagnet, intuition tells us that the ground state of this Hamiltonian should simply have all spins aligned. Consider such a state. Show that this is an eigenstate of the Hamiltonian Eq. 20.6 and find its energy.

(b) For $J<0$, the case of an antiferromagnet on a cubic lattice, one might expect that (at least for $\mathbf{B}=0$) the state where spins on alternating sites point in opposite directions might be an eigenstate. Unfortunately, this is not precisely true. Consider such a state of the system. Show that the state in question is not an eigenstate of the Hamiltonian.

Although the intuition of alternating spins on alternating sites is not perfect, it becomes reasonable for systems with large spins S. For smaller spins (like spin 1/2) one needs to consider so-called "quantum fluctuations" (which is much more advanced, so we will not do that here).

(20.2) **Frustration**

Consider the Heisenberg Hamiltonian as in Exercise 20.1 with $J < 0$, and treat the spins as classical vectors of equal length.

(a) If the system consists of only three spins arranged in a triangle (as in Fig. 20.2), show that the ground state has each spin oriented 120° from its neighbor.

(b) For an infinite triangular lattice, what does the ground state look like?

(20.3) **Spin Waves***

For the spin-S ferromagnet particularly for large S, our "classical" intuition is fairly good and we can use simple approximations to examine the excitation spectrum above the ground state.

First recall the Heisenberg equations of motion for any operator

$$i\hbar \frac{d\hat{O}}{dt} = [\hat{O}, \mathcal{H}]$$

with $\mathcal{H}$ the Hamiltonian (Eq. 20.6 with $\mathbf{S_i}$ being a spin S operator).

(a) Derive equations of motion for the spins in the Hamiltonian Eq. 20.6. Show that one obtains

$$\hbar \frac{d\mathbf{S_i}}{dt} = \mathbf{S_i} \times \left(J \sum_j \mathbf{S_j} - g\mu_B \mathbf{B} \right) \qquad (20.7)$$

where the sum is over sites j that neighbor i.

In the ferromagnetic case, particularly if S is large, we can treat the spins as not being operators, but rather as being classical variables. In the ground state, we can set all $\mathbf{S_i} = \hat{z}S$ (Assuming $\mathbf{B}$ is in the $-\hat{z}$ direction so the ground state has spins aligned in the $\hat{z}$ direction). Then to consider excited states, we can perturb around this solution by writing

$$\begin{aligned} S_i^z &= S - \mathcal{O}((\delta S)^2/S) \\ S_i^x &= \delta S_i^x \\ S_i^y &= \delta S_i^y \end{aligned}$$

where we can assume δS^x and δS^y are small compared to S. Expand the equations of motion (Eq. 20.7) for small perturbation to obtain equations of motion that are linear in δS_x and δS_y

(b) Further assume wavelike solutions

$$\begin{aligned} \delta S_j^x &= A_x e^{i\omega t - i\mathbf{k}\cdot\mathbf{r}_j} \\ \delta S_j^y &= A_y e^{i\omega t - i\mathbf{k}\cdot\mathbf{r}_j} \end{aligned}$$

This ansatz should look very familiar from our prior consideration of phonons.

Plug this form into your derived equations of motion.

▷ Show that S_i^x and S_i^y are out of phase by $\pi/2$. What does this mean?

▷ Show that the dispersion curve for "spin-waves" of a ferromagnet is given by $\hbar\omega = |F(\mathbf{k})|$ where

$$\begin{aligned} F(\mathbf{k}) &= g\mu_b|B| \\ &+ JS(6 - 2[\cos(k_x a) + \cos(k_y a) + \cos(k_z a)]) \end{aligned}$$

where we assume a cubic lattice.

▷ How might these spin waves be detected in an experiment?

(c) Assume the external magnetic field is zero. Given the spectrum you just derived, show that the specific heat due to spin wave excitations is proportional to $T^{3/2}$.

(20.4) **Small Heisenberg Models**

(a) Consider a Heisenberg model containing a chain of only two spins, so that

$$\mathcal{H} = -J\mathbf{S_1} \cdot \mathbf{S_2} \ .$$

Supposing these spins have $S = 1/2$, calculate the energy spectrum of this system. Hint: Write $2\mathbf{S_1} \cdot \mathbf{S_2} = (\mathbf{S_1} + \mathbf{S_2})^2 - \mathbf{S_1}^2 - \mathbf{S_2}^2$.

(b) Now consider three spins forming a triangle (as shown in Fig. 20.2). Again assuming these spins are $S = 1/2$, calculate the spectrum of the system. Hint: Use the same trick as in part (a)!

(c) Now consider four spins forming a tetrahedron. Again assuming these spins are $S = 1/2$, calculate the spectrum of the system.

(20.5) **One-Dimensional Ising Model with $B = 0$**

(a) Consider the one-dimensional Ising model with spin $S = 1$. We write the Hamiltonian for a chain of N spins in zero magnetic field as

$$\mathcal{H} = -J \sum_{i=1}^{N-1} \sigma_i \sigma_{i+1}$$

where each σ_i takes the value ± 1. The partition function can be written as

$$Z = \sum_{\sigma_1, \sigma_2, \ldots \sigma_N} e^{-\beta \mathcal{H}} \ .$$

Using the transformation $R_i = \sigma_i \sigma_{i+1}$ rewrite the partition function as a sum over the R variables, and hence evaluate the partition function.

▷ Show that the free energy has no cusp or discontinuity at any temperature, and hence conclude that there is no phase transition in the one-dimensional Ising model.

(b) *At a given temperature T, calculate an expression for the probability that M consecutive spins will be pointing in the same direction. How does this probability decay with M for large M? What happens as T becomes small? You may assume $N \gg M$.

(20.6) **One-Dimensional Ising Model with $B \neq 0$ ***

Consider the one-dimensional Ising model with spin $S = 1$. We write the Hamiltonian (Eq. 20.5) for a chain of N spins in magnetic field B as

$$\mathcal{H} = \sum_{i=1}^{N} \mathcal{H}_i \tag{20.8}$$

where

$$\mathcal{H}_1 = h\sigma_1$$
$$\mathcal{H}_i = -J\sigma_i\sigma_{i-1} + h\sigma_i \qquad \text{for } i > 1$$

where each σ_i takes the value ± 1 and we have defined $h = g\mu_B B$ for simplicity of notation.

Let us define a partial partition function for the first M spins (the first M terms in the Hamiltonian sum Eq. 20.8) *given that* the Mth spin is in a particular state. I.e.,

$$Z(M, \sigma_M) = \sum_{\sigma_1, \ldots, \sigma_{M-1}} e^{-\beta \sum_{i=1}^{M} \mathcal{H}_i}$$

so that the full partition function is $Z = Z(N, +1) + Z(N, -1)$.

(a) Show that these partial partition functions satisfy a recursion relation

$$Z(M, \sigma_M) = \sum_{\sigma_{M-1}} T_{\sigma_M, \sigma_{M-1}} Z(M-1, \sigma_{M-1})$$

where T is a 2 by 2 matrix, and find the matrix T. (T is known as a "transfer matrix").

(b) Write the full partition function in terms of the matrix T raised to the $(N-1)^{th}$ power.

(c) Show that the free energy per spin, in the large N limit, can be written as

$$F/N \approx -k_B T \log \lambda_+$$

where λ_+ is the larger of the two eigenvalues of the matrix T.

(d) From this free energy, derive the magnetic moment per spin for small h, and show that the susceptibility per spin is given by

$$\chi \propto \beta e^{2\beta J}$$

which matches the Curie form at high T.

(20.7) **Anisotropy Terms&**

As pointed out in margin note 12 of this chapter, for small values of spin S, anisotropy terms may be trivial.

(a) What is the smallest value of spin S for which the anisotropy term in Eq. 20.4 is nontrivial?

(b) Consider an anisotropy term of the form $S_x^4 + S_y^4$. Under what circumstances might you expect to find such a term? What is the smallest value of S for which this term is nontrivial. For this value of S, is there a simpler way to write this expression? Does the same simplification work for larger S?

(20.8) **Antiferromagnetic Spin Waves**&**

(a) Consider an antiferromagnet on a cubic lattice in zero magnetic field. Using the methods of exercise 20.3 above, derive the equations of motion

$$\hbar \frac{d\boldsymbol{\delta}\mathbf{S}_i}{dt} = \pm JS\hat{z} \times \left(\sum_j [\boldsymbol{\delta}\mathbf{S}_j + \boldsymbol{\delta}\mathbf{S}_i] \right)$$

where the $\pm$ depends on whether site i is on the even or odd sublattice, δ means a deviation from the equilibrium position in the $x-y$ direction, and the sum on j is over the neighbors of i.

(b) Instead of working with equations for δS^x and δS^y it is easier to consider a single equation of motion for $\delta S^+ = \delta S^x + i\delta S^y$. Making a wave ansatz (as in the diatomic chain, you will have two equations since the unit cell has two sites) show that the antiferromagnetic spin-wave spectrum is

$$\hbar\omega(k) = 2JS\sqrt{\sin^2(k_x a) + \sin^2(k_y a) + \sin^2(k_z a)}$$

(20.9) **Frustration in One Dimension*&**

Another way to obtain a frustrated interaction is when a neighbor interaction competes with a next neighbor interaction. Consider the Hamiltonian

$$\mathcal{H} = -J_1 \sum_i \sigma_i \sigma_{i+1} - J_2 \sum_i \sigma_i \sigma_{i+2}$$

with $J_1, J_2 < 0$ so that both interactions cannot be satisfied at the same time.

Using the method of Exercise 20.5, reduce the problem to the situation considered in Exercise 20.6 and follow the same method of solution to find the free energy as a function of temperature. From the free energy, determine the probability that two adjacent spins will be aligned. At low temperature, how does this change as a function of J_1/J_2?

21 Domains and Hysteresis

21.1 Macroscopic Effects in Ferromagnets: Domains

We might think that in a ferromagnet, all the spins in the system will align as described in the previous chapter by the Heisenberg (or Ising) model. However, in real magnets, this is frequently not the case. To understand why this is so, we imagine splitting our sample into two halves as shown in Fig. 21.1. Once we have two magnetic dipoles it is clear that they would be lower energy if one of them flipped over, as shown at the far right of Fig. 21.1 (the two north faces of these magnets repel each other[1]). This energy, the long-range dipolar force of a magnet, is not described in the Heisenberg or Ising models at all. In those models we have only included nearest-neighbor interaction between spins. As we mentioned in Section 19.2.1, the actual magnetic dipolar force between electronic spins (or orbital moments) is tiny compared to the Coulomb interaction driven "exchange" force between neighboring spins. But when you put together a whole lot of atoms (like 10^{23} of them!) to make a macroscopic magnet, the summed effect of their dipole moment can be substantial.

Of course, in an actual ferromagnet (or ferrimagnet), the material does not really break apart, but nonetheless different regions will have magnetization in different directions in order to minimize the dipolar energy. A region where the moments all point in one given direction is known as a *domain* or a *Weiss domain*.[2] The boundary of a domain,

[1] Another way to understand the dipolar force is to realize that the magnetic field will be much lower if the two magnets are antialigned with each other. Since the electromagnetic field carries energy $\frac{1}{2}\int dV|B|^2/\mu_0$, minimizing this magnetic field lowers the energy of the two dipoles.

[2] After Pierre-Ernest Weiss, one of the fathers of the study of magnets from the early 1900s.

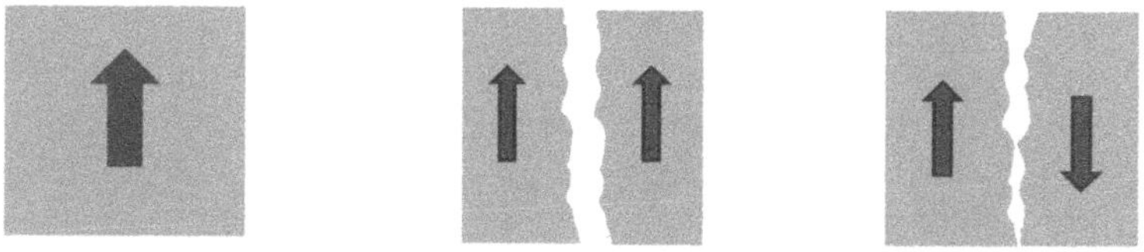

Fig. 21.1 Dipolar Forces Create Magnetic Domains. **Left:** The original ferromagnet. **Middle:** The original ferromagnet broken into two halves. **Right:** Because two dipoles next to each other are lower energy if their moments are antialigned, the two broken halves would rather line up in opposing directions to lower their energies (the piece on the right-hand side has been flipped over here). This suggests that in large ferromagnets, domains may form.

where the magnetization switches direction, is known as a *domain wall.*[3] Some possible examples of domain structures are sketched in Fig. 21.2. In the left two frames we imagine an Ising-like ferromagnet where the moment can only point up or down. The left most frame shows a magnet with net zero magnetization. Along the domain walls, the ferromagnetic Hamiltonian is "unsatisfied". In other words, spin-up atoms on one side of the domain wall have spin-down neighbors—where the local Hamiltonian says that they should want to have spin-up neighbors only. What is happening is that the system is paying an energy cost along the domain wall in order that the *global* energy associated with the long-range dipolar forces is minimized.

If we apply a small up-pointing external field to this system, we will obtain the middle picture where the up-pointing domains grow at the expense of the down-pointing domains to give an overall magnetization of the sample. In the rightmost frame of Fig. 21.2 we imagine a sample where the moment can point along any of the crystal axis directions.[4] Again in this picture the total magnetization is zero, but it has rather complicated domain structure.

[3]Domain walls can also occur in antiferromagnets. Instead of the magnetization switching directions we imagine a situation where to the left of the wall, the up-spins are on the even sites, and the down-spins are on the odd sites, whereas on the right of the domain wall the up-spins are on the odd sites and the down-spins are on the even sites. At the domain wall, two neighboring sites will be aligned rather than antialigned. Since antiferromagnets have no net magnetization, the argument that domain walls should exist in ferromagnets is not valid for antiferromagnets. In fact, it is always energetically unfavorable for domain walls to exist in antiferromagnets, although they can occur at finite temperature.

[4]See for example the Hamiltonian, Eq. 20.4, which would have moments pointing only along the coordinate axes—although that particular Hamiltonian does not have the long-range magnetic dipolar interaction written, so it would not form domains.

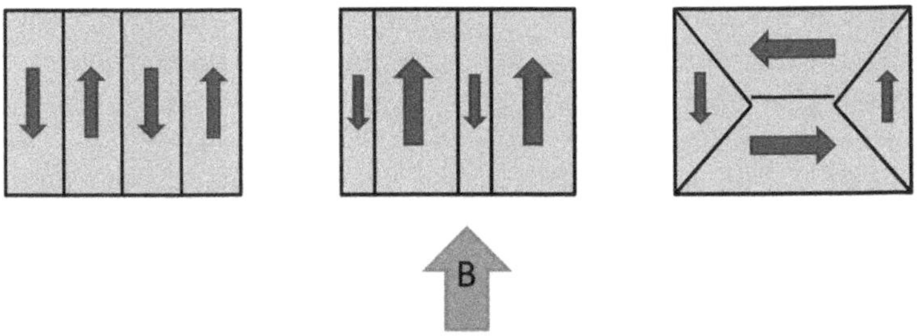

Fig. 21.2 Some possible domain structures for a ferromagnet. **Left:** An Ising-like ferromagnet where in each domain the moment can only point either up or down. **Middle:** When an external magnetic field pointing upwards is applied to this ferromagnet, it will develop a net moment by having the down-domains shrink and the up-domains expand (The local moment per atom remains constant—only the size of the domains change.) **Right:** In this ferromagnet, the moment can point in any of the (orthogonal) crystal axis directions.

21.1.1 Domain Wall Structure and the Bloch/Néel Wall

The detailed geometry of domains in a ferromagnet depends on a number of factors. First of all, it depends on the overall geometry of the sample. (For example, if the sample is a very long thin rod and the system is magnetized along the long axis, it may gain very little energy by forming domains.) It also depends on the relative energies of the neighbor interaction versus the long-range dipolar interaction: increasing the strength of the long-range dipolar forces with respect to the neighbor interaction will obviously decrease the size of domains (having no long-range dipolar forces will result in domains of infinite size). Finally, the detailed disorder in a sample can effect the shape and size of magnetic domains. For example, if the sample is polycrystalline, each domain could be a single crystallite (a single microscopic crystal)—a case which we will discuss in Section 21.2.2. In the current section we will take a closer look at the microscopic structure of the domain wall.

Our discussion of domain walls so far has assumed that the spins can only point in special directions picked out by the crystal axes—that is, the anisotropy term κ in Eq. 20.3 (or Eq. 20.4) is extremely strong. However, it often happens that this is not true—the spins would prefer to point either up or down, but there is not a huge energy penalty for pointing in other directions instead. In this case the domain wall might instead be more of a smooth rotation from spins pointing up to spins pointing down, as shown on the bottom of Fig. 21.3. This type of smooth domain wall is known as a *Bloch wall* or *Néel wall*,[5] depending on which direction the spin rotates with respect to the direction of the domain wall itself (a somewhat subtle distinction, which we will not discuss further here). The length of the domain wall (L in the figure, i.e., how many spins are pointing neither up nor down) is clearly dependent on a balance between the $-J\mathbf{S}_\mathrm{i} \cdot \mathbf{S}_\mathrm{j}$ term of Eq. 20.3 (known sometimes as the *spin stiffness*) and the κ term, the anisotropy. If κ/J is very large, then the spins must point either up or down only. In this case, the domain wall is very sharp, as depicted on the top of Fig. 21.3. On the other hand, if κ/J is small, then it costs little to point the spins in other directions, and it is more important that each spin points mostly in the direction of its neighbor. In this case, the domain wall will be very fat, as depicted on the bottom of Fig. 21.3.

[5] We have already met our heroes of magnetism—Felix Bloch and Louis Néel.

A very simple scaling argument can give us an idea of how fat the Bloch/Néel wall is. Let us say that the length of the wall is N lattice constants, so $L = Na$ is the actual length of the twist in the domain wall (see Fig. 21.3). Roughly let us imagine that the spin twists uniformly over the course of these N spins, so between each spin and its neighbor, the spin twists an angle $\delta\theta = \pi/N$. The first term $-J\mathbf{S}_i \cdot \mathbf{S}_\mathrm{j}$ in the Hamiltonian Eq. 20.3 can be rewritten in terms of the angle between the neighbors (approximating spins as classical vectors here)

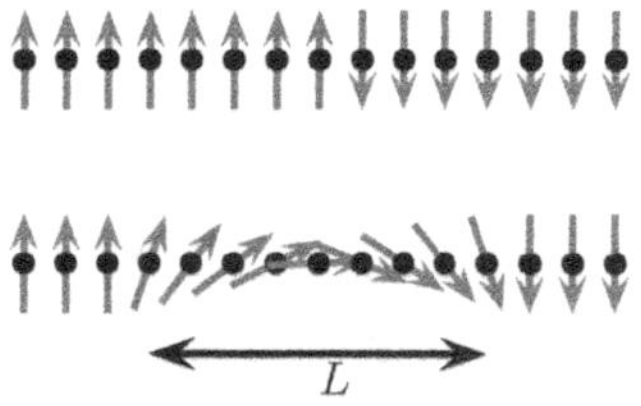

Fig. 21.3 Domain wall structures. **Top:** A sharp domain wall. This would be realized if the anisotropy energy (κ) is extremely large so the spins must point either up or down (i.e., this is a true Ising system). **Bottom:** A Bloch/Néel wall (actually this depicts a Néel wall) where the spin flips continuously from up to down over a length scale L. The anisotropy energy here is smaller so that the spin can point at intermediate angle for only a small energy penalty. By twisting slowly the domain wall will cost less spin-stiffness energy.

$$E_{one-bond} = -J\mathbf{S}_\mathrm{i} \cdot \mathbf{S}_\mathrm{j} = -JS^2 \cos(\theta_i - \theta_j) = -JS^2 \left(1 - \frac{(\delta\theta)^2}{2} + \ldots\right)$$

where we have used the fact that $\delta\theta$ is small to expand the cosine. Naturally, the energy of this term is minimized if the two neighboring spins are aligned, that is $\delta\theta = 0$. However, if they are not quite aligned there is an energy penalty of

$$\delta E_{one-bond} = JS^2(\delta\theta)^2/2 = JS^2(\pi/N)^2/2 \quad .$$

This is the energy per bond. So the energy of the domain wall due to this spin "stiffness" is

$$\frac{\delta E_{stiffness}}{A/a^2} = NJS^2(\pi/N)^2/2$$

where we have written the energy per unit area A of the domain wall, where area is measured in units of the lattice constant a.

On the other hand, in Eq. 20.3 there is a penalty proportional to κS^2 per spin when the spins are not either precisely up or down. We estimate the energy due to this term to be κS^2 per spin, or a total of

$$\frac{\delta E_{anisotropy}}{A/a^2} \approx \kappa S^2 N$$

along the length of the twist. Thus the total energy of the domain wall is

$$\frac{E_{tot}}{A/a^2} = JS^2(\pi^2/2)/N + \kappa S^2 N.$$

This can be trivially minimized, resulting in a domain wall twist having length $L = Na$ with

$$N = C_1\sqrt{(J/\kappa)} \tag{21.1}$$

and a minimum domain wall energy per unit area

$$\frac{E_{tot}^{min}}{A/a^2} = C_2 S^2 \sqrt{J\kappa}$$

where C_1 and C_2 are constants of order unity (which we will not get right here considering the crudeness our approximation, but see Exercise 21.3). As predicted, the length increases with J/κ. In many real materials the length of a domain wall can be hundreds of lattice constants.

Since domain walls cost an energy per unit area, they are energetically unfavorable. However, as mentioned at the beginning of this chapter, this energy cost is weighed against the long-range dipolar energy which tries to introduce domain walls. The more energy the domain wall costs, the larger individual domains will be (to minimize the number of domain walls). Note that if a crystal is extremely small (or, say, one considers a single crystallite within a polycrystaline sample) it can happen that the size of the crystal is much smaller than the optimum size of the domain wall twist. In this case the spins within this crystallite always stay aligned with each other.

21.2 Hysteresis in Ferromagnets

21.2.1 Disorder Pinning

We know from our experience with electromagnetism that ferromagnets show a hysteresis loop with respect to the applied magnetic field, as shown in Fig. 21.4. After a large external magnetic field is applied, when the field is returned to zero there remains a residual magnetization. We can now ask why this should be true. In short, it is because there is a large activation energy for changing the magnetization.

21.2.2 Single-Domain Crystallites

For example, let us consider the case of a ferromagnet made of many small crystallites. We determined in Section 21.1.1 that domain walls

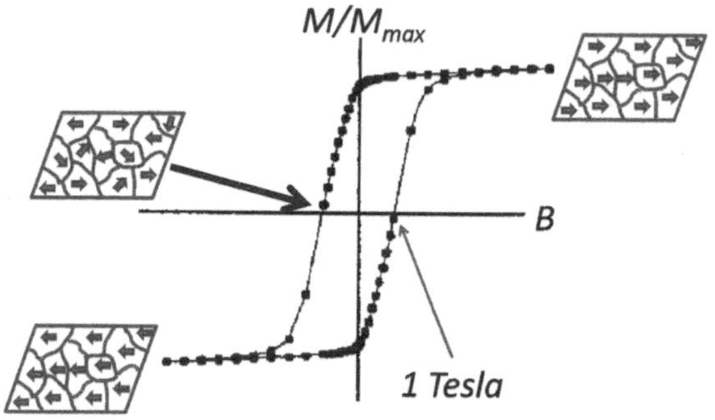

Fig. 21.4 The hysteresis loop of a ferromagnet. Hysteresis can be understood in terms of crystallites, or domains, reorienting. The data shown here is taken on samarium cobalt[6] magnetic powder. Data from Leslie-Pelecky *et al.*, *Phys. Rev. B*, **59** 457 (1999); http://prb.aps.org/abstract/PRB/v59/i1/p457_1, copyright American Physical Society. Used by permission.

are unfavorable in small enough crystallites. So if the crystallites are small enough then all of the moments in each crystallite point in a single direction. So let us imagine that all of the microscopic moments (spins or orbital moments) in this crystallite are locked with each other and point in the same direction. The energy per volume of the crystallite in an external field can be written as

$$E/V = E_0 - \mathbf{M} \cdot \mathbf{B} - \kappa'(M_z)^2$$

where here $\mathbf{M}$ is magnetization vector, and M_z is its component in the $\hat{z}$ crystal axis. Here the anisotropy term κ' stems from the anisotropy term κ in the Hamiltonian Eq. 20.3.[7] Note that we have no J term since this would just give a constant if all the moments in the crystallite are always aligned with each other.

Assuming that the external field $\mathbf{B}$ is pointing along the $\hat{z}$ axis (although we will allow it to point either up or down) we then have

$$E/V = E_0 - |M||B|\cos\theta - \kappa'|M|^2(\cos\theta)^2 \qquad (21.2)$$

where $|M|$ is the magnitude of magnetization and θ is the angle of the magnetization with respect to the $\hat{z}$ axis.

We see that this energy is a parabola in the variable $(\cos\theta)$ which ranges from $+1$ to -1. The minimum of this energy is always when the magnetization points in the same direction as the external field (which we have taken to always point in the either the $+\hat{z}$ or $-\hat{z}$ direction, corresponding to $\theta = 0$ or π). However, for small B_z the energy is not monotonic in θ. Indeed, having the magnetization point in the *opposite* direction as B is also a local minimum (because the κ' term favors pointing along the z-axis). This is shown schematically in Fig. 21.5. It is easy to show (see Exercise 21.2) that there will be a local minimum of the energy with the magnetization pointing the opposite direction as the applied field for $B < B_{crit}$ with

$$B_{crit} = 2\kappa'|M|.$$

So if the magnetization is aligned along the $-\hat{z}$ direction and a field $B < B_{crit}$ is applied in the $+\hat{z}$ direction, there is an activation barrier for the moments to flip over. Indeed, since the energy shown in Eq. 21.2

[6] Samarium cobalt magnets have particularly high permanent magnetization (neodymium magnets are the only magnets having higher magnetization). Fender Guitars makes some of their electric pickup magnets from this material.

[7] In particular, since $\mathbf{M} = -g\mu_B\mathbf{S}\rho$ with ρ the number of spins per unit volume we have $\kappa' = \kappa/[(g\mu_B)^2\rho]$. Further, we note that the $-\mathbf{M}\cdot\mathbf{B}$ term is precisely the Zeeman energy $+g\mu_B\mathbf{B}\cdot\mathbf{S}$ per unit volume.

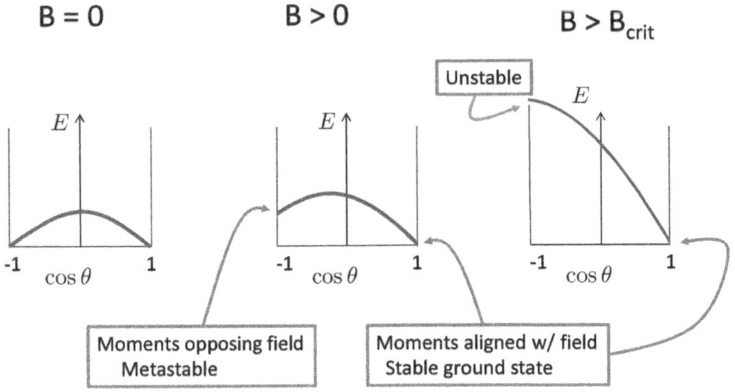

Fig. 21.5 Energy of an anisotropic ferromagnet in a magnetic field as a function of angle. **Left:** Due to the anisotropy, in zero field the energy is lowest if the spins point either in the $+\hat{z}$ or $-\hat{z}$ direction. **Middle:** When a field is applied in the $+\hat{z}$ direction the energy is lowest when the moments are aligned with the field, but there is a metastable solution with the moments pointing in the opposite direction. The moments must cross an activation barrier to flip over. **Right:** For large enough field, there is no longer a metastable solution.

is an energy *per volume*, the activation barrier can be very large.[8] As a result, the moments will not be able to flip until a large enough field ($B > B_{crit}$) is applied to lower the activation barrier, at which point the moments flip over. Clearly this type of activation barrier can result in hysteretic behavior as shown in Fig. 21.4.

[8] In principle the spins can get over the activation barrier either by being thermally activated or by quantum tunneling. However, if the activation barrier is sufficiently large (i.e., for a large crystallite) both of these are greatly suppressed.

21.2.3 Domain Pinning and Hysteresis

Even for single-crystal samples, disorder can play an extremely important role in the physics of domains. For example, a domain wall can have lower energy if it passes over a defect in the crystal. To see how this occurs let us look at a domain wall in an Ising ferromagnet as shown in Fig. 21.6. Bonds are marked with solid lines where neighboring spins are antialigned rather than aligned. In both figures the domain wall starts and ends at the same points, but on the right it follows a path through a defect in the crystal—in this case a site that is missing an atom. When it intersects the location of the missing atom, the number of antialigned bonds (marked) is lower, and therefore the energy is lower. Since this lower energy makes the domain wall stick to the missing site, we say the domain wall is *pinned* to the disorder. Even though the actual domain wall may be hundreds of lattice constants thick (as we saw in Section 21.1.1) it is easy to see that these objects still have a tendency to stick to disorder.

As mentioned at the beginning of this chapter, when a magnetic field is externally applied to a ferromagnet, the domain walls move to re-establish a new domain configuration (see the left two panels of Fig. 21.2) and therefore a new magnetization. However, when there is disorder in a sample, the domain walls can get pinned to the disorder: There is a low-energy configuration where the domain wall intersects the disorder, and there is then an activation energy to move the domain wall. This

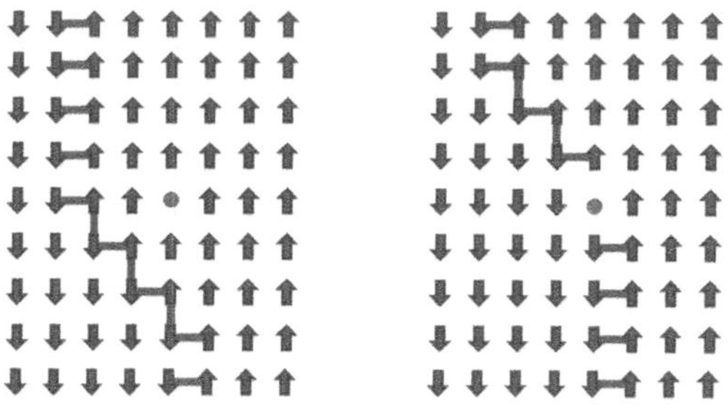

Fig. 21.6 Domain wall pinning. The energy of a domain wall is lower if the domain wall goes through the position of a defect in the crystal. Here, the dot is supposed to represent a missing spin. The drawn solid segments, where spins are antialigned, each cost energy. When the domain wall intersects the location of the missing spin, there are fewer solid segments, therefore it is a lower energy configuration (there are twelve solid segments on the left, but only ten on the right).

activation energy, analogous to what we found in Section 21.2.2, results in hysteresis of the magnet. Indeed, we can think of the hysteresis in single-domain crystallites as being very similar to the hysteresis we find in crystalline samples. The only difference is that in the case of crystallites, the domain walls always lie between the crystallites (since each crystallite has all of its moments fully aligned). But in both cases, it is the microscopic structure of the sample which prevents the domain wall from moving freely.

It is frequently the case that one wants to construct a ferromagnet which retains its magnetization extremely well—i.e., where there is strong hysteresis, and even in the absence of applied magnetic field there will be a large magnetization. This is known as a "hard" magnet (also known as a "permanent" magnet). It turns out that much of the trick of constructing hard magnets is arranging to insert appropriate disorder and microstructure to strongly pin the domain walls.

Aside: Magnetic disk drives, which store most of the world's digital information, are based on the idea of magnetic hysteresis. The disk itself is made of some ferromagnetic material. A magnetic field can be applied to magnetize a tiny region in either the up or down direction, and this magnetization remains in place so that it may be read out later.

Chapter Summary

- Although short-range interaction in a ferromagnet favors all magnetic moments to align, long-range magnetic dipolar forces favors moments to antialign. A compromise is reached with domains of aligned moments where different domains point in different directions. A very small crystal may be a single domain.
- The actual domain wall boundary may be a continuous rotation of the spin rather than a sudden flip over a single bond-length. The size of this spin structure depends on the ratio of the ferromagnetic energy to the anisotropy energy (i.e., if it is very costly to have spins point in directions between up and down then the wall will be over a single bond length).

- Domain walls are lower energy if they intersect certain types of disorder in the solid. This results in the *pinning* of domain walls—they stick to the disorder.
- In a large crystal, changes in magnetization occur by changing the size of domains. In polycrystalline samples with very small crystallites, changes in magnetization occur by flipping over individual single-domain crystallites. Both of these processes can require an activation energy (domain motion requires activation energy if domain walls are pinned) and thus result in hysteretic behavior of magnetization in ferromagnets.

References

- Hook and Hall, section 8.7
- Blundell, section 6.7
- Burns, section 15.10
- Ashcroft and Mermin, end of chapter 33

Also good (but covers material in different order from us):

- Rosenberg, chapter 12
- Kittel, chapter 12

Exercises

(21.1) **Domain Walls and Geometry**

Suppose a ferromagnet is made up of a density ρ of spins each with moment μ_B.

(a) Suppose a piece of this material forms a long circular rod of radius r and length $L \gg r$. In zero external magnetic field, if all of the moments are aligned along the L-direction of the rod, calculate the magnetic energy of this ferromagnet. (Hint: a volume of aligned magnetic dipoles is equivalent to a density of magnetic monopoles on its surface.)

(b) Suppose now the material is shaped such that $r \gg L$. What is the magnetic energy now?

(c) If a domain wall is introduced into the material, how should it be positioned to minimize the magnetic energy in the two different geometries? Estimate how much magnetic energy is saved by the introduction of the domain wall.

(d) Suppose the spins in this material are arranged in a cubic lattice, and the exchange energy between nearest neighbors is denoted J and the anisotropy energy is very large. How much energy does the domain wall cost? Comparing this energy to the magnetic energy, what should we conclude about which samples should have domain walls?

(e) Note that factors of the lattice constant a are often introduced in quoting exchange and anisotropy energies to make them into energies per unit length or unit volume. For magnetite, a common magnetic material, the exchange energy is $JS^2/a = 1.33 \times 10^{-11}$ J/m and the anisotropy energy is $\kappa S^2/a^3 = 1.35 \times 10^4$ J/m^3. Estimate the width of the domain wall and its energy per unit area. *Make sure you know the derivation of any formulas you use!*

(21.2) **Critical Field for Crystallite**

(a) Given that the energy of a crystallite in a magnetic field is given by

$$E/V = E_0 - |M||B| \cos\theta - \kappa'|M|^2(\cos\theta)^2$$

show that for $|B| < B_{crit}$ there is a local energy minimum where the magnetization points opposite the applied field, and find B_{crit}.

(b)* In part (a) we have assumed **B** is aligned with the anisotropy direction of the magnetization. Describe what can occur if these directions are not aligned.

(c) For small B, roughly how large (in energy per unit volume) is the activation barrier for the system to get from the local minimum to the global minimum.

(d) Can you make an estimate (in terms of actual numbers) of how big this activation barrier would be for a ferromagnetic crystal of magnetite that is a sphere of radius 1 nm ? You may use the parameters given in Exercise 21.1.e.

(21.3) **Exact Domain Wall Solution***

The approximation used in Section 21.1.1 of the energy of the anisotropy (κ) term is annoyingly crude. To be more precise, we should instead write $\kappa S^2(\sin\theta_i)^2$ and then sum over all spins i. Although this makes things more complicated, it is still possible to solve the problem so long as the spin twists slowly so that we can replace the finite difference $\delta\theta$ with a derivative, and replace the sum over sites with an integral. In this case, we can write the energy of the domain wall as

$$E = \int \frac{dx}{a}\left\{\frac{JS^2a^2}{2}\left(\frac{d\theta(x)}{dx}\right)^2 + \kappa S^2[\sin\theta(x)]^2\right\}$$

with a the lattice constant.

(a) Using calculus of variations show that this energy is minimized when

$$(Ja^2/\kappa)d^2\theta/dx^2 - \sin(2\theta) = 0$$

(b) Verify that this differential equation has the solution

$$\theta(x) = 2\tan^{-1}\left(\exp\left[\sqrt{2}(x/a)\sqrt{\frac{\kappa}{J}}\right]\right)$$

thus demonstrating the same $L \sim \sqrt{J/\kappa}$ scaling.

(c) Show that the total energy of the domain wall becomes $E_{tot}/(A/a^2) = 2\sqrt{2}S^2\sqrt{J\kappa}$.

(21.4) **Superparamagnetism&**

Consider a system of magnetite spheres as discussed in Exercise 21.2.d. If the temperature is larger than the activation temperature discussed in Exercise 21.2 there will be no hystersis, and the system will not maintain a magnetization in zero applied field. Estimate the magnetic susceptibility for this system as a function of temperature assuming the system temperature is above the activation temperature. Compare your result to the susceptibility of a typical paramagnet. Hint: the entire crystalite acts as one big spin now. (The enhanced paramagnetism of small ferromagnetic particles is often known as "superparamagnetism").

Mean Field Theory

22

Given a Hamiltonian for a magnetic system, we are left with the theoretical task of how to predict its magnetization as a function of temperature (and possibly as a function of external magnetic field). Certainly at low temperature the spins will be maximally ordered, and at high temperature the spins will thermally fluctuate and will be disordered. But calculating the magnetization as a function of temperature and applied magnetic field is typically a very hard task. Except for a few very simple exactly solvable models (like the Ising model in one dimension, see Exercises 20.5 and 20.6) we must resort to approximations. The most important and probably the simplest such approximation is known as "mean field theory". Generally a "mean field theory" is any method that approximates some non-constant quantity by an average.[1] Although mean field theories come in many varieties, there is a particularly simple and useful variety of mean field theory known as "molecular field theory" or "Weiss mean field theory",[2] which we will now discuss in depth.

Molecular or Weiss mean field theory generally proceeds in two steps:

- First, one examines one site (or one unit cell, or some small region) and treats it exactly. Any object outside of this site (or unit cell or small region) is approximated as an expectation (an average or a *mean*).
- The second step is to impose self-consistency. Every site (or unit cell, or small region) in the entire system should look the same. So the one site we treated exactly should have the same average as all of the others.

This procedure is extremely general and can be applied to problems ranging from magnetism to liquid crystals to fluid mechanics. We will demonstrate the procedure as it applies to ferromagnetism. In Exercise 22.5 we consider how mean field theory can be applied to antiferromagnets as well (further generalizations should then be obvious).

[1] In Chapter 2 we already saw another example of mean field theory, when we considered the Boltzmann and Einstein models of specific heat of solids. There we considered each atom to be in a harmonic well formed by all of its neighbors. The single atom was treated exactly, whereas the neighboring atoms were treated only approximately in that their positions were essentially averaged in order to simply form the potential well—and nothing further was said of the neighbors. Another example in similar spirit was given in margin note 3 of Chapter 18 where an alloy of Al and Ga with As is replaced by some averaged atom Al_xGa_{1-x} and is still considered a periodic crystal.

[2] The same Pierre-Ernest Weiss for whom Weiss domains are named.

22.1 Mean Field Equations for the Ferromagnetic Ising Model

As an example, let us consider the spin-1/2 Ising model

$$\mathcal{H} = -\frac{1}{2}\sum_{\langle i,j\rangle} J\sigma_i\sigma_j + g\mu_B B \sum_j \sigma_j$$

where $J > 0$, and here $\sigma = \pm 1/2$ is the z-component of the spin and the magnetic field B is applied in the $\hat{z}$ direction (and as usual μ_B is the Bohr magneton). For a macroscopic system, this is a statistical mechanical system with 10^{23} degrees of freedom, where all the degrees of freedom are now coupled to each other. In other words, it looks like a hard problem!

To implement mean field theory, we focus in on one site of the problem, say, site i. The Hamiltonian for this site can be written as

$$\mathcal{H}_i = \left(g\mu_B B - J \sum_j \sigma_j \right) \sigma_i$$

where the sum is over sites j that neighbor i. We think of the term in brackets as being caused by some effective magnetic field seen by the spin on site i, thus we define $B_{eff,i}$ such that

$$g\mu_B B_{eff,i} = g\mu_B B - J \sum_j \sigma_j$$

with again j neighboring i. Now $B_{eff,i}$ is not a constant, but is rather an operator since it contains the variables σ_j which can take several values. However, the first principle of mean field theory is that we should simply take an average of all quantities that are not site i. Thus we write the Hamiltonian of site i as $\mathcal{H}_i = g\mu_B \langle B_{eff} \rangle \sigma_i$. This is precisely the same Hamiltonian we considered when we studied paramagnetism in Eq. 19.6, and it is easily solvable. In short, one writes the partition function

$$Z_i = e^{-\beta g \mu_B \langle B_{eff} \rangle /2} + e^{\beta g \mu_B \langle B_{eff} \rangle /2} .$$

From this we can derive the expectation of the spin on site i (compare Eq. 19.8)

$$\langle \sigma_i \rangle = -\frac{1}{2} \tanh \left(\beta g \mu_B \langle B_{eff} \rangle /2 \right). \tag{22.1}$$

However, we can also write that

$$g\mu_B \langle B_{eff} \rangle = g\mu_B B - J \sum_j \langle \sigma_j \rangle.$$

The second step of the mean field approach is to set $\langle \sigma \rangle$ to be equal on all sites of the lattice, so we obtain

$$g\mu_B \langle B_{eff} \rangle = g\mu_B B - Jz \langle \sigma \rangle \tag{22.2}$$

where z is the number of neighbors j of site i (this is known as the *coordination number* of the lattice, and this factor has replaced the sum on j). Further, again assuming that $\langle \sigma \rangle$ is the same on all lattice sites, from Eqs. 22.1 and 22.2, we obtain the self-consistency equation for $\langle \sigma \rangle$ given by

$$\langle \sigma \rangle = -\frac{1}{2} \tanh \left(\beta \left[g\mu_B B - Jz \langle \sigma \rangle \right] /2 \right). \tag{22.3}$$

The expected moment per site will correspondingly be given by[3]

$$m = -g\mu_B \langle \sigma \rangle. \tag{22.4}$$

[3]Recall that the spin points opposite the moment! Ben Franklin, why do you torture us so? (See margin note 15 of Section 4.3.)

22.2 Solution of Self-Consistency Equation

The self-consistency equation, Eq. 22.3, is still complicated to solve. One approach is to find the solution graphically. For simplicity, let us set the external magnetic field B to zero. We then have the self-consistency equation

$$\langle\sigma\rangle = \frac{1}{2}\tanh\left(\frac{\beta J z}{2}\langle\sigma\rangle\right). \tag{22.5}$$

We then choose a value of the parameter $\beta J z/2$. Let us start by choosing a value $\beta J z/2 = 1$, which is somewhat small value, i.e., a high temperature. Then, in the top of Fig. 22.1 we plot both the right-hand side of Eq. 22.5 and the left-hand side of Eq. 22.5 both as a function of $\langle\sigma\rangle$. Note that the left-hand side is $\langle\sigma\rangle$ so we are just plotting the straight line $y = x$. We see that there is only a single point where the two curves meet, i.e., where the left side equals the right side. This point, in this case is $\langle\sigma\rangle = 0$. From this we conclude that for this value temperature, within mean field approximation, there is no magnetization in zero field.

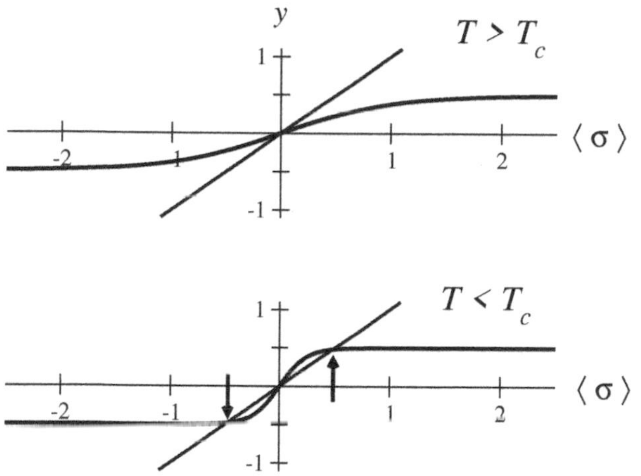

Fig. 22.1 Graphical solution of the mean field self-consistency equations. **Top:** Relatively high temperature $\beta J z/2 = 1$. The smooth line is the tanh of Eq. 22.5. The straight line is just the line $y = x$. Eq. 22.5 is satisfied only where the two curves cross—i.e., at $\langle\sigma\rangle = 0$, meaning that at this temperature, within the mean field approximation, there is no magnetization. **Bottom:** Relatively low temperature $\beta J z/2 = 4$. Here, the curves cross at three possible values ($\langle\sigma\rangle = 0$ and $\langle\sigma\rangle \approx \pm.479$). The fact that there is a solution of the self-consistency equations with non-zero magnetization tells us that the system is ferromagnetic (the zero magnetization solution is non-physical, see Exercise 22.2).

Let us now reduce the temperature substantially to $\beta J z/2 = 4$. Analogously, in the bottom of Fig. 22.1 we plot both the right-hand side of Eq. 22.5 as a function of $\langle\sigma\rangle$ and the left-hand side of Eq. 22.5 (again the straight line is just $y = x$). Here, however, we see there are three possible self-consistent solutions to the equations. There is the solution at $\langle\sigma\rangle = 0$ as well as two solutions marked with arrows in the figure at $\langle\sigma\rangle \approx \pm.479$. The two non-zero solutions tell us that at low temperature this system can have non-zero magnetization even in the absence of applied field—i.e., it is ferromagnetic.

The fact that we have possible solutions with the magnetization pointing in both directions is quite natural. The Ising ferromagnet can be polarized either spin-up or spin-down. However, the fact that there is also a self-consistent solution with zero magnetization at the same temperature seems a bit puzzling. We will see in Exercise 22.2 that when there are three solutions, the zero magnetization solution is actually a solution of maximal free energy not minimal free energy, and therefore should be discarded.[4]

[4] In particular we show that our self-consistency equations are analogous to when we find the minimum of a function by differentiation—we may also find maxima as well.

Thus the picture that arises is that at high temperature the system has zero magnetization (and we will see in the next subsection that it is paramagnetic) whereas at low temperature a non-zero magnetization develops and the system becomes ferromagnetic.[5] The transition between these two behaviors occurs at a temperature known as T_c, which stands for *critical temperature*[6] or *Curie temperature.*[7] It is clear from Fig. 22.1 that the behavior changes from one solution to three solutions precisely when the straight line is tangent to the tanh curve, i.e., when the slope of the tanh is unity. This tangency condition thus determines the critical temperature. Expanding the tanh for small argument, we obtain the tangency condition

[5] It is quite typical that at high temperatures a ferromagnet will turn into a paramagnet, unless something else happens first—like the crystal melts.

[6] Strictly speaking it should only be called a critical temperature if the transition is second order, i.e., if the magnetization turns on continuously at this transition. For the Ising model, this is in fact true, but for some magnetic systems it is not true.

[7] Named for Pierre again.

$$1 = \frac{1}{2}\left(\frac{\beta_c J z}{2}\right)$$

or equivalently the critical temperature is

$$k_B T_c = \frac{Jz}{4}.$$

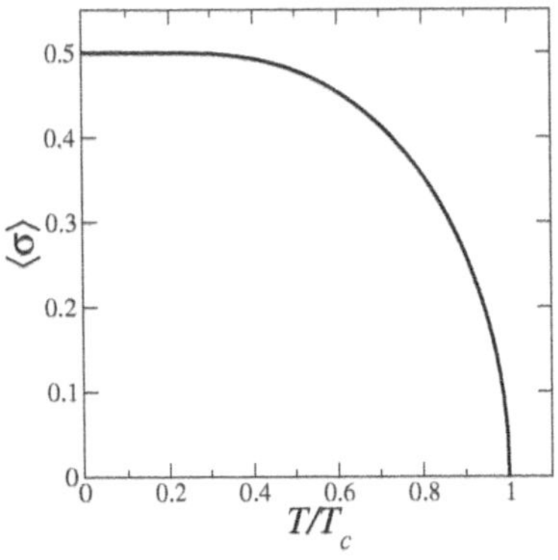

Fig. 22.2 Magnetization as a function of temperature. The plot shows the magnitude of the moment per site in units of $g\mu_B$ as a function of temperature in the mean field approximation of the spin-1/2 Ising model, with zero external magnetic field applied.

Using the graphical technique described in this section, one can in principle solve the self-consistency equations (Eq. 22.5) at any temperature (although there is no nice analytic expression, it can always be solved numerically). The results are shown in Fig. 22.2. Note that at low enough temperature, all of the spins are fully aligned ($\langle\sigma\rangle = 1/2$ which is the maximum possible for a spin-1/2). In Fig. 22.3 we show a comparison of the prediction of mean field theory to real experimental measurement of magnetization as a function of temperature. The agreement is typically quite good (although not exact). One can also, in principle, solve the self-consistency equation (Eq. 22.3) with finite magnetic field **B** as well.

22.2.1 Paramagnetic Susceptibility

At high temperature there will be zero magnetization in zero externally applied field. However, at finite field, we will have a finite magnetization. Let us apply a small magnetic field and solve the self-consistency equations Eq. 22.3. Since the applied field is small, we can assume that the induced $\langle\sigma\rangle$ is also small. Thus we can expand the tanh in Eq. 22.3 to obtain

$$\langle\sigma\rangle = \frac{1}{2}\left(\beta\left[Jz\langle\sigma\rangle - g\mu_B B\right]/2\right).$$

Rearranging this then gives

$$\langle\sigma\rangle = -\frac{\frac{1}{4}(\beta g\mu_B)B}{1-\frac{1}{4}\beta Jz} = -\frac{\frac{1}{4}(g\mu_B)B}{k_B(T-T_c)}$$

which is valid only so long as $\langle\sigma\rangle$ remains small. The moment per site is then given by (see Eq. 22.4) $m = -g\mu_B\langle\sigma\rangle$, which divided by the volume of a unit cell gives the magnetization M. Thus we find that the susceptibility is

$$\chi = \mu_0 \frac{\partial M}{\partial B} = \frac{\frac{1}{4}\rho(g\mu_B)^2\mu_0}{k_B(T-T_c)} = \frac{\chi_{Curie}}{1-T_c/T} \qquad (22.6)$$

where ρ is the number of spins per unit volume and χ_{Curie} is the pure Curie susceptibility of a system of (non-interacting) spin-1/2 particles (compare Eq. 19.9). Eq. 22.6 is known as the *Curie–Weiss law*. Thus, we see that a ferromagnet above its critical temperature is roughly a paramagnet with an enhanced susceptibility. Note that the susceptibility diverges at the transition temperature when the system becomes ferromagnetic.[8]

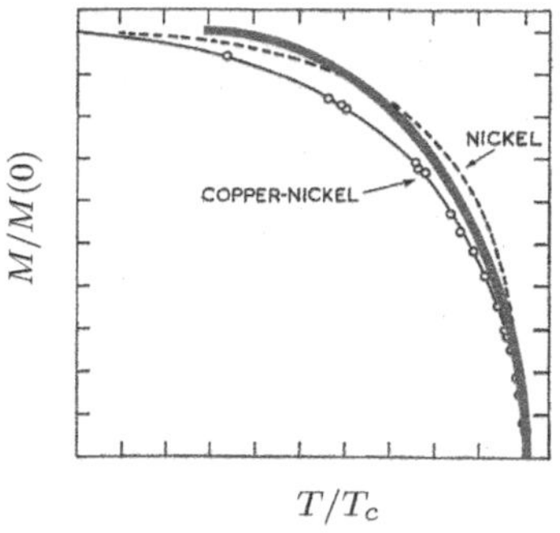

Fig. 22.3 Comparison of mean field prediction to experimental measurement of magnetization as a function of temperature. The heavy line is the mean field prediction for spin-1/2 Ising model (exactly the same as Fig. 22.2). The dotted line is experimentally measured magnetization of nickel, and the small points are a nickel–copper alloy. The vertical axis is the magnetization divided by magnetization at zero temperature. Data from Sucksmith *et al., Rev. Mod. Phys.*, **25**, 34 (1953); http://rmp.aps.org/abstract/RMP/v25/i1/p34_1 Copyright American Physical Society. Reprinted by permission.

22.2.2 Further Thoughts

In Exercise 22.5 we will also study the antiferromagnet. In this case, we divide the system into two sublattices—representing the two sites in a unit cell. In that example we will want to treat one spin of each sublattice exactly, but as in the ferromagnetic case each spin sees only the average field from its neighbors. One can generalize even further to consider very complicated unit cells.

Aside: It is worth noting that the result of solving the antiferromagnetic Ising model gives

$$\chi = \frac{\chi_{Curie}}{1+T_c/T}$$

compared to Eq. 22.6. It is this difference in susceptibility that pointed the way to the discovery of antiferromagnets (see Section 20.1.2).

We see that in both the ferromagnetic and antiferromagnetic case, at temperatures much larger than the critical temperature (much larger than the exchange energy scale J), the system behaves like a pure free spin Curie paramagnet. In Section 19.6.3 we asked where we might find free spins so that a Curie paramagnet might be realized. In fact, now we discover that any ferromagnet or antiferromagnet (or ferrimagnet for that matter) will appear to be free spins at temperatures high enough compared to the exchange energy. Indeed, it is almost always the case that when one thinks that one is observing free spins, at low enough energy scales one discovers that in fact the spins are coupled to each other!

The principle of mean field theory is quite general and can be applied to a vast variety of difficult problems in physics. No matter what the problem, the principle remains the same—isolate some small part of the system to treat exactly, average everything outside of that small system, then demand self-consistency: the average of the small system should look like the average of the rest of the system.

While the mean field approach is merely an approximation, it is frequently a very good approximation for capturing a variety of physical phenomena. Furthermore, many of its shortcomings can be systematically improved by considering successively more corrections to the initial mean field approach.[9]

[8] This divergence is in fact physical. As the temperature is reduced towards T_c, the divergence tells us that it takes a smaller and smaller applied B field to create some fixed magnetization M. This actually makes sense since once the temperature is below T_c, the magnetization will be non-zero even in the absence of any applied B.

[9] The motivated student might want to think about various ways to improve mean field theory systematically. One approach is discussed in Exercise 22.6.

Chapter Summary

- Understand the mean field theory calculation for ferromagnets. Understand how you would generalize this to any model of antiferromagnets, ferrimagnets, different spins, anisotropic models, etc., etc.
- For the ferromagnet the important results of mean field theory include:
  - a finite-temperature phase transition from a low-temperature ferromagnetic phase to a high-temperature paramagnetic phase at a transition temperature known as the Curie temperature.
  - Above the Curie temperature the paramagnetic susceptibility is $\chi = \chi_{Curie}/(1 - T_c/T)$, where χ_{Curie} is the susceptibility of the corresponding model where the ferromagnetic coupling between sites is turned off.
  - Below T_c the magnetic moment turns on, and increases to saturation at the lowest temperature.

References on Mean Field Theory

- Ibach and Luth, chapter 8 (particularly 8.6, 8.7)
- Hook and Hall, chapter 8 (particularly 8.3, 8.4)
- Kittel, beginning of chapter 12
- Burns, section 15.5
- Ashcroft and Mermin, chapter 33
- Blundell, chapter 5

Exercises

(22.1) ‡ **Weiss Mean Field Theory of a Ferromagnet**
Consider the spin-1/2 ferromagnetic Heisenberg Hamiltonian on the cubic lattice:

$$\mathcal{H} = -\frac{J}{2}\sum_{\langle i,j\rangle} \mathbf{S}_i \cdot \mathbf{S}_j + g\mu_B \mathbf{B} \cdot \sum_i \mathbf{S}_i \qquad (22.7)$$

Here, $J > 0$, with the sum indicated with $\langle i,j\rangle$ means summing over i and j being neighboring sites of the cubic lattice, and $\mathbf{B}$ is the externally applied magnetic field, which we will assume is in the $\hat{z}$ direction for simplicity. The factor of 1/2 out front is included so that each pair of spins is counted only once. Each site i is assumed to have a spin $\mathbf{S}_i$ of spin $S = 1/2$. Here μ_B is the conventional Bohr magneton defined to be positive. The fact that the final term has a + sign out front is from the fact that the electron charge is negative, therefore the magnetic moment opposes the spin direction. If one were to assume that these were nuclear spins the sign would be reversed (and the magnitude would be much smaller due to the larger nuclear mass).

(a) Focus your attention on one particular spin $\mathbf{S}_i$, and write down an effective Hamiltonian for this spin, treating all other variables $\mathbf{S}_j$ with $j \neq i$ as expectations $\langle \mathbf{S}_j \rangle$ rather than operators.

(b) Calculate $\langle \mathbf{S}_i \rangle$ in terms of the temperature and the fixed variables $\langle \mathbf{S}_j \rangle$ to obtain a mean-field self-consistency equation. Write the magnetization $M = |\mathbf{M}|$ in terms of $\langle \mathbf{S} \rangle$ and the density of spins.

(c) At high temperature, find the susceptibility $\chi = dM/dH = \mu_0 dM/dB$ in this approximation.

(d) Find the critical temperature in this approximation.

▷ Write the susceptibility in terms of this critical temperature.

(e) Show graphically that in zero external field ($\mathbf{B} = 0$), below the critical temperature, there are solutions of the self-consistency equation with $M \neq 0$.

(f) Repeat parts (a)–(d) but now assuming there is an $S = 1$ spin on each site (meaning that S_z takes the values $-1, 0, +1$).

(g)*& See Exercise 22.7.

(22.2) **Bragg-Williams Approximation**

This exercise provides a different approach to obtaining the Weiss mean-field equations. For simplicity we will again assume spin 1/2 variables on each site.

Assume there are N lattice sites in the system. Let the average spin value be $\langle S_i^z \rangle = s$. Thus the probability of a spin being an up spin is $P_\uparrow = 1/2 + s$ whereas the probability of a spin being a down spin is $P_\downarrow = 1/2 - s$. The total number of up spins or down spins is then $NP_\uparrow$ and $NP_\downarrow$ respectively.

(a) Consider first a case where sites do not interact with each other. In the micro-canonical ensemble, we can count the number of configurations (microstates) which have the given number of spin-ups and spin-downs (determined by s). Using $S = k_B \ln \Omega$, calculate the entropy of the system in the large N limit.

(b) Assuming all sites have independent probabilities $P_\uparrow$ and $P_\downarrow$ of pointing up and down respectively, calculate the probability that two neighboring sites will point in the same direction and the probability that two neighboring sites will point in opposite directions.

▷ Use this result to calculate an approximation to the expectation of the Hamiltonian. Note: This is not an exact result, as in reality, sites that are next to each other will have a tendency to have the same spin because that will lower their energies, but we have ignored this effect here.

(c) Putting together the results of (a) and (b) above, derive the approximation to the free energy

$$
\begin{aligned}
F &= E - TS \\
&= Nk_BT\left[(\frac{1}{2}+s)\log(\frac{1}{2}+s)\right. \\
&\quad \left. + (\frac{1}{2}-s)\log(\frac{1}{2}-s)\right] \\
&\quad + g\mu_B B_z N s - JNzs^2/2
\end{aligned}
$$

where z is the number of neighbors each spin has, and we have assumed the external field $\mathbf{B}$ to be in the $\hat{z}$ direction. (Again we assume the spin is electron spin so that the energy of a spin interacting with the external field is $+g\mu_b \mathbf{B} \cdot \mathbf{S}$.)

(d) Extremize this expression with respect to the variable s to obtain the Weiss mean field equations.

▷ Below the critical temperature note that there are three solutions of the mean field equations.

▷ By examining the second derivative of F with respect to s, show that the $s = 0$ solution is actually a maximum of the free energy rather than a minimum.

▷ Sketch $F(s)$ both above and below the critical temperature for $B = 0$. At non-zero B?

(22.3) **Spin S Mean Field Theory**

Using the result of Exercise 19.7 use mean field theory to calculate the critical temperature for a spin S ferromagnet with a given g-factor g, having coordination number z and nearest-neighbor exchange coupling J_{ex}. (It may be useful to re-solve Exercise 19.7 if you don't remember how this is done.)

(22.4) **Low-Temperature Mean Field Theory**

Consider the $S = 1/2$ ferromagnet mean field calculation from Exercise 22.1. At zero temperature, the magnet is fully polarized.

(a) Calculate the magnetization in the very low temperature limit. Show that the deviation from fully polarized becomes exponentially small as T goes to zero.

(b)* Now consider a spin S ferromagnet. Determine the magnetization in the low T limit.

(c)* In fact this exponential behavior is not observed experimentally! The reason for this has to do with spinwaves, which are explored in Exercise 20.3, but are not included in mean field theory. Using some results from that exercise, determine (roughly) the low-temperature behavior of the magnetization of a ferromagnet.

(22.5) **Mean Field Theory for the Antiferromagnet**

For this Exercise we use the molecular field (Weiss mean field) approximation for the spin-1/2 *antiferromagnetic* model on a three-dimensional cubic lattice. The full Hamiltonian is exactly that of Eq. 22.7, except that now we have $J < 0$, so neighboring spins want to point in opposite directions (compared to a ferromagnet where $J > 0$ and neighboring spins want to point in the same direction). For simplicity let us assume that the external field points in the $\hat{z}$ direction.

At mean field level, the ordered ground state of this Hamiltonian will have alternating spins pointing up and down respectively. Let us call the sublattices of alternating sites, sublattice A and sublattice B respectively (i.e, A sites have lattice coordinates (i, j, k) with $i + j + k$ odd whereas B sites have lattice coordinates with $i + j + k$ even).

In mean field theory the interaction between neighboring spins is replaced by an interaction with an average spin. Let $s_A = \langle S^z \rangle_A$ be the average value of the spins on sublattice A, and $s_B = \langle S^z \rangle_B$ be the average value of the spins on sublattice B (we assume that these are also oriented in the $\pm\hat{z}$ direction).

(a) Write the mean field Hamiltonian for a single site on sublattice A and the mean field Hamiltonian for a single site on sublattice B.

(b) Derive the mean-field self-consistency equations

$$\begin{aligned} s_A &= \frac{1}{2}\tanh(\beta[JZs_B - g\mu_B B]/2) \\ s_B &= \frac{1}{2}\tanh(\beta[JZs_A - g\mu_B B]/2) \end{aligned}$$

with $\beta = 1/(k_B T)$. Recall that $J < 0$.

(c) Let $B = 0$. Reduce the two self-consistency equations to a single self-consistency equation. (Hint: Use symmetry to simplify! Try plotting s_A versus s_B.)

(d) Assume $s_{A,B}$ are small near the critical point and expand the self-consistency equations. Derive the critical temperature T_c below which the system is antiferromagnetic (i.e., $s_{A,B}$ become non-zero).

(e) How does one detect antiferromagnetism experimentally?

(f) In this mean-field approximation, the magnetic susceptibility can be written as

$$\chi = -(\rho/2)g\mu_0\mu_B \lim_{B\to 0} \frac{\partial(s_A + s_B)}{\partial B}$$

with ρ the density of sites (why the factor of 1/2 out front?).

▷ Derive this susceptibility for $T > T_c$ and write it in terms of T_c.

▷ Compare your result with the analogous result for a ferromagnet (Exercise 22.1). In fact, it was this type of measurement that first suggested the existence of antiferromagnets!

(g)** For $T < T_c$ show that

$$\chi = \frac{(\rho/4)\mu_0(g\mu_B)^2(1-(2s)^2)}{k_B T + k_B T_c(1-(2s)^2)}$$

with s the staggered moment (ie, $s(T) = |s_A(T)| = |s_B(T)|$). (See the result of Exercise 22.7 below).

▷ Compare this low T result with that of part f.

▷ Give a sketch of the susceptibility at all T.

(22.6) **Correction to Mean Field***

Consider the spin-1/2 Ising ferromagnet on a cubic lattice in d dimensions. When we consider mean field theory, we treat exactly a single spin σ_i and the $z = 2d$ neighbors on each side will be considered to have an average spin $\langle\sigma\rangle$. The critical temperature you calculate should be $k_B T_c = Jz/4$.

To improve on mean field theory, we can instead treat a block of two connected spins σ_i and $\sigma_{i'}$ where the neighbors outside of this block are assumed to have the average spin $\langle\sigma\rangle$. Each of the spins in the block has $2d - 1$ such averaged neighbors. Use this improved mean field theory to write a new equation for the critical temperature (it will be a transcendental equation). Is this improved estimate of the critical temperature higher or lower than that calculated in the more simple mean-field model?

(22.7) **Differential Susceptibility*&**

One can define a differential susceptibility $\chi = \mu_0 dM/dB$ even when the magnetization or magnetic field is nonzero. Consider the spin 1/2 ferromagnet as defined in Exercise 22.1. Using the mean field self-consistency equations, show that the differential susceptibility is always given by

$$\chi = \frac{(\rho/4)\mu_0(g\mu_B)^2(1-(2s)^2)}{k_B T - k_B T_c(1-(2s)^2)}$$

where $s = \langle\sigma\rangle$ is the spin per site, and ρ the density of sites. You will need to use implicit differentiation, as well as some hyperbolic trigonometry identities. See also the result of Exercise 22.5 above.

Magnetism from Interactions: The Hubbard Model

23

So far we have only discussed ferromagnetism in the context of isolated spins on a lattice that align due to their interactions with each other. However, in fact many materials have magnetism where the magnetic moments, the aligned spins, are not pinned down, but rather can wander through the system. This phenomenon is known as *itinerant ferromagnetism*.[1] For example, it is easy to imagine a free electron gas where the number of up spins is different from the number of down spins. However, for completely free electrons it is always lower energy to have the same number of up and down spins than to have the numbers differ.[2] So how does it happen that electrons can decide, even in absence of external magnetic field, to polarize their spins? The culprit is the strong Coulomb interaction between electrons. On the other hand, we will see that antiferromagnetism can also be caused by strong interaction between electrons as well!

The Hubbard model[3] is an attempt to understand the magnetism that arises from interactions between electrons. It is certainly the most important model of interacting electrons in modern condensed matter physics. Using this model we will see how interactions can produce both ferro- and antiferromagnetism (this was alluded to in Section 19.2.1).

The model is relatively simple to describe.[4] First we write a tight binding model for a band of electrons as we did in Chapter 11 with hopping parameter t. (We can choose to do this in one, two, or three dimensions as we see fit[5]). We will call this Hamiltonian H_0. As we derived in Chapter 11 (and should be easy to derive in two and three dimensions now) the full bandwidth of the band is $4dt$ in d dimensions. We can add as many electrons as we like to this band. Let us define the number of electrons in the band per site to be called the doping, x (so

[1] Itinerant means traveling from place to place without a home (from Latin iter, or itiner, meaning journey or road—in case anyone cares). Most of the ferromagnets that we are familiar with, such as iron, are itinerant.

[2] The total energy of having N electrons spin-up in a system is proportional to $NE_F \sim N(N/V)^{2/d}$, where d is the dimensionality of the system (you should be able to prove this easily). We can write $E = CN^{1+a}$ with $a > 0$ and C some constant. For $N_\uparrow$ up-spins and $N_\downarrow$ down-spins, we have a total energy $E = CN_\uparrow^{1+a} + CN_\downarrow^{1+a} = C(N_\uparrow^{1+a} + (N-N_\uparrow)^{1+a})$ where N is the total number of electrons. Setting $dE/dN_\uparrow = 0$ immediately gives $N_\uparrow = N/2$ as the minimum energy configuration.

[4] The reason most introductory books do not cover the Hubbard model is that the model is conventionally introduced using so-called "second quantized" notation—that is, using field-theoretic methods. We will avoid this approach, but as a result, we cannot delve too deeply into the physics of the model. Even so, this chapter should probably be considered to be more advanced material than the rest of the book.

[5] In one dimension, the Hubbard model is exactly solvable.

[3] John Hubbard, a British physicist, wrote down this model in 1963, and it quickly became an extremely important example in the attempt to understand interacting electrons. Despite the success of the model, Hubbard, who died relatively young in 1980, did not live to see how important his model became. In 1986, when the phenomenon of "high-temperature superconductivity" was discovered by Bednorz and Müller (resulting in a Nobel Prize the following year), the community quickly came to believe that an understanding of this phenomenon would only come from studying the Hubbard model. (It is a shame that we do not have space to discuss superconductivity in this book.) Over the next two decades the Hubbard model took on the status of being *the* most important question in condensed matter physics. Its complete solution remains elusive despite the tens of thousands of papers written on the subject.

that $x/2$ is the fraction of k-states in the band which are filled being that there are two spin states). As long as we do not fill all of the states in the band ($x < 2$), in the absence of interactions, this partially filled tight binding band is a metal. Finally we include the Hubbard interaction

$$H_{interaction} = \sum_i U\, n_{i\uparrow}\, n_{i\downarrow} \tag{23.1}$$

where here $n_{i\uparrow}$ is the number of electrons with spin-up on site i, and $n_{i\downarrow}$ is the number of electrons on site i with spin-down, and $U > 0$ is an energy known as the repulsive Hubbard interaction energy. This term gives an energy penalty of U whenever two electrons sit on the same site of the lattice (Pauli exclusion prevents two electrons with the same spin from being at the same site). This short-ranged interaction term is an approximate representation of the Coulomb interaction between electrons. The full Hubbard model Hamiltonian is given by the sum of the kinetic and interaction pieces

$$H = H_0 + H_{interaction} \quad .$$

23.1 Itinerant Ferromagnetism

Why should this on-site interaction create magnetism? Imagine for a moment that all of the electrons in the system had the same spin state (a so-called "spin-polarized" configuration). If this were true, by the Pauli exclusion principle, no two electrons could ever sit on the same site. In this case, the expectation of the Hubbard interaction term would be zero

$$\langle \text{Polarized Spins} | H_{interaction} | \text{Polarized Spins} \rangle = 0$$

which is the lowest possible energy that this interaction term could have. On the other hand, if we filled the band with only one spin species, then the Fermi energy (and hence the kinetic energy of the system) would be much higher than if the electrons could be distributed between the two possible spin states. Thus, it appears that there will be some competition between the potential and kinetic energy that decides whether the spins align or not.

23.1.1 Hubbard Ferromagnetism Mean Field Theory

To try to decide quantitatively whether spins will align or not we start by writing

$$U\, n_{i\uparrow}\, n_{i\downarrow} = \frac{U}{4}(n_{i\uparrow} + n_{i\downarrow})^2 - \frac{U}{4}(n_{i\uparrow} - n_{i\downarrow})^2 \quad .$$

Now we make the approximation of treating all operators $n_{i,\uparrow}$ and $n_{i,\downarrow}$ as their expectations.

$$U\, n_{i\uparrow}\, n_{i\downarrow} \approx \frac{U}{4}\langle n_{i\uparrow} + n_{i\downarrow}\rangle^2 - \frac{U}{4}\langle n_{i\uparrow} - n_{i\downarrow}\rangle^2 \quad .$$

[6]This is a slightly different type of mean field theory from that encountered in Chapter 22. Previously we considered some local degree of freedom (some local spin) which we treated exactly, and replaced all other spins by their average. Here, we are going to treat the kinetic energy as if the electrons were non-interacting, and we replace the operators in the potential energy term by their averages.

This approximation is a type of mean-field theory,[6] similar to that we encountered in the previous Chapter 22: we replace operators by their expectations. The expectation $\langle n_{i\uparrow} + n_{i,\downarrow}\rangle$ in the first term is just the average number of electrons on site i which is just the average number of particles per site,[7] which is equal to the doping x, which we keep fixed.

Correspondingly, the second expectation, $\langle n_{i\uparrow} - n_{i\downarrow}\rangle$, is related to the magnetization of the system. In particular, since each electron carries[8] a magnetic moment of μ_B, the magnetization[9] is

$$M = (\mu_B/v)\langle n_{i\downarrow} - n_{i\uparrow}\rangle$$

with v the volume of the unit cell. Thus, within this approximation the expectation of the energy of the Hubbard interaction is given by

$$\langle H_{interaction}\rangle \approx (V/v)(U/4)\left(x^2 - (Mv/\mu_B)^2\right) \qquad (23.2)$$

where V/v is the number of sites in the system. Thus, as expected, increasing the magnetization M decreases the expectation of the interaction energy. To determine if the spins actually polarize we need to weigh this potential energy gain against the kinetic energy cost.

[7] This assumes that the system remains homogeneous—that is, that all sites have the same average number of electrons.

[8] We have assumed an electron g-factor of $g = 2$ and an electron spin of 1/2. Everywhere else in this chapter the symbol g will only be used for density of states.

[9] Recall that magnetization is moment per unit volume.

23.1.2 Stoner Criterion[10]

Here we calculate the kinetic energy cost of polarizing the spins in our model and we balance this against the potential energy gain. We will recognize this calculation as being almost identical to the calculation we did way back in Section 4.3 when we studied Pauli paramagnetism (but we repeat it here for clarity).

Consider a system (at zero temperature for simplicity) with the same number of spin-up and spin-down electrons. Let $g(E_F)$ be the total density of states at the Fermi surface per unit volume (for both spins put together). Now, let us flip over a small number of spins so that the spin-up and spin-down Fermi surfaces have slightly different energies[11]

$$\begin{aligned} E_{F,\uparrow} &= E_F + \delta\epsilon/2 \\ E_{F,\downarrow} &= E_F - \delta\epsilon/2. \end{aligned}$$

The difference in the number density of up and down electrons is then

$$\rho_\uparrow - \rho_\downarrow = \int_0^{E_F+\delta\epsilon/2} dE\, \frac{g(E)}{2} - \int_0^{E_F-\delta\epsilon/2} dE\, \frac{g(E)}{2}$$

where we have used the fact that the density of states per unit volume for either the spin-up or spin-down species is $\frac{g(E)}{2}$.

Although we could carry forward at this point and try to perform the integrals generally for arbitrary $\delta\epsilon$ (see Exercise 23.1) it is enough for our present discussion to consider the simpler case of very small $\delta\epsilon$. In this case, we have

$$\rho_\uparrow - \rho_\downarrow = \delta\epsilon \frac{g(E_F)}{2}.$$

[10] This has nothing to do with the length of your dreadlocks or the number of Grateful Dead shows you have been to (I've been to six shows ... I think).

[11] If we were being very careful we would adjust E_F to keep the overall electron density $\rho_\uparrow + \rho_\downarrow$ fixed as we change $\delta\epsilon$. For small $\delta\epsilon$ we find that E_F remains unchanged as we change $\delta\epsilon$, but this is not true for larger $\delta\epsilon$.

The difference in the number of up and down electrons is related to the magnetization of the system by[8]

$$M = \mu_B(\rho_\downarrow - \rho_\uparrow) = -\mu_B \delta\epsilon \frac{g(E_F)}{2}.$$

The kinetic energy per unit volume is a bit more tricky. We write

$$\begin{aligned}
K &= \int_0^{E_F+\delta\epsilon/2} dE\, E\, \frac{g(E)}{2} + \int_0^{E_F-\delta\epsilon/2} dE\, E\, \frac{g(E)}{2} \\
&= 2\int_0^{E_F} dE\, E\, \frac{g(E)}{2} + \int_{E_F}^{E_F+\delta\epsilon/2} dE\, E\, \frac{g(E)}{2} - \int_{E_F-\delta\epsilon/2}^{E_F} dE\, E\, \frac{g(E)}{2} \\
&\approx K_{M=0} + \frac{g(E_F)}{2}\left[\left(\frac{(E_F+\delta\epsilon/2)^2}{2} - \frac{E_F^2}{2}\right) - \left(\frac{E_F^2}{2} - \frac{(E_F-\delta\epsilon/2)^2}{2}\right)\right] \\
&= K_{M=0} + \frac{g(E_F)}{2}(\delta\epsilon/2)^2 \\
&= K_{M=0} + \frac{g(E_F)}{2}\left(\frac{M}{\mu_B g(E_F)}\right)^2 \qquad (23.4)
\end{aligned}$$

where $K_{M=0}$ is the kinetic energy per unit volume for a system with no net magnetization (equal numbers of spin-up and spin-down electrons).

We can now add this result to Eq. 23.2 to give the total energy of the system per unit volume

$$E_{tot} = E_{M=0} + \left(\frac{M}{\mu_B}\right)^2 \left[\frac{1}{2g(E_F)} - \frac{vU}{4}\right]$$

with v the volume of the unit cell. We thus see that for

$$U > \frac{2}{g(E_F)v}$$

the energy of the system is lowered by increasing the magnetization from zero. This condition for itinerant ferromagnetism is known as the *Stoner criterion*.[12,13]

Aside: We did a lot of work to arrive at Eq. 23.4. In fact, we could have almost written it down with no work at all based on the calculation of the Pauli susceptibility we did back in Section 4.3. Recall first that when an external magnetic field is applied in the up direction to a system, there is an energy induced from the coupling of the spins to the field which is given by $\mu_B(\rho_\uparrow - \rho_\downarrow)B = -MB$ (with positive M being defined in the same direction as positive B so that having the two aligned is low energy). Also recall in Section 4.3 that we derived the (Pauli) susceptibility of an electron system is

$$\chi_{Pauli} = \mu_0 \mu_B^2 g(E_F)$$

which means that when a magnetic field B is applied, a magnetization $\chi_{Pauli}B/\mu_0$ is induced. Thus we can immediately conclude that the energy of such a system in an external field *must* be of the form

$$E(M) = \frac{M^2 \mu_0}{2\chi_{Pauli}} - MB + \text{Constant}.$$

[12] Edmund Stoner was a British physicist who, among other things, figured out the Pauli exclusion principle in 1924 a year before Pauli. However, Stoner's work focused on the spectra, and behavior, of atoms, and he was not bold enough to declare that exclusion was a fundamental property of electrons. Stoner was diagnosed with diabetes in 1919 at 20 years of age and grew progressively weaker for the next eight years. In 1927, insulin treatment became available, saving his life. He died in 1969.

[13] Although this type of calculation of the Stoner criterion has been gospel for half a century, like many things, the truth can be somewhat more complicated. For example, recent numerical work has shown that ferromagnetism never occurs at low electron density in the Hubbard model! However, numerics on repulsive fermions not confined to hop on a lattice do show ferromagnetism.

To see that this is correct, we minimize the energy with respect to M at a given B, and we discover that this properly gives us $M = \chi_{Pauli} B/\mu_0$. Thus, at zero applied B, the energy should be

$$E(M) - \text{Constant} = \frac{M^2 \mu_0}{2\chi_{Pauli}} = \frac{M^2}{2\mu_B^2 g(E_F)}$$

exactly as we found in Eq. 23.4!

23.2 Mott Antiferromagnetism

In fact, the Hubbard model is far more complex than the above mean-field calculation would lead one to believe. Let us now consider the case where the doping is such that there is exactly one electron per site of the lattice. For non-interacting electrons, this would be a half-filled band, and hence a conductor. However, if we turn on the Hubbard interaction with a large U, the system becomes an insulator. To see this, imagine one electron sitting on every site. In order for an electron to move, it must hop to a neighboring site which is already occupied. This process therefore costs energy U, and if U is large enough, the hopping cannot happen. This is precisely the physics of the Mott insulator which we discussed in Section 16.4.

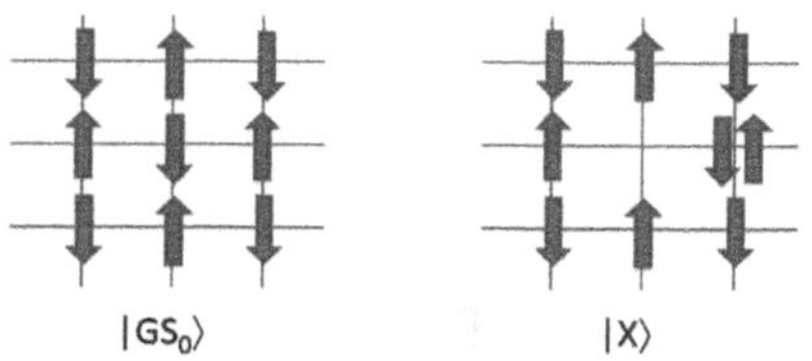

Fig. 23.1 Spin configurations of the half-filled Hubbard model. **Left:** The proposed antiferromagnetic ground state in the limit that t is very small. **Right:** A higher-energy state in the limit of small t which can occur by an electron from one site hopping onto a neighboring site. The energy penalty for double occupancy is U.

With one immobile electron on each site we can now ask which way the spins align (in the absence of external field). For a square or cubic lattice, there are two obvious options: either the spins want to be aligned with their neighbors or they want to be antialigned with their neighbors (ferromagnetism or antiferromagnetism). It turns out that antiferromagnetism is favored! To see this, consider the antiferromagnetic state $|GS_0\rangle$ shown on the left of Fig. 23.1. In the absence of hopping this state is an eigenstate with zero energy (as is any other state where there is precisely one electron on each site). We then consider adding the hopping perturbatively. Because the hopping Hamiltonian allows an electron to hop from site to site (with hopping amplitude $-t$), the electron can make a "virtual" hop to a neighboring site, as shown in the right of Fig. 23.1. The state on the right $|X\rangle$ is of higher energy (in the absence of hopping it has energy U because of the double occupancy). Using second-order perturbation theory we obtain

$$E = E(|GS_0\rangle) + \sum_X \frac{|\langle X|H_{hop}|GS_0\rangle|^2}{E_{GS_0} - E_X} = E(|GS_0\rangle) - \frac{Nz|t|^2}{U}$$

In the first line the sum is over all $|X\rangle$ states that can be reached in a single hop from the state $|GS_0\rangle$. In the second line we have counted the number of such terms to be Nz, where z is the coordination number (number of nearest neighbors) and N is the number of sites. Further, we have inserted $-t$ for the amplitude of hopping from one site to the next. Note that if the spins were all aligned, no virtual intermediate state $|X\rangle$ could exist, since it would violate the Pauli exclusion principle (hopping of electrons conserves spin state, so spins cannot flip over during a hop, so there is strictly no double occupancy). Thus we conclude that the antiferromagnetic state has its energy lowered compared to the ferromagnetic state in the limit of large U in a Mott insulating phase.

Admittedly this argument appears a bit handwaving (it is correct though!). To make the argument more precise, one should be much more careful about how one represents states with multiple electrons. This typically requires field theoretic techniques. A very simple example of how this is done (without more advanced techniques) is presented in the appendix to this chapter.

Nonetheless, the general physics of why the antiferromagnetic Mott insulator state should be lower energy than its ferromagnetic counterpart can be understood qualitatively without resorting to the more precise arguments. On each site one can think of an electron as being confined by the interaction with its neighbors. In the ferromagnetic case, the electron cannot make any excursions to neighboring sites because of the Pauli exclusion principle (these states are occupied). However, in the antiferromagnetic case the electron can make excursions, and even though when the electron wanders onto neighboring sites the energy is higher, there is nonetheless some amplitude for this to happen.[14] Allowing the electron wavefunction to spread out always lowers its energy.[15]

Indeed, in general a Mott insulator (on a square or cubic lattice) is typically an antiferromagnet (unless some other interesting physics overwhelms this tendency). It is generally believed that there is a substantial range of t, U, and doping x where the ground state is antiferromagnetic. Indeed, many real materials are thought to be examples of antiferromagnetic Mott insulators. Interestingly, it turns out that in the limit of very very strong on-site interaction $U \to \infty$, adding even a single additional hole to the half-filled Mott insulator will turn the Mott antiferromagnet into a ferromagnet! This rather surprising result, due to Nagaoka and Thouless[16] (one of the few key results about the Hubbard model which is rigorously proven), shows the complexity of this model.

[14] Similar to when a particle is in a potential well $V(x)$, there is some amplitude to find the electron at a position such that $V(x)$ is very large.

[15] By increasing Δx we can decrease Δp and thus lower the kinetic energy of the particle, as per the Heisenberg uncertainty principle.

[16] David Thouless, born in Scotland, is one of the most important names in modern condensed matter physics. He won a Nobel prize in 2016 for applications of ideas of topology to condensed matter physics. Yosuki Nagaoka was a prominent Japanese theorist.

Chapter Summary

- Hubbard model includes tight-binding hopping t and on-site "Hubbard" interaction U
- For partially filled band, the repulsive interaction (if strong enough) makes the system an (itinerant) ferromagnet: aligned spins can have lower energy because they do not double occupy sites, and therefore are lower energy with respect to U although it costs higher kinetic energy to align all the spins.
- For a half-filled band, the repulsive interaction makes the Mott insulator antiferromagnetic: virtual hopping lowers the energy of antialigned neighboring spins.

References on Hubbard Model

Unfortunately there are essentially no references that I know of that are readable without background in field theory and second quantization.

23.3 Appendix: Hubbard Model for the Hydrogen Molecule

Since the perturbative calculation showing antiferromagnetism is very hand-waving, I thought it useful to do a real (but very simple) calculation showing how, in principle, these calculations are done more properly. Again, please consider this material to be more advanced, but if you are confused about the discussion of antiferromagnetism in the Hubbard model, this appendix might be enlightening to read.

The calculation given here will address the Hubbard model for the hydrogen molecule. Here we consider two nuclei A and B near each other, with a total of two electrons—and we consider only the lowest spatial orbital (the s-orbital) for each atom.[17] There are then four possible states which an electron can be in:

$$A\uparrow \quad A\downarrow \quad B\uparrow \quad B\downarrow.$$

To indicate that we have put electron 1 in, say the $A\uparrow$ state, we write the wavefunction

$$|A\uparrow\rangle \quad \longleftrightarrow \quad \varphi_{A\uparrow}(1)$$

where φ is the wavefunction, and (1) is shorthand for the position $\mathbf{r}_1$ as well as the spin σ_1 coordinate.

For a two-electron state we are only allowed to write wavefunctions that are overall antisymmetric. So given two single electron orbitals α and β (α and β take values in the four possible orbitals $A\uparrow, A\downarrow, B\uparrow, B\downarrow$) we write so-called Slater determinants to describe the antisymmetric two-particle wavefunctions

$$|\alpha;\beta\rangle = \frac{1}{\sqrt{2}}\det\begin{vmatrix} \alpha(1) & \beta(1) \\ \alpha(2) & \beta(2) \end{vmatrix} = (\alpha(1)\beta(2) - \beta(1)\alpha(2))/\sqrt{2} = -|\beta;\alpha\rangle.$$

Note that this Slater determinant can be generalized to write a fully antisymmetric wavefunction for any number of electrons. If the two orbitals are the same, then the wavefunction vanishes (as it must by Pauli exclusion).

For our proposed model of the Hydrogen molecule, we thus have six possible states for the two electrons

$$\begin{aligned}
|A\uparrow;A\downarrow\rangle &= -|A\downarrow;A\uparrow\rangle \\
|A\uparrow;B\uparrow\rangle &= -|B\uparrow;A\uparrow\rangle \\
|A\uparrow;B\downarrow\rangle &= -|B\downarrow;A\uparrow\rangle \\
|A\downarrow B\uparrow\rangle &= -|B\uparrow;A\downarrow\rangle \\
|A\downarrow;B\downarrow\rangle &= -|B\downarrow;A\downarrow\rangle \\
|B\uparrow;B\downarrow\rangle &= -|B\downarrow;B\uparrow\rangle
\end{aligned}$$

The Hubbard interaction energy (Eq. 23.1) is diagonal in this basis—it simply gives an energy penalty U when there are two electrons on the same site. We thus have

$$\langle A\uparrow;A\downarrow|H_{interaction}|A\uparrow;A\downarrow\rangle = \langle B\uparrow;B\downarrow|H_{interaction}|B\uparrow;B\downarrow\rangle = U$$

[17] This technique can in principle be used for any number of electrons in any number of orbitals, although exact solution becomes difficult as the Schroedinger matrix becomes very high-dimensional and hard to diagonalize exactly, necessitating sophisticated approximation methods.

and all other matrix elements are zero.

To evaluate the hopping term we refer back to where we introduced tight binding in Section 6.2.2 and Chapter 11. Analogous to that discussion, we see that the hopping term with amplitude $-t$ (with dimensions of energy) turns an $A\uparrow$ orbital into a $B\uparrow$ orbital or vice versa, and similarly turns a $A\downarrow$ into a $B\downarrow$ and vice versa (the hopping does not change the spin). Thus, for example, we have

$$\langle A\downarrow; B\uparrow |H_{hop}| A\downarrow; A\uparrow\rangle = -t$$

where here the hopping term turned the B into an A. Note that this implies similarly that

$$\langle A\downarrow; B\uparrow |H_{hop}| A\uparrow; A\downarrow\rangle = t$$

since $|A\downarrow; A\uparrow\rangle = -|A\uparrow; A\downarrow\rangle$.

Since there are six possible basis states, our Hamiltonian can be expressed as a six by six matrix. We thus write our Schroedinger equation as

$$\begin{pmatrix} U & 0 & -t & t & 0 & 0 \\ 0 & 0 & 0 & 0 & 0 & 0 \\ -t & 0 & 0 & 0 & 0 & -t \\ t & 0 & 0 & 0 & 0 & t \\ 0 & 0 & 0 & 0 & 0 & 0 \\ 0 & 0 & -t & t & 0 & U \end{pmatrix} \begin{pmatrix} \psi_{A\uparrow A\downarrow} \\ \psi_{A\uparrow B\uparrow} \\ \psi_{A\uparrow B\downarrow} \\ \psi_{A\downarrow B\uparrow} \\ \psi_{A\downarrow B\downarrow} \\ \psi_{B\uparrow B\downarrow} \end{pmatrix} = E \begin{pmatrix} \psi_{A\uparrow A\downarrow} \\ \psi_{A\uparrow B\uparrow} \\ \psi_{A\uparrow B\downarrow} \\ \psi_{A\downarrow B\uparrow} \\ \psi_{A\downarrow B\downarrow} \\ \psi_{B\uparrow B\downarrow} \end{pmatrix}$$

where here we mean that the full wavefunction is the sum

$$\begin{aligned} |\Psi\rangle &= \psi_{A\uparrow A\downarrow}|A\uparrow; A\downarrow\rangle + \psi_{A\uparrow B\uparrow}|A\uparrow; B\uparrow\rangle + \psi_{A\uparrow B\downarrow}|A\uparrow; B\downarrow\rangle \\ &+ \psi_{A\downarrow B\uparrow}|A\downarrow; B\uparrow\rangle + \psi_{A\downarrow B\downarrow}|A\downarrow; B\downarrow\rangle + \psi_{B\uparrow B\downarrow}|B\uparrow; B\downarrow\rangle. \end{aligned}$$

We note immediately that the Hamiltonian is block diagonal. We have eigenstates

$$|A\uparrow; B\uparrow\rangle \qquad |A\downarrow; B\downarrow\rangle$$

both with energy $E = 0$ (hopping is not allowed and there is no double occupancy, so no Hubbard interaction either). The remaining four by four Schroedinger equation is then

$$\begin{pmatrix} U & t & -t & 0 \\ t & 0 & 0 & t \\ -t & 0 & 0 & -t \\ 0 & t & -t & U \end{pmatrix} \begin{pmatrix} \psi_{A\uparrow A\downarrow} \\ \psi_{A\uparrow B\downarrow} \\ \psi_{A\downarrow B\uparrow} \\ \psi_{B\uparrow B\downarrow} \end{pmatrix} = E \begin{pmatrix} \psi_{A\uparrow A\downarrow} \\ \psi_{A\uparrow B\downarrow} \\ \psi_{A\downarrow B\uparrow} \\ \psi_{B\uparrow; B\downarrow} \end{pmatrix}$$

We find one more eigenvector $\propto (0,1,1,0)$ with energy $E = 0$ corresponding to the state[18]

$$\frac{1}{\sqrt{2}}\left(|A\uparrow; B\downarrow\rangle + |A\downarrow; B\uparrow\rangle\right).$$

A second eigenstate has energy U and has a wavefunction

$$\frac{1}{\sqrt{2}}\left(|A\uparrow; A\downarrow\rangle - |B\uparrow; B\downarrow\rangle\right).$$

[18] The three states with $E = 0$ are in fact the $S_z = -1, 0, 1$ states of $S = 1$. Since the Hamiltonian is rotationally invariant, these all have the same energy.

The remaining two eigenstates are more complicated, and have energies $\frac{1}{2}\left(U \pm \sqrt{U^2+16t^2}\right)$. The ground state, always has energy

$$E_{ground} = \frac{1}{2}\left(U - \sqrt{U^2+16t^2}\right).$$

In the limit of t/U becoming zero, the ground-state wavefunction becomes very close to

$$\frac{1}{\sqrt{2}}\left(|A\uparrow; B\downarrow\rangle - |A\downarrow; B\uparrow\rangle\right) + \mathcal{O}(t/U) \qquad (23.5)$$

with amplitudes of order t/U for the two electrons to be on the same site. In this limit the energy goes to

$$E_{ground} = -4t^2/U$$

which is almost in agreement with our perturbative calculation—the prefactor differs from that mentioned in the above calculation by a factor of 2. The reason for this discrepancy is that the ground state is not just $\uparrow$ on one site and $\downarrow$ on the other, but rather a superposition between the two. This superposition can be thought of as a (covalent) chemical bond (containing two electrons) between the two atoms.

In the opposite limit, $U/t \to 0$ the ground-state wavefunction for a single electron is the symmetric superposition $(|A\rangle + |B\rangle)/\sqrt{2}$ (see Section 6.2.2) assuming $t > 0$. This is the so-called "bonding" orbital. So the ground state for two electrons is just the filling of this bonding orbital with both spins—resulting in

$$\begin{aligned}
& \frac{|A\uparrow\rangle + |B\uparrow\rangle}{\sqrt{2}} \otimes \frac{|A\downarrow\rangle + |B\downarrow\rangle}{\sqrt{2}} \\
= & \frac{1}{2}\left(|A\uparrow; A\downarrow\rangle + |A\uparrow; B\downarrow\rangle + |B\uparrow; A\downarrow\rangle + |B\uparrow; B\downarrow\rangle\right) \\
= & \frac{1}{2}\left(|A\uparrow; A\downarrow\rangle + |A\uparrow; B\downarrow\rangle - |A\downarrow; B\uparrow\rangle + |B\uparrow; B\downarrow\rangle\right).
\end{aligned}$$

Note that eliminating the double occupancy states (simply crossing them out)[19] yields precisely the same result as Eq. 23.5. Thus, as the interaction is turned on it simply suppresses the double occupancy in this case.

[19] Eliminating doubly occupied orbitals by hand is known as *Gutzwiller projection* (after Martin Gutzwiller) and is an extremely powerful approximation tool for strongly interacting systems.

Exercises

(23.1) **Itinerant Ferromagnetism***

(a.i) Review 1: For a three-dimensional tight binding model on a cubic lattice, calculate the effective mass in terms of the hopping matrix element t between nearest neighbors and the lattice constant a.

(a.ii) Review 2: Assuming the density n of electrons in this tight binding band is very low, one can view the electrons as being free electrons with this effective mass m^*. For a system of spinless electrons show that the total energy per unit volume (at zero temperature) is given by

$$E/V = nE_{min} + Cn^{5/3}$$

where E_{min} is the energy of the bottom of the band.

▷ Calculate the constant C.

(b) Let the density of spin-up electrons be $n_\uparrow$ and the density of spin-down electrons be $n_\downarrow$. We can write these as

$$n_\uparrow = (n/2)(1+\alpha) \quad (23.6)$$
$$n_\downarrow = (n/2)(1-\alpha) \quad (23.7)$$

where the total net magnetization of the system is given by

$$M = -\mu_B n\alpha.$$

Using the result of part (a), fixing the total density of electrons in the system n,

▷ calculate the total energy of the system per unit volume as a function of α.

▷ Expand your result to fourth order in α.

▷ Show that $\alpha = 0$ gives the lowest possible energy.

▷ Argue that this remains true to all orders in α

(c) Now consider adding a Hubbard interaction term

$$H_{Hubbard} = U\sum_i N_\uparrow^i N_\downarrow^i$$

with $U \geqslant 0$ where N_σ^i is the number of electrons of spin σ on site i.

Calculate the expectation value of this interaction term given that the up and down electrons form Fermi seas with densities $n_\uparrow$ and $n_\downarrow$ as given by Eqns. 23.6 and 23.7.

▷ Write this energy in terms of α.

(d) Adding together the kinetic energy calculated in part b with the interaction energy calculated in part c, determine the value of U for which it is favorable for α to become non-zero.

▷ For values of U not too much bigger than this value, calculate the magnetization as a function of U.

▷ Explain why this calculation is only an approximation.

(e) Consider now a two-dimensional tight binding model on a square lattice with a Hubbard interaction. How does this alter the result of part (d)?

(23.2) **Antiferromagnetism in the Hubbard Model***

Consider a tight binding model with hopping t and a strong Hubbard interaction

$$H_{Hubbard} = U\sum_i N_\uparrow^i N_\downarrow^i \ .$$

(a) If there is one electron per site, if the interaction term U is very strong, explain qualitatively why the system must be an insulator.

(b) On a square lattice, with one electron per site, and large U, use second-order perturbation theory to determine the energy difference between the ferromagnetic state and the antiferromagnetic state. Which one is lower energy?

(23.3) **Nagaoka-Thouless Ferromagnetism**&**

(a) Consider a Hubbard model with one electron per site on a square lattice in the limit where $U \to \infty$. Referring back to Exercise 23.2, what is the energy difference between the ferromagnetic and antiferromagnetic arrangement of spins?

(b) Assume that there are $N = L^2$ sites in a finite sized system with periodic boundaries. Let us remove a single electron from the system, so the density is $n = 1 - 1/N$ electrons per site. In the ferromagnetic case (all spins aligned) what is the ground state energy with the single removed electron. I.e., how much kinetic energy did the single removed electron have?. Equivalently, what is the kinetic energy of the hole that has been inserted into the filled spin-polarized band?

(c) Now consider starting with a Mott antiferromagnet (one electron per site with alternating spin-up and spin-down). If we remove a single electron we can consider the effect of the hole hopping around in the antiferromagnetic background. What happens to the antiferromagnet if the hole hops in a closed loop?

(d) When the hole in an antiferromagnetic background hops in a loop, why can the wavefunction not interfere with itself constructively? Why does lack of constructive self-interference increase the total energy compared to the case with polarized spins?

(e) Conclude that ferromagnetism is favored in this case. This is not a rigorous proof, but it is essentially correct.

(f) For finite but very large U explain why this effect will not persist to infinite N.

Sample Exam and Solutions

A

The current Oxford syllabus covers this entire book with the exception of the chapter on device physics and the final chapter on interactions and magnetism.

Numbers in brackets [ ] indicate the number of points (or "marks", as is said in the UK) expected to be allocated for the given part of the question. Give yourself three hours to do all four questions (or give yourself 90 minutes to do any two, as is done in Oxford).

EXAM

Question 1. Write down a formula for the structure factor $S(hkl)$, and find the condition for reflections to be missing in the diffraction pattern from *any* crystal whose lattice is all-face-centered. [5]

Silicon crystallizes in a cubic structure whose lattice is face-centered with a basis [000] and $[\frac{1}{4}\frac{1}{4}\frac{1}{4}]$.

(1) Write out the fractional coordinates of *all* the atoms in the conventional unit cell of silicon.

(2) Sketch a plan diagram of the silicon structure viewed down the [001] axis, taking care to mark the axes and heights of the atoms within the unit cell.

(3) Write down the fractional coordinates of a center of inversion symmetry in the structure. [Center of inversion symmetry is when for every atom at x, y, z there is an equivalent atom at $-x, -y, -z$.]

(4) Show that reflections for which $h + k + l = 4n + 2$ have zero intensity in a diffraction pattern. Careful measurements of the (222) reflection nevertheless do show a small peak in intensity: suggest a possible explanation for this. [11]

How many acoustic and optic branches are to be found in the phonon dispersion diagram of silicon? How many *distinct* branches would you expect along a (100) direction? [4]

Question 2. Explain what is meant by the *Brillouin zone* and *Wigner–Seitz construction.* [4]

Consider a body-centered cubic crystal. Taking into account any missing reciprocal lattice points, draw a diagram of the $hk0$ section of the reciprocal lattice, marking (hkl)-indices for each lattice point. Use the Wigner–Seitz construction to draw the first Brillouin zone on your diagram. [7]

Consider a linear chain of atoms with only nearest-neighbour interactions with alternating force constants C_1 and C_2 between atoms of mass M. Let a be the *repeat* distance in this chain. Let the displacement of atoms be denoted by u_s and v_s, where s is an index for each pair of atoms. Show that the equations of motion are given by

$$\begin{aligned} M\frac{\mathrm{d}^2 u_s}{\mathrm{d}t^2} &= C_1\left(v_s - u_s\right) + C_2\left(v_{s-1} - u_s\right), \\ M\frac{\mathrm{d}^2 v_s}{\mathrm{d}t^2} &= C_2\left(u_{s+1} - v_s\right) + C_1\left(u_s - v_s\right). \end{aligned}$$

Show that

$$\omega^2 = \alpha \pm \sqrt{\alpha^2 - \beta \sin^2 \frac{ka}{2}}$$

where ω is the frequency of vibration for a wave-vector k. Hence find expressions for α and β in terms of C_1, C_2 and M. What is ω at the Brillouin zone boundary? [9]

Question 3. Describe which types of material are likely to show *diamagnetism* and *paramagnetism.* [3]

An insulating material contains N ions per unit volume where each ion has spin $S = 1/2$ and Landé g-factor g (the orbital angular momentum $L = 0$). Obtain an expression for the magnetisation of the system as a function of temperature T and applied magnetic field B. Under what conditions does your expression lead to Curie's law, which is given by $\chi = \alpha N/T$, where χ is the magnetic susceptibility, T is the temperature, and α is a constant of proportionality? Express α in terms of physical constants and g. [5]

Calculate the entropy of the system as a function of temperature and magnetic field. [Entropy S is related to the Helmholtz free energy F by $S = -\left(\frac{\partial F}{\partial T}\right)_V$.] Evaluate the high temperature limit and explain what you expect the low temperature limit to be. A crystal of this material is cooled to a temperature T_i in a magnetic field B_i. The magnetic field is then altered to a new value B_f under adiabatic conditions. What is the final value of the temperature? [9]

The process described above can be used as a method to cool down the material and is known as adiabatic demagnetisation. Discuss the factors which will limit the minimum temperature which might be achieved in such a process. [3]

Question 4. Show that the density of states per unit volume in the conduction band of a semiconductor is given by $g(E) = \alpha\sqrt{E - E_c}$, where E is the electron energy, E_c is the energy at the bottom of the conduction band. Express α in terms of the electron effective mass m_e^* and $\hbar$. [4]

Show that the density of electrons n in the conduction band of the semiconductor at temperature T is given by

$$n = AT^{3/2} \exp\left(\frac{\mu - E_c}{k_B T}\right),$$

where μ is the chemical potential. Express A in terms of m_e^* and physical constants. Write down a similar expression for the density of holes p in the valence band and hence derive an expression for the number density n_i of electrons in an intrinsic semiconductor. [10]

$$\left[\text{You may need: } \int_0^\infty x^{1/2} e^{-x} dx = \frac{\sqrt{\pi}}{2}\right]$$

The density of minority carriers (holes) in a sample of n-type germanium is found to be $2 \times 10^{14}\text{m}^{-3}$ at room temperature ($T = 300$K). Calculate the density of donor ions in this sample. Briefly explain how the band-gaps of germanium could be measured. [6]
[The (indirect) band-gap of germanium is 0.661eV at $T = 300$K. Assume that the electron and hole effective masses in germanium are $m_e^* = 0.22m_e$ and $m_h^* = 0.34m_e$ respectively.]

SOLUTIONS

Question 1

The structure factor is

$$S(hkl) = \sum_d f_d(\mathbf{G}_{hkl}) e^{i\mathbf{G}_{hkl}\cdot\mathbf{x}_d}$$

where the sum is over atom positions $\mathbf{x}_d$ within the unit cell. Here we assume a conventional unit cell, and the sum is over all d within the conventional cell.

The form factor f is given by

$$f_d(\mathbf{G}_{hkl}) \sim \int \mathbf{dr} V(\mathbf{r}) e^{i\mathbf{G}_{hkl}\cdot\mathbf{r}}$$

with V the scattering potential. Note that the integral is over all space.

Note that the question asked about *any* face-centered crystal. So most generally we have the lattice times an arbitrary basis. We write the position of any atom as $\mathbf{x}_d = \mathbf{R} + \mathbf{y}_d$ with $\mathbf{R}$ the lattice point and $\mathbf{y}_d$ the displacement. Then we obtain

$$S(hkl) = \sum_d \sum_{\mathbf{R}} f_d(\mathbf{G}_{hkl}) e^{i\mathbf{G}_{hkl}\cdot(\mathbf{R}+\mathbf{y}_d)}$$

where here the sum over d is a sum over the atoms in a primitive unit cell and the sum over $\mathbf{R}$ is the sum over all lattice points within the conventional unit cell. This can then be factorized as

$$\begin{aligned} S(hkl) &= \left[\sum_{\mathbf{R}} e^{i\mathbf{G}_{hkl}\cdot\mathbf{R}}\right]\left[\sum_d f_d(\mathbf{G}_{hkl}) e^{i\mathbf{G}_{hkl}\cdot\mathbf{y}_d}\right] \\ &= S_{lattice}(hkl) \times S_{basis}(hkl) \end{aligned}$$

Thus if $S_{lattice}$ vanishes for some (hkl) then the full S does too.

Now since the question asks about *any* face-centered crystal, it could be referring to a general orthorhombic crystal! So let us define orthogonal primitive lattice vectors $\mathbf{a_1}, \mathbf{a_2}, \mathbf{a_3}$ not necessarily the same length (resulting in orthogonal reciprocal primitive lattice vectors $\mathbf{b_1}, \mathbf{b_2}, \mathbf{b_3}$). The coordinates of the lattice points within a conventional unit cell can be written as

$$\begin{aligned} \mathbf{R}_1 &= [0,0,0] \\ \mathbf{R}_2 &= [0,1/2,1/2] \\ \mathbf{R}_3 &= [1/2,0,1/2] \\ \mathbf{R}_4 &= [1/2,1/2,0] \end{aligned}$$

where the fractional coordinates refer to fractions of the primitive lattice vectors. We also have

$$\mathbf{G}_{hkl} = h\mathbf{b_1} + k\mathbf{b_2} + l\mathbf{b_3}$$

Thus giving

$$S_{lattice}(hkl) = \sum_{\mathbf{R}} e^{i\mathbf{G}_{hkl}\cdot\mathbf{R}} = 1 + e^{i\pi(k+l)} + e^{i\pi(h+l)} + e^{i\pi(k+h)}$$

where we used here that $\mathbf{a_i} \cdot \mathbf{b_j} = 2\pi\delta_{ij}$. It is easy to see that this is non-zero only when all three (hkl) are even or all are odd.

(1) The coordinates are:

$$\begin{array}{llll} [0,0,0] & [0,1/2,1/2] & [1/2,0,1/2] & [1/2,1/2,0] \\ [1/4,1/4,1/4] & [1/4,3/4,3/4] & [3/4,1/4,3/4] & [3/4,3/4,1/4] \end{array}$$

(2) Plan view. Heights are marked in units of a. Unmarked points are at height 0 and a.

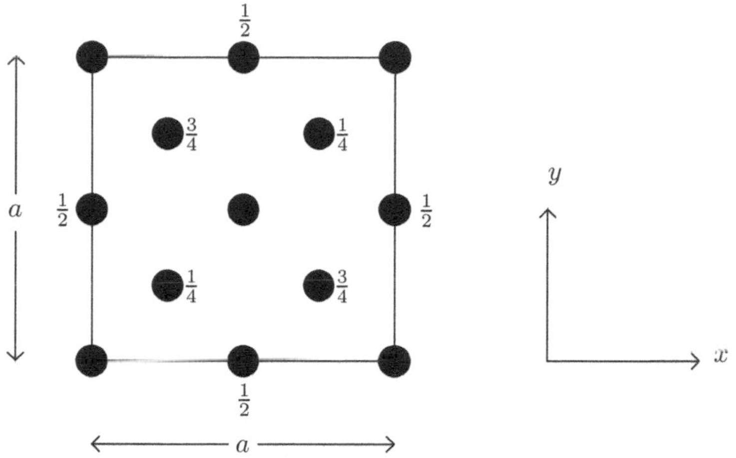

(3) If the center of inversion has position $[UVW]$ for each atom at position $[uvw]$ there must also be an atom at the inverted position $[2U-u, 2V-v, 2W-w]$. We have an inversion center at $[1/8, 1/8, 1/8]$. Inversion maps

$$[0,0,0] \leftrightarrow [1/4,1/4,1/4] \tag{A.1}$$
$$[0,1/2,1/2] \leftrightarrow [1/4,-1/4,-1/4] = [1/4,3/4,3/4] \tag{A.2}$$
$$[1/2,0,1/2] \leftrightarrow [-1/4,1/4,-1/4] = [3/4,1/4,3/4] \tag{A.3}$$
$$[1/2,1/2,0] \leftrightarrow [-1/4,-1/4,1/4] = [3/4,3/4,1/4] \tag{A.4}$$

(4) The basis is $[000]$ and $[1/4,1/4,1/4]$ so

$$S_{basis} = f_{Si}(1 + e^{i(\pi/2)(h+k+l)})$$

Thus for $h + k + l = 4n + 2$ this vanishes.

This final piece of part (4) is rather obscure (and is not expected to be solved by many students). There are two possible reasons for observing a (222) reflection peak. Reason (1) is multiple scattering. In a single scatter the probe particle can scatter by (111), and with a second scattering by (111) one observes a resulting (222) reflection. A second possible reason for observing a (222) peak is the realization that the two silicon atoms at [000] and [1/4,1/4,1/4] are actually inequivalent, in that they experience precisely inverted environments. As a result of being inequivalent, they have slightly different form factors. So we should really write

$$S_{basis} = f_{Si[000]} + f_{Si[1/4,1/4/1/4]} e^{i(\pi/2)(h+k+l)}$$

and there will not be precise destructive interference.

Students are likely to say that there may be some deviation of the basis atom [1/4,1/4,1/4] from that precise position, and this may result in observing a (222) peak as well. While this should be worth partial credit, the issue with this solution is that it is hard to come up with a mechanism for such deviation. For example, uniaxial pressure, or even shear stress, does not generally change the position of this basis atom in terms of the primitive lattice vectors.

Since there are two atoms in the basis, there should be six phonon modes. Three of these are optical, and three are acoustic (one longitudinal, and two transverse). Along the (100) direction, the two transverse modes have the same energy though (presumably this is what the question means by asking if the modes are "distinct"). This final part is difficult. To see that these modes have to be degenerate, note that a 90-degree rotation around the (100) axis is identical to a translation by [1/4,1/4,1/4]. So there is one longitudinal optical mode, two (degenerate) transverse optical modes, one longitudinal acoustic mode, and two (degenerate) transverse acoustic modes. Another way to see this is to realize that a wave traveling in the (100) direction should have the same energy as a wave traveling in the opposite direction—and this inversion of direction is also equivalent to a 90-degree rotation.

Question 2

A *Brillouin zone* is a unit cell of the reciprocal lattice.

A *Wigner–Seitz Cell* of a point $\mathbf{R_0}$ of a lattice is the set of all points that are closer to $\mathbf{R_0}$ than to any other lattice point. (Note: the Wigner–Seitz cell is primitive and has the same symmetries as the lattice.)

The Wigner–Seitz construction is a method of finding the Wigner–Seitz cell. To do this, one finds the perpendicular bisectors between $\mathbf{R_0}$ and each nearby lattice point. The area around $\mathbf{R_0}$ bounded by these perpendicular bisectors is the Wigner–Seitz cell.

Note that we will work with the conventional unit cell to define the Miller indices. So for the conventional cell,

$$(hkl) = (2\pi/a)(h\hat{x} + k\hat{y} + l\hat{z})$$

The short way to find the reciprocal lattice points is just to remember that, due to the selection rules, the missing Miller indices are those for which $h+k+l$ is odd. The $(hk0)$ cut through the reciprocal lattice then looks as follows:

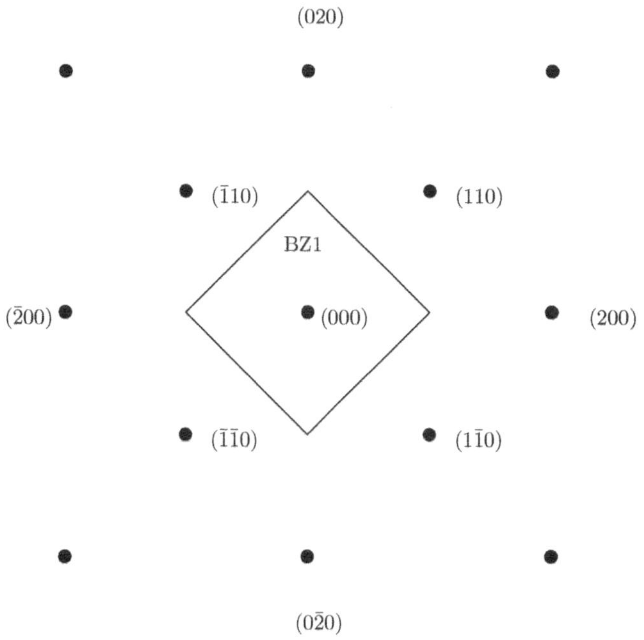

One can also solve this problem a harder way. The bcc lattice has primitive lattice vectors

$$\begin{aligned}\mathbf{a_1} &= a[100]\\ \mathbf{a_2} &= a[010]\\ \mathbf{a_3} &= a[1/2,1/2,1/2].\end{aligned}$$

The reciprocal lattice has primitive lattice vectors

$$\begin{aligned}\mathbf{b_1} &= (2\pi/a)(1,0,-1)\\ \mathbf{b_2} &= (2\pi/a)(0,1,-1)\\ \mathbf{b_3} &= (2\pi/a)(0,0,2).\end{aligned}$$

We can check that this is correct, either by using the formula

$$\mathbf{b_i} = \frac{\mathbf{a_j} \times \mathbf{a_k}}{\mathbf{a_1} \cdot (\mathbf{a_2} \times \mathbf{a_3})}$$

with i, j, k cyclic. Or (easier) we can just confirm that

$$\mathbf{a_i} \cdot \mathbf{b_j} = 2\pi\delta_{ij} \quad .$$

We can take linear combinations of these primitive lattice vectors to get the more usual primitive lattice vectors of an fcc lattice.

$$\begin{aligned}\mathbf{b_1}' = \mathbf{b_1} + \mathbf{b_3} &= (4\pi/a)(1/2,0,1/2)\\ \mathbf{b_2}' = \mathbf{b_2} + \mathbf{b_3} &= (4\pi/a)(0,1/2,1/2)\\ \mathbf{b_3}' = \mathbf{b_1} + \mathbf{b_2} + \mathbf{b_3} &= (4\pi/a)(1/2,1/2,0).\end{aligned}$$

Note that the lattice constant here is $4\pi/a$. The previous picture is precisely a cut through such an fcc lattice. It is potentially more obvious to use another set of primitive lattice vectors

$$\begin{aligned}
\mathbf{b_1}'' = \mathbf{b_1} - \mathbf{b_2} &= (2\pi/a)(1,1,0) \\
\mathbf{b_2}'' = \mathbf{b_1} + \mathbf{b_2} + \mathbf{b_3} &= (2\pi/a)(1,-1,0) \\
\mathbf{b_3}'' &= (2\pi/a)(0,0,2).
\end{aligned}$$

The equations of motion

$$\begin{aligned}
M\frac{\mathrm{d}^2u_s}{\mathrm{d}t^2} &= C_1\left(v_s - u_s\right) + C_2\left(v_{s-1} - u_s\right) \\
M\frac{\mathrm{d}^2v_s}{\mathrm{d}t^2} &= C_2\left(u_{s+1} - v_s\right) + C_1\left(u_s - v_s\right)
\end{aligned}$$

are nothing more than $F = ma$. (Not sure what is expected to "show" this: perhaps one writes down a Lagrangian, and writes Euler-Lagrange equations, or differentiates the energies to get a force.)

We propose a wave ansatz

$$\begin{aligned}
u_s &= Ae^{iksa-i\omega t} \\
v_s &= Be^{iksa-i\omega t}.
\end{aligned}$$

Plugging into our equations of motion and doing some cancellation we get

$$\begin{aligned}
M(-i\omega)^2A &= C_1(B-A) + C_2(e^{-ika}B - A) \\
M(-i\omega)^2B &= C_2(e^{ika}A - B) + C_1(A-B)
\end{aligned}$$

This can be rewritten as a matrix equation

$$M\omega^2\begin{pmatrix} A \\ B \end{pmatrix} = \begin{pmatrix} C_1 + C_2 & -C_1 - C_2e^{-ika} \\ -C1 - C_2e^{ika} & C_1 + C_2 \end{pmatrix}\begin{pmatrix} A \\ B \end{pmatrix}$$

which has the secular determinant equation

$$(C_1 + C_2 - M\omega^2)^2 - |C_1 + C_2e^{ika}|^2 = 0.$$

This can be rewritten as

$$\begin{aligned}
M\omega^2 &= C_1 + C_2 \pm |C_1 + C_2e^{ika}| \\
&= C_1 + C_2 \pm \sqrt{C_1^2 + C_2^2 + 2C_1C_2\cos(ka)} \\
&= C_1 + C_2 \pm \sqrt{(C_1+C_2)^2 + 2C_1C_2(\cos(ka) - 1)} \\
&= C_1 + C_2 \pm \sqrt{(C_1+C_2)^2 - 4C_1C_2\sin^2(ka/2)}
\end{aligned}$$

so

$$\omega^2 = \alpha \pm \sqrt{\alpha^2 - \beta\sin^2\frac{ka}{2}}$$

where

$$\alpha = \frac{C_1 + C_2}{M} \qquad \beta = 4C_1C_2/M^2$$

At the Brillouin zone boundary, $k = \pi/a$ and $\sin(ka/2) = 1$. Thus we obtain

$$\begin{aligned} M\omega^2 &= C_1 + C_2 \pm \sqrt{(C_1+C_2)^2 - 4C_1C_2} \\ &= C_1 + C_2 \pm \sqrt{(C_1-C_2)^2} \\ &= 2C_1 \quad \text{or} \quad 2C_2 \end{aligned}$$

Thus we have

$$\omega = \sqrt{2C_1/M} \quad \text{or} \quad \sqrt{2C_2/M}$$

Question 3

Paramagnetism is typical of:

- The canonical example of a paramagnet is a systems of non-interacting spins (Curie law). This is typical of atoms with localized moments, which arises from non-filled shells such that $J \neq 0$. Rare earth ions are a good example of this. (Mott insulator physics can localize the electrons on individual sites.)
- Ferromagnets and antiferromagnets above their critical temperatures are typically paramagnets.
- Free-electron gas and free-electron-like metals (like group I metals, sodium and potassium) have much weaker (Pauli) paramagnetism. Note that this Pauli paramagnetism can be outweighed by other diamagnetic contributions, particularly when one counts the Larmor diamagnetism of the core electrons.
- Van Vleck paramagnets (advanced material) have localized moment $J = 0$ but have low-energy excitations with $J \neq 0$ such that at second order in perturbation theory there is weak paramagnetism.

Diamagnetism is typical of:

- Atoms with $J = 0$ and no conduction electrons that can cause Pauli paramagnetism and no low-energy excitations that can cause Van Vleck paramagnetism. This is typical of filled-shell configurations where $J = L = S = 0$, such as noble gases.
- Inert molecules with filled shells of molecular orbitals (again with $J = L = S = 0$). For example, N_2.
- Metals can be diamagnets if the Larmor (and Landau, advanced material) diamagnetism is stronger than the Pauli paramagnetism (copper is an example of this).
- Superconductors are perfect diamagnets (not covered in this book).

For a single isolated spin 1/2 we can calculate the partition function

$$Z = e^{\beta g \mu_B B/2} + e^{-\beta g \mu_B B/2}$$

with μ_B the Bohr magneton. The expectation of the moment (per spin) is then

$$m = -d\log Z/d(B\beta) = (g\mu_B/2)\tanh(\beta g\mu_B B/2).$$

For small B this is

$$m = (g\mu_B/2)^2(B/k_BT)$$

Assuming the solid is made of independent non-interacting spins, the total magnetization is

$$M = N(g\mu_B/2)^2(B/k_BT)$$

for N spins per unit volume. Thus the susceptibility is

$$\chi = dM/dH = \mu_0 dM/dB = N\mu_0(g\mu_B/2)^2(1/k_BT),$$

thus giving the constant

$$\alpha = \mu_0(g\mu_B/2)^2/k_B$$

To determine the entropy,

$$Z = [2\cosh(\beta g\mu_B B/2)]^N$$

so

$$F = -k_BTN\ln[2\cosh(\beta g\mu_B B/2)]$$

and

$$\begin{aligned} S &= -\partial F/\partial T \\ &= k_BN\ln[2\cosh(\beta g\mu_B B/2)] - k_BN(\beta g\mu_B B/2)\tanh(\beta g\mu_B B/2). \end{aligned}$$

In the high-temperature limit β is small, so the second term vanishes and the cosh goes to 1. We thus obtain

$$S = k_BN\ln 2.$$

This is expected, being that we have two states per spin which are equally likely at high temperature.

At low temperature the system freezes into a single configuration, hence we expect zero entropy in agreement with the third law. We can check this with a real calculation. At low temperature β is large. Using $2\cosh(\beta g\mu_B/2) \to \exp(\beta g\mu_B/2)$, and the tanh goes to unity, we then have

$$S = k_BN\ln\exp(\beta g\mu_B B/2) - k_BN(\beta g\mu_B B/2) = 0$$

Note that S is a function of B/T only (not a function of B and T separately. Thus if we are to make any adiabatic changes, in order to keep S fixed we must keep the ratio of B/T fixed. So

$$T_f = T_i(B_f/B_i)$$

Two key considerations in performing adiabatic demagnetization in practice:

(1) Adiabatic demagnetization works only as well as the system is isolated from the environment. In truth the system will have some weak coupling to an environment, so that heat can leak into the system (limiting how cold the system can get). This is essentially an issue of how well you can insulate your system.

(2) At low enough temperature, the system will not be an ideal system of non-interacting spins. (In fact, in order that the system forms an ensemble at a real temperature there must be some mechanism by which the spins interact to exchange energy and to allow individual spins to flip over!) For interacting spins, the entropy is certainly not purely a function of B/T. This does not mean that adiabatic demagnetization does not work at all; it just works less well once the spins begin to order. Suppose at some $T < T_c$ the spins are ordered even at $B = 0$ and one therefore has a very small entropy S_0. This S_0 may be exponentially small if T is much less than T_c. Now unless you begin the experiment in a huge B field so that the initial entropy is smaller than S_0, you cannot cool past this temperature (and it requires a huge initial B field in order to obtain an initial entropy which is exponentially small).

Question 4

The energy of an electron in the conduction band is (assuming we are at energies that are not too much higher than the bottom of the band)

$$E = E_c + \frac{\hbar^2(\mathbf{k} - \mathbf{k_0})^2}{2m_e^*}$$

where $\mathbf{k_0}$ is the location of the minimum of the conduction band in the Brillouin zone. We are assuming that there is a single "valley" being considered (i.e., there is only a single $\mathbf{k_0}$ where the band energy is minimum), and we assume the effective mass is isotropic.

Let us imagine we fill up a ball of radius q around the point $\mathbf{k_0}$, we have

$$E - E_c = \hbar^2 q^2/(2m_e^*)$$

or

$$q = \sqrt{2m_e^*(E - E_c)}/\hbar.$$

The number of states per unit volume in k-space is $2V/(2\pi)^3$, where the factor 2 out front is for spins. So given that we are filling a ball of radius q, we have

$$n = \left[(4/3)\pi q^3\right](2/(2\pi)^3) = q^3/(3\pi^2).$$

The density of states per unit volume is

$$\begin{aligned} g(E) &= (dn/dq)(dq/dE) = (q^2/\pi^2)/(\hbar^2 q/m_e^*) \\ &= q m_e^*/(\hbar^2\pi^2) = \alpha\sqrt{E - E_c} \end{aligned}$$

where

$$\alpha = \frac{\sqrt{2}(m_e^*)^{3/2}}{\hbar^3\pi^2}.$$

Again assuming no valley degeneracy and isotropic effective mass. The density of electrons in the conduction band is

$$n = \int_{E_c}^{\infty} dE g(E) n_F(\beta(E-\mu)).$$

One must further assume that the chemical potential is not too close to the bottom of the conduction band so that the Fermi function may be replaced by a Boltzmann factor.

$$\begin{aligned} n &= \int_{E_c}^{\infty} dE g(E) e^{-\beta(E-\mu)} \\ &= \alpha e^{-\beta(E_c-\mu)} \int_{E_c}^{\infty} dE (E-E_c)^{1/2} e^{-\beta(E-E_c)} \\ &= \alpha e^{-\beta(E_c-\mu)} \beta^{-3/2} \int_0^{\infty} dx\, x^{1/2}\, e^{-x} \\ &= \alpha e^{-\beta(E_c-\mu)} (k_B T)^{3/2} \frac{\sqrt{\pi}}{2} \end{aligned}$$

so

$$n = AT^{3/2} \exp\left(\frac{\mu - E_c}{k_B T}\right),$$

with

$$A = k_B^{3/2} \alpha \frac{\sqrt{\pi}}{2} = \left(\frac{k_B m_e^*}{\pi\hbar^2}\right)^{3/2} \frac{1}{\sqrt{2}}.$$

For density of holes in the valence band, one can really write down the result by symmetry (just turning the energy upside-down at the chemical potential) that we should have

$$p = AT^{3/2} \exp\left(\frac{E_v - \mu}{k_B T}\right),$$

where E_v is the top of the valence band and

$$A = \left(\frac{k_B m_h^*}{\pi\hbar^2}\right)^{3/2} \frac{1}{\sqrt{2}}$$

with m_h^* is the hole effective mass.

To fill out a few of the details of this calculation, the density of states per unit volume near the top of the valence band is given by

$$g(E) = \alpha_v \sqrt{E_v - E}$$

with E_v the top of the valence band and

$$\alpha_v = \frac{\sqrt{2}(m_h^*)^{3/2}}{\hbar^3\pi^2}$$

with m_h^* the hole effective mass. The density of holes in the valence band is then

$$p = \int_{-\infty}^{E_v} dE g(E) \left[1 - n_F(\beta(E - \mu))\right].$$

We replace $1 - n_F$ by a Boltzmann factor to obtain

$$p = \int_{-\infty}^{E_v} dE g(E) e^{\beta(E-\mu)}$$

very similar manipulation now obtains the result.

Multiplying n by p, we obtain the law of mass action

$$np = \frac{1}{2}\left(\frac{k_B T}{\pi \hbar^2}\right)^3 (m_e^* m_h^*)^{3/2} e^{-\beta E_{gap}}$$

where $E_{gap} = E_c - E_v$ is the band gap. For an intrinsic semiconductor $n = p$, so we obtain

$$n_i = \sqrt{np} = \frac{1}{\sqrt{2}}\left(\frac{k_B T}{\pi \hbar^2}\right)^{3/2} (m_e^* m_h^*)^{3/4} e^{-\beta E_{gap}/2}$$

A direct application of the law of mass action just derived. With $T = 300$ K, $E_{gap} = .661$ eV, $m_h^* = .34 m_e$, and $m_e^* = .22 m_e$ we obtain.

$$np = 1 \times 10^{38} \text{m}^{-6}$$

with $p = 2 \times 10^{14} \text{m}^{-3}$ we then obtain

$$n = 5 \times 10^{23} \text{m}^{-3}$$

Since

$$n - p = \text{density of donors ions} - \text{density of acceptors ions}$$

and with p very small, assuming there are no acceptor ions in the sample, we conclude that n is very close to the density of donor ions.

Measuring the band gap: for an intrinsic sample, probably the easiest way to measure the gap is by measuring the density of carriers (by measuring Hall coefficient) as a function of temperature. This will change roughly as $e^{-\beta E_{gap}/2}$. If the sample is extrinsic (such as the doped sample mentioned here) then there are two approaches. One can raise the temperature of the sample until it becomes intrinsic (i.e., the intrinsic densities are greater than the dopant densities) and then follow the same scheme as for the intrinsic case (for the sample in this question this would be above about 800 K). Or at any temperature one could look at the optical absorption spectrum. Even though the gap is indirect, there should still be a small step in the absorption at the indirect gap energy.

Note that throughout this problem we have assumed the temperature is high enough so that we are above the freeze-out temperature for any impurities. This is a reasonable assumption at room temperature.

List of Other Good Books

B

The following are general purpose references that cover a broad range of the topics found in this book.

- **Solid State Physics, 2nd ed**
  J. R. Hook and H. E. Hall, Wiley
  This is frequently the book that students like the most. It is a first introduction to the subject and is much more introductory than Ashcroft and Mermin.
- **States of Matter**
  D. L. Goodstein, Dover
  Chapter 3 of this book is a very brief but well written and easy to read overview of much of what is covered in my book (not all, certainly). The book is also published by Dover, which means it is super-cheap in paperback. Warning: It uses cgs units rather than SI units, which is a bit annoying.
- **Solid State Physics**
  N. W. Ashcroft and N. D. Mermin, Holt-Sanders
  This is the standard complete introduction to solid state physics. It has many many chapters on topics not covered here, and goes into great depth on almost everything. It may be overwhelming to read this because of information overload, but it has good explanations of almost everything. Warning: Uses cgs units.
- **The Solid State, 3ed**
  H. M. Rosenberg, Oxford University Press
  This slightly more advanced book was written a few decades ago to cover what was the solid state course at Oxford at that time. Some parts of the course have since changed, but other parts are well covered in this book.
- **Solid-State Physics, 4ed**
  H. Ibach and H. Luth, Springer-Verlag
  Another very popular book. It is more advanced than Ashcroft and Mermin (much more than Hook and Hall) and has quite a bit of information in it. Some modern topics are covered well.
- **Introduction to Solid State Physics, 8ed**
  C. Kittel,[1] Wiley
  This is a classic text. It gets mixed reviews by being unclear on some matters. It is somewhat more complete than Hooke and Hall, less so than Ashcroft and Mermin. Its selection of topics and organization may seem a bit strange in the modern era.

[1] Kittel happens to be my dissertation-supervisor's dissertation-supervisor's dissertation-supervisor's dissertation-supervisor, for whatever that is worth.

- **Solid State Physics**
  G. Burns, Academic
  Another more advanced book. Some of its descriptions are short but very good. The typesetting is neolithic.
- **Fundamentals of Solid State Physics**
  J. R. Christman, Wiley
  Slightly more advanced book, with many many good problems in it (some with solutions). The ordering of topics is not to my liking, but otherwise it is very useful.

The following are good references for specific topics (but should not be considered general references for solid state physics):

- **The Structure of Crystals**
  M. A. Glazer, Bristol
  This is a very nice, very very short book that tells you almost everything you would want to know about crystal structure. It only does a little bit on reciprocal space and diffraction, but gets the most important pieces.
- **The Basics of Crystallography and Diffraction, 3ed**
  C. Hammond, Oxford University Press
  This book has historically been part of the Oxford syllabus, particularly for scattering theory and crystal structure. I don't like it much, but it would probably be very useful if you were actually doing diffraction experiments.
- **Structure and Dynamics**
  M. T. Dove, Oxford University Press
  This is a more advanced book that covers scattering and crystal structure in particular. It is used in the Oxford condensed matter fourth year masters option.
- **Magnetism in Condensed Matter**
  S. Blundell, Oxford University Press
  Well written advanced material on magnetism. It is used in the Oxford condensed matter fourth year masters option.
- **Band Theory and Electronic Properties of Solids**
  J. Singleton, Oxford University Press
  More advanced material on electrons in solids and band structure. Also used in the Oxford condensed matter fourth-year masters option.
- **Semiconductor Devices: Physics and Technology**
  S. M. Sze, Wiley
  This is an excellent first text for those who want to know some more details of semiconductor device physics.
- **Principles of Condensed Matter Physics**
  P. M. Chaikin and T. C. Lubensky, Cambridge
  A book that covers condensed matter physics much more broadly than solid state. Some of it is fairly advanced.

- **The Chemical Bond, 2ed**
  J. N. Murrell, S. F. A. Kettle, and J. M. Tedder, Wiley
  If you feel you need more basic information about chemical bonding, this is a good place to start. It is probably designed for chemists, but it should be easily readable by physicists.
- **The Nature of the Chemical Bond and the Structure of Molecules and Crystals**
  L. Pauling, Cornell
  If you want to really learn about chemistry, this is a classic written by the master. The first few chapters are very readable and are still of interest to physicists.

Indices

This book has two indices.[2]

In the index of people, Nobel laureates are marked with *. There are well over fifty of them. To be fair, a few of the Nobel laureates we have mentioned here (such as Fredrick Sanger) are mentioned offhandedly in these notes, but have little to do with the content of this book. On the other hand, there many more Nobel laureates who won their prizes for work in condensed matter physics who we simply did not have space to mention! At any rate, the total count of Nobel laureates is easily over fifty (and quite a few random people got into the index as well).

A few people whose names are mentioned did not end up in the index because the use of their name is so common that it is not worth indexing them as people. A few examples are Coulomb's law, Fourier transform, Taylor expansion, Hamiltonian, Jacobian, and so forth. But then again, I did index Schroedinger equation and Fermi statistics under Schroedinger and Fermi respectively. So I'm not completely consistent. So sue me.

[2] Making it a tensor. har har.

Index of People

Index of Topics